T0226238

# Lecture Notes in Artificial Intelligence    11110

Subseries of Lecture Notes in Computer Science

More information about this series at http://www.springer.com/series/1244

Jacques Fleuriot · Dongming Wang
Jacques Calmet (Eds.)

# Artificial Intelligence and Symbolic Computation

13th International Conference, AISC 2018
Suzhou, China, September 16–19, 2018
Proceedings

Springer

*Editors*
Jacques Fleuriot
University of Edinburgh
Edinburgh
UK

Jacques Calmet
Karlsruhe Institute of Technology
Karlsruhe
Germany

Dongming Wang
Beihang University
Beijing
China

ISSN 0302-9743      ISSN 1611-3349   (electronic)
Lecture Notes in Artificial Intelligence
ISBN 978-3-319-99956-2      ISBN 978-3-319-99957-9   (eBook)
https://doi.org/10.1007/978-3-319-99957-9

Library of Congress Control Number: 2018952478

LNCS Sublibrary: SL7 – Artificial Intelligence

This Springer imprint is published by the registered company Springer Nature Switzerland AG
The registered company address is: Gewerbestrasse 11, 6330 Cham, Switzerland

# Preface

This volume contains the invited and contributed papers for AISC 2018, the 13th International Conference on Artificial Intelligence and Symbolic Computation, held during September 16–19, 2018, in Suzhou, China.

For the past 26 years or so, AISC has considered artificial intelligence (AI) and symbolic computation as two significant approaches to problem solving, especially in mathematics. As AI has gained renewed interest, especially on the (non-symbolic) machine-learning front, it is particularly timely to re-emphasize how the two fields intersect with each other in a significant number of areas with respect to symbols. Thus, the AISC conference series is an important forum when it comes to ensuring that ideas, theoretical insights, methods, and results from traditional AI can be discussed and showcased while fostering new links with other areas of AI such as probabilistic reasoning and deep learning. The papers in this volume hint at these opportunities, with (a non-exhaustive list of) topics that include: traditional domains such as theorem proving, SAT solving, heuristic (numerical) problem solving and intelligent knowledge management; probabilistic modeling and reasoning for word detection in Chinese texts and proof automation; the understanding of neural models; and the analysis of crowdsourcing. As AI assumes a transformative role in society, aside from its long-standing role in promoting the synergies between the field and symbolic computation, AISC may be in a unique position with regard to the investigation of areas such as explainable AI, which many agree will require novel research in symbolic representation and reasoning. It is our hope that the community will ensure this long-running conference series can not only perdure but gain new momentum.

For this conference, original research contributions were solicited in areas encompassing AI, symbolic computation, and their interactions. Two special tracks on "Intelligent Documents" and "Collective Intelligence" were also announced. The 18 accepted papers, together with two invited ones, make up the proceedings published in this LNAI volume. Each paper received was reviewed by three members of the Program Committee with subreviewers, and the acceptance was based on the evaluation with respect to relevance and significance. The conference program featured three invited talks by Chee K. Yap, Alan Bundy, and Zhi-Hua Zhou for the main track, two invited tutorials by James H. Davenport and Ilias S. Kotsireas, and three invited talks by Xiaoyu Chen, Cezary Kaliszyk, and Guoliang Li for the special tracks.

We thank all the authors of submitted papers, the members of the Program Committee and external reviewers, the invited speakers and the organizers, and we acknowledge the support of the Suzhou Institute of Beihang University, which contributed to the success of the conference.

July 2018

Jacques Fleuriot
Dongming Wang
Jacques Calmet

# Organization

## General Chairs

Jacques Calmet      Karlsruhe Institute of Technology, Germany
Dongming Wang      Beihang University, China and CNRS, France

## Program Chair

Jacques Fleuriot      University of Edinburgh, UK

## Program Committee

| | |
|---|---|
| Jesús Aransay | University of La Rioja, Spain |
| Yves Bertot | Inria, France |
| Francisco Botana | University of Vigo at Pontevedra, Spain |
| Krysia Broda | Imperial College, UK |
| Mnacho Echenim | University of Grenoble, France |
| Matthew England | Coventry University, UK |
| Xiao-Shan Gao | Chinese Academy of Sciences, China |
| Tetsuo Ida | University of Tsukuba, Japan |
| Paul Jackson | University of Edinburgh, UK |
| Predrag Janičić | University of Belgrade, Serbia |
| Deepak Kapur | University of New Mexico, USA |
| Michael Kohlhase | FAU Erlangen-Nuremberg, Germany |
| Ekaterina Komendantskaya | Heriot-Watt University, UK |
| Robert Y. Lewis | Vrije University Amsterdam, The Netherlands |
| Xinjun Mao | National University of Defense Technology, China |
| Chenqi Mou | Beihang University, China and LIP6-UPMC, France |
| Julien Narboux | University of Strasbourg, France |
| Petros Papapanagiotou | University of Edinburgh, UK |
| Tomás Recio | University of Cantabria, Spain |
| Jose-Luis Ruiz-Reina | University of Seville, Spain |
| Carolyn Talcott | SRI International, USA |
| Laurent Thery | Inria, France |
| Yongxin Tong | Beihang University, China |
| Josef Urban | Czech Technical University in Prague, Czech Republic |
| Dongming Wang | Beihang University, China and CNRS, France |
| Wolfgang Windsteiger | Johannes Kepler University, Austria |
| Ye Yuan | Northeastern University, China |
| Zimu Zhou | ETH Zurich, Switzerland |

## Local Arrangements

| | |
|---|---|
| Wenjun Wu | Beihang University, China |
| Xiaoyu Chen | Beihang University, China |

## Publicity

| | |
|---|---|
| Xiaohong Jia | Chinese Academy of Sciences, China |

# Contents

## Intelligent Documents and Collective Intelligence

# Invited Presentations

# Automated Reasoning in the Age of the Internet

Alan Bundy[✉], Kwabena Nuamah, and Christopher Lucas

School of Informatics, University of Edinburgh, Edinburgh, UK
{A.Bundy,k.nuamah,c.lucas}@ed.ac.uk

**Abstract.** The internet hosts a vast store of information that we cannot and should not ignore. It's not enough just to retrieve facts. To make full use of the internet we must also infer new information from old. This is an exciting new opportunity for automated reasoning, but it also presents new kinds of research challenge.

- There are a huge number of potential axioms from which to infer new theorems. Methods of choosing appropriate axioms are needed.
- Information is stored on the Internet in diverse forms, e.g., graph and relational databases, JSON (JavaScript Object Notation), CSV (Comma-Separated Values) files, and many others. Some contain errors and others are incomplete: lacking vital contextual details such as time and units of measurements.
- Information retrieved from the Internet must be automatically curated into a common format before we can apply inference to it. Such a representation must be flexible enough to represent a wide diversity of knowledge formats, as well as supporting the diverse kinds of inference we propose.
- We can employ forms of inference that are novel in automated reasoning, such as using regression to form new functions from sets of number pairs, and then extrapolation to predict new pairs.
- Information is of mixed quality and accuracy, so introduces uncertainty into the theorems inferred. Some inference operations, such as regression, also introduce uncertainty. Uncertainty estimates need to be inherited during inference and reported to users in an intelligible form.

We will report on the FRANK (Formally know as RIF: Rich Inference Framework. We changed the name as the RIF acronym is already in use, standing for Requirements Interchange Format.) system that explores this new research direction.

**Keywords:** Query answering · Prediction · Automated reasoning
World Wide Web

This work has been funded by a University of Edinburgh studentship for the second author and Huawei grant HIRP O20170511.

J. Fleuriot et al. (Eds.): AISC 2018, LNAI 11110, pp. 3–18, 2018.
https://doi.org/10.1007/978-3-319-99957-9_1

# 1   Introduction

We describe the FRANK (Functional Reasoning Acquires New Knowledge) system. FRANK applies inference to knowledge sources on the World Wide Web to derive estimates of new information and reliably assigns an uncertainty to it. It applies deductive, arithmetic and statistical reasoning to the results of information retrieval. We call this *rich inference*. An earlier description appeared in [9]. FRANK's main focus is on estimating the values of numeric attributes, but it sometimes returns qualitative answers, e.g., the query "Which country will have the largest population in Africa in 2021?" returns the name of the African country with the maximum estimated population.

Our hypothesis is:

*A combination of information retrieval with deductive, arithmetic and statistical reasoning can be used accurately to estimate novel information and to assign a reliable uncertainty estimate to it.*

To address the issues raised in the abstract above, we have adopted the following techniques:

- The knowledge required to answer a query is retrieved from a wide variety of different knowledge sources on the Web. We employ APIs for each of the common knowledge formats in order to match the knowledge sought to the knowledge sources from which we retrieve it.
- This knowledge is then dynamically curated into a common format and stored in a query-specific ontology. This enables our inference operations to combine knowledge from diverse sources. Our common format is *alists*, i.e., sets of attribute/value pairs (see Definition 1). Alists can also be interpreted as $n$-ary, typed, logical relations (and sometimes also as functions), where $n + 1$ is the size of the set, the compulsory *Predicate* attribute's value is the predicate of the relation and the other attribute names are the types (see Definition 1). These pairs are both extracted from the particular knowledge item, e.g., the *Subject*, *Predicate* and *Object* attributes, and also augmented with attribute values from the source itself, e.g., the *Time*, *Units* and *Uncertainty* attributes. Alists provide the flexibility we need to cope with relations of diverse type signatures.
- Queries are represented as conjunctions of alists. Some of their attributes' values will be logical variables, whose value is unknown when the query is posed and which it is intended will be instantiated to a concrete value as a side effect of inference. Some of the variables in the query alist will be instantiated and returned as the answer to the query.
- FRANK's inference constructs a search tree with both AND and OR branches. Nodes are labelled with (sub-)goals represented as alists; the root node is labelled with the original query. Arcs are labelled with inference rules that enable a parent alist to be inferred from its child alists, i.e., inference is backwards from the root query to the leaf facts. If the search is successful, then the search tree will contain a proof as a subtree, which will contain

only AND branches. This proof tree will provide just those inference steps required to prove the query. During this proof, the query's variables will be instantiated to provide the required answer.

– The variables associated with the leaf alists of the proof tree are instantiated by matching them to facts stored in knowledge sources. The values of variables in parent alists are calculated by applying arithmetic aggregation operations to some of the variables in their child alists. The variables whose instantiated values are projected from child to parent are distinguished as *projection variables* (see Definition 2).

– Projected numeric values are assumed to have a Gaussian distribution and are returned as a mean and standard deviation. The mean is regarded as the answer and the standard deviation as an error bar on this answer. Aggregation operations are applied to both mean and standard deviation as they are inherited from leaf to root. Leaf nodes are assigned uncertainty values associated with the knowledge source from which they are taken. Knowledge sources are initially assigned default uncertainties, but these uncertainties are incrementally adjusted by a Bayesian process which compares the compatibility of rival sources of the same knowledge. Some inference operations also add additional uncertainty that is inherent in their nature, e.g., regression/extrapolation.

## 2   Alists: A Common Knowledge Format

Each node of the FRANK search tree is labelled by an association list or *alist*, which is a set of attribute/value pairs[1]. For example, the assertion that the population of the UK in 2011 is 63,182,000 people is represented by:

$$\{\langle Subject, UK\rangle, \langle Predicate, Population\rangle, \langle Object, 63,182,000\rangle, \langle Time, 2011\rangle\} \tag{1}$$

Alists enable FRANK to represent relations of any arity and with whatever types of arguments are required by the application. For example, alist (1) represents the ternary relation:

$$Population(UK, 63,182,000, 2011)$$

where the type signature of *Population* is:

$$Population : Subject \times Object \times Time \mapsto Bool$$

So, alists can be seen just as a syntax for typed logical formula and deduction with them as a logical inference process. All the different knowledge formats used in the knowledge sources accessed by FRANK can be curated into alists.

---

[1] See https://en.wikipedia.org/wiki/Association_list accessed on 5.6.18. Alists are not lists but sets, but the 'alist' terminology has, unfortunately, become standard.

## 2.1  Definition of Simple Alists

We can formalise a simple alist as follows:

**Definition 1 (Simple Alist).** *A simple* alist *is a set of pairs* $\{\langle A_i, a_i \rangle | 1 \le i \le n\}$, *where each* $A_i$ *is an attribute and* $a_i$ *is its value. This will sometimes be written as* $\{\langle A_1, a_1 \rangle, \ldots, \langle A_n, a_n \rangle\}$ *or abbreviated as* $\mathcal{A}$.

- *We will use the notation* $\mathcal{A}(t)$ *to indicate that* $\mathcal{A}$ *contains a distinguished term* $t$ *at some unspecified redex.*
- *We will use the notation* $\mathcal{A}[A] = a$, *when* $\langle A, a \rangle \in \mathcal{A}$, *i.e., that* $a$ *is the value of attribute* $A$ *in* $\mathcal{A}$.
- *We will use the notation* $\mathcal{A}[\boldsymbol{b}/\boldsymbol{a}]$ *to indicate that the values* $\boldsymbol{a}$ *of some attributes* $\boldsymbol{A}$ *are pairwise replaced by* $\boldsymbol{b}$ *in an alist* $\mathcal{A}$.
- *One attribute must be* Predicate. *This allows the alternative representation of:*

$$\{\langle Predicate, P, \rangle, \langle A_1, a_1 \rangle, \ldots, \langle A_n, a_n \rangle\}$$

*as* $P(a_1, \ldots, a_n)$ *where* $P : A_1 \times \ldots \times A_n \mapsto Bool$.

Typical attributes are *Subject, Object, Predicate, Time*, etc. Values can be names, numbers, functions, etc. *Object* values are often numbers, but not exclusively so.

## 2.2  Variables in Alists

A (sub-)goal alist usually has some attribute values that are variables. During proof search, these variables may be instantiated. Variables in leaf alists are instantiated by being matched against facts stored in knowledge sources. Projection variables are instantiated to values that are passed from child alists to their parents. Each alist has an aggregation operation attribute with a function value $h$, say. This function $h$ is applied to the projection variables of the child alists to instantiate the projection variable of the parent. This aggregation operation is associated with the inference rule on the AND branch connecting the parent to its children. The aggregation operation enables each alist to be regarded, not just as a relation, but also as a function from projection variables of the children to the projection variable of the parent.

The various variables appearing in an alist are defined as follows:

**Definition 2 (Projection, Auxiliary and Operand Variables).** *Let* $\mathcal{A}$ *be an alist.*

- *Its* projection variables *are the variables whose values are to be projected from it to its parents. They are prefixed with a* ?, *e.g.,* ?$x$ *denotes a projection variable. In general, an alist may have several projection variables, so we use vector notation to denote them all, e.g.,* ?$\boldsymbol{x}$.

- *Its* auxiliary variables *are the variables whose values are used locally within* $\mathcal{A}$, *but are not projected to its parents. They are prefixed with a* \$, *e.g.,* \$x *denotes an auxiliary variable. In general, an alist may have several auxiliary variables, so we use vector notation to denote them all, e.g.,* **\$x**.
- *Its* operand variables *are the variables that are used as arguments for* $\mathcal{A}$*'s* aggregation operation $h$. *An operand can be either a projection or an auxiliary variable but must exist as an attribute value in* $\mathcal{A}$.

By distinguishing projection variables, we can also view alists as *functions*, which return the value(s) of the projection variable(s) as their results. This view of alists is crucial in formalising the propagation of projection variables (see Sect. 5) and the treatment of nested queries (see Sect. 2.3).

A query or (sub-)goal is represented as an alist containing projection variables, e.g., if we want to ask what the population of the UK was in 2011, then the query would be:

$$\{\langle Subject, UK\rangle, \langle Predicate, Population\rangle, \langle Object, ?p\rangle, \langle Time, 2011\rangle\} \qquad (2)$$

where $?p$ is a projection variable which will be projected up.

### 2.3 Nested Queries and Alists

Some queries are *nested*, e.g., "What was the GDP in 2010 of the country predicted to have the largest total population in Europe in 2018?". FRANK's initial formalisation of nested queries is to represent them as *compound alists*, i.e., alists which have alists as some of their values. If alists are viewed only as relations, then nesting one relation inside of another would be a syntax error. It does make syntactic sense, however, if the nested alists are given their *functional* interpretation. That is, the inner alist returns the values of its projected variables as the value of an attribute of the outer alist. We can represent the situation abstractly as:

$$\{\ldots, \langle Attribute_1, \{\ldots \langle Attribute_2, ?x\rangle, \ldots\}\rangle, \ldots\} \qquad (3)$$

where the projection variable value $?x$ of the attribute $Attribute_2$ of the inner alist becomes the value of the attribute $Attribute_1$ of the outer alist.

For FRANK's inference system to apply, however, such compound alists need to be normalised into conjunctions of simple alists. In the case of compound alist (3), normalisation gives the following conjunction of two simple alists.

$$\{\ldots \langle Attribute_2, ?x\rangle, \ldots\} \wedge \{\ldots, \langle Attribute_1, \$x\rangle, \ldots\}$$

Note that $x$ does not necessarily become a projection variable of the outer alist, so we have used $\$x$ here, rather than $?x$.

## 3 Curation and Enrichment

Curation is a bridge between the diverse knowledge source formats and the target common format used by FRANK. The leaf alists in the search tree are sub-goals that must be translated into the format used by the knowledge source being queried and then matched to the knowledge in that source. Matching instantiates variables in the sub-goal alist. FRANK incorporates APIs for each of the knowledge formats used by the knowledge sources that it queries. It also incorporates information retrieval procedures for each type of knowledge source, e.g., SQL, SPARQL, JSON, OWL.

FRANK uses a variety of KBs for (1) finding synonyms of terms in lookup decompositions, (2) finding sub-parts of geographical entities in geospatial decompositions and (3) retrieving facts about entities. KBs used include Wordnet [7], Geonames [11], Wikidata [12], ConceptNet [6], Google Knowledge Graph [10], and the World Bank's datasets on country development indicators[2].

Some knowledge formats have restricted functionality, e.g., representing only unary or binary relations, e.g., only a predicate between a subject and an object. The leaf alist, however, may represent an $n$-ary relation for $n > 2$, and some of these additional attributes may contain variables that must be instantiated, e.g., units, time and uncertainty. The additional fields can often be found as global properties of the knowledge source, e.g., a car manufacturer may express all dimensions as centimetres, census data will record the year of the census, FRANK will have a record of the uncertainty it currently assigns to each knowledge source. These global properties enable variables in these additional attributes to be given values.

## 4    Search and Proof Trees

FRANK's inference can be represented as an AND/OR search tree. The OR branches represent the different ways in which FRANK may attempt to prove a sub-goal. Only one of these branches needs to succeed in order for the sub-goal to be proved. The AND branches represent the different child sub-goals that all need to be proved in order to prove the parent sub-goal. An example search tree is given in Fig. 1.

- Each node in the search tree is a box labelled by a truncated representation of its alist. The arcs between nodes represent inference operations. AND branching is represented by a circular line connecting the branches. OR branches have no such line.
- The first word in each alist is the aggregation operation. For instance, LOOKUP returns the value to which projection variable(s) have been instantiated by information retrieval; VALUE returns the value of its unique child's alists' projection variable(s); MAX returns the maximum value of the children's projection variables; REGRESS returns the result of extrapolating, to a new $x$ value, a function formed by regression on the children's $\langle x, y \rangle$ pairs.
- FRANK always tries direct look-up first. Only if this fails does it apply an inference operation. $\otimes$ represents a failure, e.g., look-up failed because no matching fact could be found in any knowledge source.

---

[2] https://data.worldbank.org/.

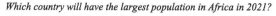

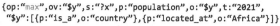

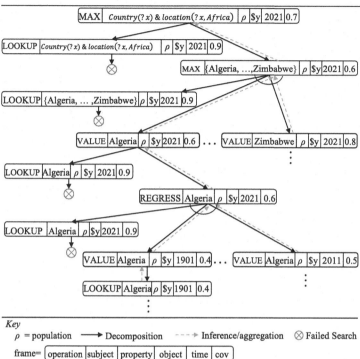

**Fig. 1.** FRANK's Search Tree for the query "Which country will have the largest population in Africa in 2021?"

- The main inference operations used are geospatial and temporal decomposition. Geospatial decomposition breaks a *Subject* into parts, applies the query to each part and then combines the results, e.g., by summing them or taking the maximum. Temporal decomposition applies the query to different (often older) time values, applies regression to form a function and then applies that function to the original time.
- A successful search tree contains a proof sub-tree. This is indicated by the dotted arc lines in Fig. 1.
- In Fig. 1, after failure to find the query's answer by direct look-up, geospatial decomposition is applied to apply the query directly to each African country and then to return the country whose population is the maximum. Direct look-up of each country's population in 2021 fails, so temporal decomposition is applied to census data for each country from the years 1901 to 2011. Regression is applied to this data to form a graph, which is then extrapolated to 2021. Since the AND branching rates are quite high, ellipsis has been used to compact the search tree to readable dimensions.

## 5   Inference and Aggregation

FRANK's current inference operations are information retrieval and the geospatial and temporal decomposition rules, which have been described in Sect. 4 above. Plans to extend these are outlined in Sect. 10.1. A unique property of FRANK's inference is its combination of deductive reasoning with statistical reasoning. In particular, it forms functions by regression, which provides the ability to reason about functions: second-order deduction, such as calculus.

An aggregation operation is associated with each application of an inference operation. Aggregation propagates the values of instantiated projection variables from child alists to parent alists, and so back to the root alist, where it becomes the mean value of the answer to the original query.

**Geospatial Decomposition:** Depending on the query, the values of the children's projection variables can be aggregated by various arithmetic operations, such as finding: the maximum or minimum; the mean, median or mode; the sum or product; or the number of children. If there are only two children, then we can also find whether the first is equal to, greater than or less than the second.

**Temporal Decomposition:** Each of the children's alists returns a $\langle x, y \rangle$ pair. Regression is applied to these values to form a function $f$. This $f$ is extrapolated or interpolated to a new value of $x$ by applying this function to it and returning the corresponding $f(x)$ as the parent's projection variable value.

## 6   Uncertainty

It is important that some measure of uncertainty is associated with the results returned by FRANK. Knowledge obtained from the Web is of variable quality, depending on the reliability of the source. Moreover, some of the inference operations we use, e.g., regression, contribute additional uncertainty. FRANK must keep track of this uncertainty and report it to user, so that they know how much to trust the result. We propose error bars as the best measure of uncertainty to assign to the kind of numerical results estimated by FRANK.

– Probabilities do not work. For instance, the probability of $\exists ?p. \; Population$ $(UK, ?p, 2025)$ is 1, i.e., it is certain that the UK will have some population count in 2025. What we need to know is the accuracy of the value FRANK assigns to $?p$. Due to the inherently vagueness of population counts, the probability that any one value of $?p$ is absolutely correct is essentially 0—or, more accurately, the question is inherently meaningless unless we know what value to assign to people who are in the process of dying, being born or in a vegetative state, etc., and what instant in time the census was taken.
– What we really need is to give a *range* for the answer. Error bars are a well known way of expressing such ranges, that many people will have seen on graphs, etc. They are also standard in numerical science.

Gaussian distributions (also known as bell curves) are ubiquitous in many numerical estimates. They can be defined by two measures: the *mean*, which gives an average of the distribution and the *standard deviation*, which describes the spread of the distribution, so is ideal to express the error bars. We have, therefore, adopted Gaussians as our distribution of uncertainty.

We return the mean value as our estimate of the value of a numerical projection function. The width of the error bar then gives a measure of the uncertainty associated with the mean. We use two different ways to express error bars. Firstly, we can use the standard deviation, which gives an absolute measure of the range of values of the projection variable that fall within the standard deviation. Secondly, we use the *coefficient of variation* (CoV). This is the mean divided by the standard deviation. It gives a relative measure of the range. For instance, we could turn the CoV into a percentage by multiplying it by 100, and then say that the mean was, say, within 5% of the correct[3] value. The CoV is ideal for propagating the uncertainty from leaf to root nodes. That's because the projection variables vary from node to node of the proof tree. So, the standard deviations are not comparable, but the CoVs are, so can be combined [2]. To report the final uncertainty back to the user, though, the standard deviation is sometimes preferable. It can be readily calculated by multiplying the CoV by the mean. For more details about the use of uncertainty in FRANK, see [8].

Note that this measure of uncertainty only applies to real numbered values. We are looking into measures of qualitative uncertainty as future work (see Sect. 10.2).

## 7   Interface

FRANK has a simple natural language interface. This enables users to type queries and receive answers in a restricted grammar of English via a GUI. The natural language processing employs the spaCy: off-the-shelf NLP library [5]. The grammar restrictions are to ensure that the query can be represented as an alist. A snapshot of FRANK's GUI is given in Fig. 2.

- The question is typed in the query box at the top.
- This query is then translated into alist form, which is displayed in abbreviated form in the dark box immediately below the query.
- FRANK's answer of 30,034,356.64 is then displayed below this with a standard deviation of $\pm 32119461.051857124$ as the error bar.
- The instantiated root alist is given below this. Note that the uncertainty value here is the CoV, not the standard deviation, which only appears in the final answer.
- At the bottom is FRANK's search tree, in which the nodes labels are given as numbers to save clutter, but these can be unpacked by clicking on them.

---

[3] Assuming that the correct value lies within one standard deviation. Since the potential range is infinite, this is a compromise between being informative and reasonable accurate. One could, instead, use two or more standard deviations.

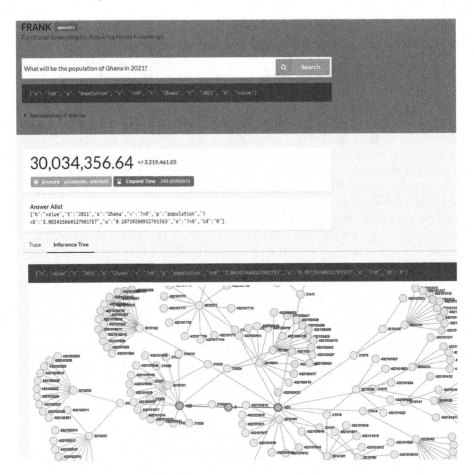

**Fig. 2.** FRANK's GUI for the query "What will be the population of Ghana in 2021?"

The search tree has a zoom option, so that the user can get an overview or examine one part in more detail.

This interface is currently in an early stage of development. This will include giving appropriate feedback to users who ask queries outwith FRANK's grammar. Eventually, we plan to deliver this interface as an open web service.

## 8   Evaluation

We have evaluated our hypothesis that:

*A combination of information retrieval with deductive, arithmetic and statistical reasoning can be used accurately to estimate novel information and to assign a reliable uncertainty estimate to it.*

Our evaluation has two parts. Firstly, we want to know how accurately FRANK has estimated the answer. For instances, is the estimated answer within one standard deviation of the true answer? Secondly, we want to know how accurate our uncertainty estimates are. For instance, are the true errors proportional to the estimated errors. For both parts of this evaluation, we need to know the true answers. We do this by a 'leave one out' methodology. That is, our queries are of known values, but we prevented FRANK from looking the values up directly, forcing it to estimate them from other known values. We compared FRANK's success rate with two comparator query answering systems: Google search and Wolfram|Alpha. These comparators were not prevented from direct look-up[4].

We randomly generated a set of 100 queries using property terms related to the country indicators in the World Bank data-set. We used 60 of these queries as a training set during the development of FRANK and used the remaining 40 for the test set. These 40 were grouped into four types:

**Retrieval:** Queries whose answers were found by direct look-up. FRANK was not prevented from direct look-up for these queries.

**Inference Queries:** Simple queries where several facts needed to be combined by inference but where regression was not needed.

**Nested Queries:** Compound queries that had to be normalised, but where regression was not needed.

**Prediction:** Queries for which regression and extrapolation/interpolation were required.

Table 1 shows a favourable comparison of FRANK's percentage success rate to two popular query answering systems: Google Search and Wolfram|Alpha[5], that also use the World Bank's dataset. A result is counted as a success if it is within one standard deviation of the true answer. FRANK performs better than both its two comparators on all four query types but, as might be expected, it did especially well when predication was required, since no prediction answers were pre-stored.

**Table 1.** Evaluation results by query types, showing the percentage of queries answered successfully

| Queries | Google search (%) | Wolfram|Alpha(%) | FRANK (%) |
|---|---|---|---|
| Retrieval | 70 | 80 | 90 |
| Aggregation queries | 20 | 70 | 80 |
| Nested queries | - | 50 | 80 |
| Prediction | 10 | 20 | 70 |
| Average % | 25 | 55 | 80 |

---

[4] Mainly because we couldn't do so, so they did have an advantage over FRANK.

[5] https://www.wolframalpha.com.

Figure 3 is a scatter plot to compare actual error to estimated error. On the $y$ axis is the ratios between (a) the absolute difference between the true and estimated values and (b) the true value. On the $x$ axis is the estimated error represented by the CoV. Ideally, this scatter plot would approximate a straight line, showing that actual and estimated error were proportional. The dotted line is the best fit straight line to these points. This is a fair fit to the data.

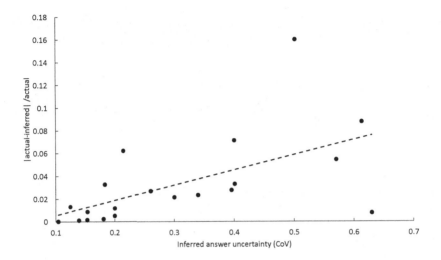

**Fig. 3.** Comparison of estimated error against actual error

## 9   Related Work

We have found nothing quite like FRANK to compare it to. The best fit is probably the first author's previous work on the GORT system [1]. This system solved guesstimation problems, where an approximate answer was required to a numeric problem, e.g., "how many cars, parked bumper to bumper, would be needed to reach from Edinburgh to Glasgow?". It also searched the Web for facts and inferred new information from it, but its inference operations were limited to simple arithmetic and its error bars just showed the range of different answers these methods had found.

Table 1 showed a favourable comparison of FRANK's performance to two other popular query answering systems. We plan further such comparisons, but inference-based query answering systems, e.g., [4], have gone out of fashion since they cannot operate at web-scale. So we did not find a lot of modern systems against which to compare FRANK. IBM's Watson [3] was too finely tuned to solving questions in the game show, Jeopardy!. As a result, it could not be directly compared to FRANK.

Currently, the field of information retrieval[6] is focused on extracting known information from the Web. Its main research challenge is interpreting queries in natural language (mostly, but not exclusively, English). FRANK's simple NL interface is described in Sect. 7, but this is not the focus of our research.

Given our focus on the inference of new information from old, Table 2 gives a comparison of FRANK to traditional work in the automation of reasoning.

**Table 2.** Comparison of FRANK and automated reasoning

| FRANK | Automated reasoning |
|---|---|
| Lots of uncertain information | A few, certain axioms |
| Lots of facts, few rules | More rules than facts |
| Diverse formats | Uniform format |
| Diverse inference operations | Deductive inference |
| Depleted information | All information present |
| Killer app: query answering | Killer app: formal verification |

## 10   Future Work

FRANK is still under active development and we have plans to extend it in several directions.

### 10.1   Generalising Decomposition Rules

FRANK currently uses only two decomposition rules: temporal and geospatial, but there is the potential for many more. The general form of a decomposition rules is given in Definition 3.

**Definition 3 (Decomposition Rule).** *A decomposition rule is an implication of the form:*

$$Decompose(\mathcal{A}, \tau) = [\mathcal{A}_j | 1 \leq j \leq m] \wedge \bigwedge_{j=1}^{m} \mathcal{A}_j[?\boldsymbol{x}]$$
$$\implies \mathcal{A}[h(\epsilon?\boldsymbol{x}.\, \mathcal{A}_1(?\boldsymbol{x}), \ldots, \epsilon?\boldsymbol{x}.\, \mathcal{A}_m(?\boldsymbol{x}))/\boldsymbol{z}]$$

*where:*

- *$[\mathcal{A}_j | 1 \leq j \leq m]$ is a form of list composition, that we have invented, which is analogous to set comprehension (as used in Definition 1 for instance).*

---

[6] https://en.wikipedia.org/wiki/Information_retrieval (accessed 4.7.18).

- **$h$** *is the inference operation that takes the values $\epsilon?x$. $\mathcal{A}_j(?x)$ assigned to the projection variables $?x$ of the child alists $\mathcal{A}_j$ and calculates the value $h(\epsilon?x\ \mathcal{A}_1(?x), \ldots, \epsilon?x\ \mathcal{A}_m(?x))$ of the operands $z$ of the parent alist $\mathcal{A}$.*
- *Decompose is a function that takes the parent alist $\mathcal{A}$ and the type of decomposition $\tau$ and returns a list of $m$ child alists $\mathcal{A}_j$. A list, rather than a set, is required here, as the order of the arguments to $h$ must be specified. Vectors would also work.*
- *Note that the implication is from left to right: the values of the projection variables of the child alists determine the values of the operands of the parent alist. But FRANK works backwards to build the proof tree from the goal alist to the leaf node alists, whose projection variable values are then looked up on the Web.*

Different decomposition rules can be generated by varying the definition of *Decompose*. For instance, Geospatial decomposition uses the *partOf* hierarchies in various KBs to partition the value $s$ of the *Subject* attribute in $\mathcal{A}$. Currently, FRANK only uses this for breaking geographical regions into parts. It could equally well be applied to break a product into its components, e.g., to identify the most costly component.

Similarly, *isa* hierarchies could be used to identify the sub-types of an *Object*. This would be useful, say, to find the cheapest laptop meeting some minimal conditions on speed, memory capacity, etc.

We are currently exploring the space of potential decomposition rules and the applications they make possible.

## 10.2   Qualitative Uncertainty

CoVs provide a good method of assigning uncertainty to real-valued query answers and intermediate values used in their calculation. We plan to extend FRANK to non-numeric queries. Currently, FRANK is limited to non-numeric queries that involve only numeric calculations during aggregation, but with a final non-numeric answer, e.g., returning those members of a set that attain either a maximum or a minimum value on a particular numeric attribute. For these, we can use the CoV associated with the calculation that this value is indeed the extreme one. For instance, if the question is: "Which country will have the largest population in Africa in 2021?", then, although the answer will be a particular African country, we can assign to that answer the CoV associated with the calculation that its population is a maximum among the set of all African countries.

In this case, all the aggregation operations involved in the proof tree were arithmetic ones. We want to investigate how uncertainty values might be aggregated for the values of non-numeric projection variables. We will probably need a new uncertainty measure, as CoVs are associated with Gaussian distributions, which are fundamentally numeric. We will need to combine these new uncertainty measure with CoVs. We then need to associate appropriate aggregation operations to apply to non-numeric projection variables.

## 11   Conclusion

We have described the FRANK query answering system, which draws inferences from information on the Web to discover new information, including make predictions. FRANK is focused on numerical questions.

- The Web contains a huge and rapidly growing source of information. Despite the inherent uncertainty in this information, it is a source we can't afford to ignore.
- Merely retrieving known facts from the Web is to neglect most of its potential. We must infer new information from old. This is a job for automated reasoning.
- But this job raises a new range of challenges for the automated reasoning field.
  - It is necessary to locate the axioms needed from this huge store. FRANK's top-down proof search identifies the kind of axioms it needed, so that information retrieval can be used to find them.
  - The information we need is stored in a diverse number of formats. In order for automated reasoning to combine information in diverse formats, these must all be curated into a common format. FRANK uses alists, as they assimilate all the other formats.
  - Some source formats are overly restrictive, e.g., only allowing unary or binary predicates. Additional attribute values are often needed, e.g., time and units. Curation must also include finding these additional attribute values, so that they can match values in goal alists.
  - The inherent uncertainty in both knowledge sources and some inference methods must be inherited back up through the proof tree to provide the user with an uncertainty estimation for the answer that FRANK returns. FRANK propagates coefficients of variation: the standard deviation of the answer normalised by the mean. CoVs provide an error bar on the answer, which is returned as the mean. The propagated CoV is converted back to a standard deviation for the final answer, as this provides a numeric range in which the true answer is likely to fall.
- A user friendly interface is required for users to pose queries and receive answers. FRANK allows uses to pose questions in a restricted grammar of English.
- With these new challenges come exciting new opportunities.
  - Information retrieval is freed from simple factoid look-up, and can infer new information—even making predictions.
  - Inference can combine deduction, arithmetic and statistics. FRANK's evaluation shows that this combination of inference methods can both accurately estimate novel information and assign a reliable uncertainty estimate to it.

# References

1. Bundy, A., Sasnauskas, G., Chan, M.: Solving guesstimation problems using the semantic web: four lessons from an application. Semant. Web, 1–14, 10 October 2013
2. Clifford, A.A.: Multivariate Error Analysis: A Handbook of Error Propagation and Calculation in Many-Parameter Systems. Applied Science Publishers, London (1973)
3. Ferrucci, D., et al.: Building Watson: an overview of the DeepQA project. AI Mag. **31**(3), 59–79 (2010)
4. Green, C.C., Raphael, B.: The use of theorem-proving techniques in question-answering systems. In: Proceedings of the 1968 23rd ACM National Conference, pp. 169–181. ACM (1968)
5. Honnibal, M., Johnson, M.: An improved non-monotonic transition system for dependency parsing. In: Proceedings of the 2015 Conference on Empirical Methods in Natural Language Processing, pp. 1373–1378. Association for Computational Linguistics, Lisbon, September 2015. https://aclweb.org/anthology/D/D15/D15-1162
6. Liu, H., Singh, P.: ConceptNet - a practical commonsense reasoning tool-kit. BT Technol. J. **22**(4), 211–226 (2004)
7. Miller, G.A.: WordNet: a lexical database for English. Commun. ACM **38**(11), 39–41 (1995)
8. Nuamah, K., Bundy, A.: Calculating error bars on inferences from web data. In: Intelligent Systems Conference. IEEE (2018)
9. Nuamah, K., Bundy, A., Lucas, C.: Functional inferences over heterogeneous data. In: Ortiz, M., Schlobach, S. (eds.) RR 2016. LNCS, vol. 9898, pp. 159–166. Springer, Cham (2016). https://doi.org/10.1007/978-3-319-45276-0_12
10. Singhal, A.: Introducing the knowledge graph: things, not strings. Official Google Blog, May 2012
11. Vatant, B., Wick, M.: Geonames ontology (2012)
12. Vrandečić, D., Krötzsch, M.: Wikidata: a free collaborative knowledgebase. Commun. ACM **57**(10), 78–85 (2014)

# Methodologies of Symbolic Computation

James Davenport$^{(\boxtimes)}$ (iD)

University of Bath, Bath, UK
J.H.Davenport@bath.ac.uk
http://staff.bath.ac.uk/masjhd

**Abstract.** The methodologies of computer algebra are about making algebra (in the broad sense) *algorithmic*, and *efficient* as well. There are ingenious algorithms, even in the obvious settings, and also mechanisms where problems are translated into other (generally smaller) settings, solved there, and translated back. Much of the efficiency of modern systems comes from these translations. One of the major challenges is *sparsity*, and the complexity of algorithms in the sparse setting is often unknown, as many problems are NP-hard, or much worse.

In view of this, it is argued that the traditional complexity-theoretic method of measuring progress has its limits, and computer algebra should look to the work of the SAT community, with its large families of benchmarks and serious contests, for lessons.

**Keywords:** Computer algebra · Benchmarking

## 1 Introduction

Symbolic Computation (also called Computer Algebra) is exactly what it says: getting computers to do algebra. Having said that, this simple phrase needs elaboration.

### 1.1 "Computers"

The moment a mathematician examines a modern computer, the mathematician observes two mismatches, symbolised as $\mathtt{int} \neq \mathbf{Z}$ and $\mathtt{float} \neq \mathbf{R}$ (we might as well have said $\mathtt{double}$: the issue is with any finite precision). Both are immediate from the fact that the computer types have a finite number of values, and the mathematical sets an infinite number. It is superficially tempting to hope that, though finite, there are "enough" values, but it is very rapidly seen that both are insufficient for many non-trivial calculations, either for the answer or, much more common, for intermediate results.

Slightly less obvious is the fact that $\mathtt{double}$ does not obey the laws of arithmetic: $(1 + 10^{20}) - 10^{20} \xrightarrow{\mathtt{double}} 0$ whereas $1 + (10^{20} - 10^{20}) \xrightarrow{\mathtt{double}} 1$. Hence the use of floating-point arithmetic in computer algebra is relatively rare, except in the area of lattice reduction [29], and even here substantial care is needed.

Some other references to floating point in computer algebra are these: [23, 33, 34, 36].

© Springer Nature Switzerland AG 2018
J. Fleuriot et al. (Eds.): AISC 2018, LNAI 11110, pp. 19–33, 2018.
https://doi.org/10.1007/978-3-319-99957-9_2

## 1.2 "to do"

While human beings know a variety of mathematical algorithms, e.g. long multiplication or finding the solutions of a quadratic, much manual algebraic manipulation is not algorithmic, proceeding via a series of possibilities. Examples of this are the factorisation of polynomials, even in one variable. Factoring a quadratic is algorithmic, factoring a cubic $ax^3 + bx^2 + cx + d$ reduces to looking for a linear factor $a' \pm d'$ where $a'$ is a factor of $a$ and $d'$ a factor of $d$. But most people, even most professional mathematicians, will resort to a bunch of heuristics when given a quartic such as $x^4 + 2x^3 - x^2 + 2x + 1$, and may fail to spot the factor of $x^2 + 3x + 1$. For higher degrees, such as $x^5 - 2x^4 + 8x^3 + 3x^2 + 6x - 4 = (x^3 - 3x^2 + 10x - 4) \cdot (x^2 + x + 1)$, the problem is probably out of range for most people.

The mathematician knows various tests, such as the Schönemann–Eisenstein test [17] which can prove *some* polynomials irreducible, and shifting the polynomial by $x \mapsto x + a$ for suitable $a$ can prove *more* irreducible polynomials to be so, but not all [37]. Hence the output of the typical mathematician would be

three-valued: $\begin{cases} \text{factors} \\ \text{irreducible} \\ \text{don't know} \end{cases}$ .

## 1.3 "Algebra"

Many problems of mathematics are obviously algebra: solving systems of equations, or factoring polynomials for example. Ever since Descartes, many problems of geometry can be expressed in terms of algebra, and indeed "Algebraic Geometry" is a flourishing subject. "Geometric Theorem Proving" is a flourishing branch of computer algebra, even though one has to be careful and insert "non-trivial" or "non-degenerate" in many standard theorems for them to be true [41].

It is also possible to regard much of traditional "calculus" as algebra. We define differentiation (denoted by postfix $'$) as a linear operator (thereby converting the "sum rule" and "product rule" from theorems to axioms) from $K[x]$ to itself with $k' = 0$ for $k \in K$ and $x' = 1$. The elements of $K$ are called *constants*. The operator $'$ then extends uniquely to $K(x)$ and its algebraic closure. We then *define* the logarithm by $(\log \theta)' = \frac{\theta'}{\theta}$ (which only defines it up to an additive constant) and exponentials by $(\exp \theta)' = \theta' \exp \theta$ etc. The "calculus" question "integrate $f$" (where $dx$ is implicit in our definition of differentiation) then becomes the "algebra" question "find $F$ with $F' = f$". So where should we look for $F$? This question also illuminates the difference between two answers to the old chestnut "integrate $e^{-x^2}$" to which the answers are either "it has no integral" or "it's continuous, so has an integral". More formally, these answers are "there is no $F$ in the class of formulae generated from $\mathbf{Q}(x)$ by algebraic closure, adding logarithm and exponentials (in any order) such that $F' = e^{-x^2}$" and "define $F$ such that $F' = e^{-x^2}$ and then $F$ is clearly the answer".

Hence the output of the typical mathematician would be three-valued:

$$\begin{cases} \text{I've found } F \\ \text{I can't find } F \text{ (but it might exist)}. \\ \text{Someone has proved there's no } F \end{cases}$$

# 2 "Straightforward" Methods

As we have said, we do know some complete algorithms, e.g. multiplying two polynomials. We know more, but when we come to program them, and therefore run them on larger examples than we do by hand, we find some surprises.

## 2.1 Greatest Common Divisors

We can compute the greatest common divisor of two polynomials (over $\mathbf{Q}(x)$, say) by the obvious extension of Euclid's algorithm: keep dividing the larger polynomial by the smaller, and replacing it by the remainder.

But this algorithm is more costly than might be expected in practice: consider the following two polynomials[1]:

$$A(x) = x^8 + x^6 - 3x^4 - 3x^3 + 8x^2 + 2x - 5; \tag{1}$$
$$B(x) = 3x^6 + 5x^4 - 4x^2 - 9x - 21. \tag{2}$$

The first elimination gives $A - \left(\frac{x^2}{3} - \frac{2}{9}\right)B$, that is

$$\frac{-5}{9}x^4 + \frac{127}{9}x^2 - \frac{29}{3}, \tag{3}$$

and the subsequent eliminations give

$$\frac{50157}{25}x^2 - 9x - \frac{35847}{25}$$
$$\frac{93060801700}{1557792607653}x + \frac{23315940650}{173088067517}$$

and, finally,

$$\frac{76103000073384 7895048691}{8660312813046 7228900}.$$

Since this is a number, it follows that no *polynomial* can divide both $A$ and $B$, i.e. that $\gcd(A, B) = 1$.

Just clearing fractions gives an algorithm that ends up with

$$7436622422540486538114177255855890572956445312500.$$

It turns out that a careful analysis can predict cancellations, and produce the *subresultant g.c.d. algorithm* [14] which ends with 1954124052188—more reasonable, but still larger than we'd expect. Modern computer algebra does better: see Sect. 4.

---

[1] This analysis is mostly taken from [9,10], but with one change (originally an error, but it makes the point better): $-21$ instead of $+21$ for the trailing coefficient of $B$.

## 2.2   Gaussian Elimination

It is easier to see the meaning of "can predict" in this setting. Consider a matrix $M$ whose elements are $m_{i,j}$. Assuming that $m_{1,1} \neq 0$ we can use row 1 to eliminate the rest of column 1 to get a matrix $M^{(1)}$ whose elements are

$$m_{i,j}^{(1)} = \frac{m_{1,1}m_{i,j} - m_{i,1}m_{1,j}}{m_{1,1}}, \tag{4}$$

or, if we eliminate fractions,

$$m^{(1')} = m_{1,1}m_{i,j} - m_{i,1}m_{1,j}. \tag{5}$$

Assuming that $m_{2,2}^{(1')} \neq 0$ we can use row 2 to eliminate the rest of column 2 to get a matrix $M^{(2')}$ whose elements are

$$m_{i,j}^{(2)} = \frac{m_{2,2}^{(1')}m_{i,j}^{(1')} - m_{i,2}^{(1')}m_{2,j}^{(1')}}{m_{1,1}^{(1')}}, \tag{6}$$

or, if we eliminate fractions,

$$m_{i,j}^{(2')} = m_{2,2}^{(1')}m_{i,j}^{(1')} - m_{i,2}^{(1')}m_{2,j}^{(1')}. \tag{7}$$

Substituting (5) into (7) to get an expression for $m_{i,j}^{(2')}$ in terms of the original $m_{i,j}$ gives a quartic. This is slightly odd, as $m_{3,3}^{(2')}$ ought to be related to the determinant of the $3 \times 3$ top-left corner, which is a cubic, and in fact what we get if we substitute (4) into (6) is $\begin{vmatrix} m_{1,1} & m_{1,2} & m_{1,j} \\ m_{2,1} & m_{2,2} & m_{2,j} \\ m_{i,1} & m_{i,2} & m_{1,j} \end{vmatrix} \Big/ \begin{vmatrix} m_{1,1} & m_{1,2} \\ m_{2,1} & m_{2,2} \end{vmatrix}$, a $\frac{\text{cubic}}{\text{quadratic}}$ quotient of determinants, and the $m_{1,1}$ has cancelled.

This might seem like a minor improvement from quartic to cubic, but in fact these cascade, and, after eliminating $n$ columns, the expressions are $m_{i,j}^{(n)} = \frac{\text{degree } n+1}{\text{degree } n}$, while $m_{i,j}^{(n')}$ has degree $2^n$. We do not need to work with fractions, though: we can indeed work with (5), (7) and their analogues, as it can be proved [5,20] that $m_{n-1,n-1}^{((n-1)')}$ always divides the $m_{i,j}^{(n')}$, and the adjusted $m_{i,j}^{(n')}$ have degree $n+1$.

## 2.3   Greatest Common Divisors via Matrices

The Euclidean algorithm of Sect. 2.1 applied to $f = \sum_{i=0}^{n} a_i x^i$, $g = \sum_{j=0}^{m} b_j x^j$ can be viewed as Gaussian elimination in the Sylvester matrix

$$\mathrm{Syl}(f,g) = \begin{pmatrix} a_n & a_{n-1} & \cdots & a_1 & a_0 & 0 & 0 & \cdots & 0 \\ 0 & a_n & a_{n-1} & \cdots & a_1 & a_0 & 0 & \cdots & 0 \\ \vdots & \ddots & \ddots & \ddots & \cdots & & \ddots & \ddots & \vdots \\ 0 & \cdots & 0 & a_n & a_{n-1} & \cdots & a_1 & a_0 & 0 \\ 0 & \cdots & 0 & 0 & a_n & a_{n-1} & \cdots & a_1 & a_0 \\ b_m & b_{m-1} & \cdots & b_1 & b_0 & 0 & 0 & \cdots & 0 \\ 0 & b_m & b_{m-1} & \cdots & b_1 & b_0 & 0 & \cdots & 0 \\ \vdots & \ddots & \ddots & \ddots & \cdots & & \ddots & \ddots & \vdots \\ 0 & \cdots & 0 & b_m & b_{m-1} & \cdots & b_1 & b_0 & 0 \\ 0 & \cdots & 0 & 0 & b_m & b_{m-1} & \cdots & b_1 & b_0 \end{pmatrix}$$

where there are $m$ lines constructed with the $a_i$, $n$ lines constructed with the $b_i$. Complications arise when, as in deducing (3) from (1) and (2), the degree drops more than expected, which is equivalent to needing to pivot in the Gaussian elimination formulation, which is why the sub-resultant algorithm is more complicated than just "divide by the previous leading coefficient", which would be the obvious translation of Sect. 2.2.

# 3    Less Straightforward Algorithms

Computer algebra has been very successful at finding algorithms where none were previously known (or at least made explicit).

## 3.1    Gröbner Bases

We saw in Sect. 2.2 how to solve linear equations by a smarter version of Gaussian elimination. What about nonlinear (but polynomial) equations? We can sometimes apply linear algebra: given the three equations

$$x^2 - y = 0 \qquad x^2 - z = 0 \qquad y + z = 0,$$

we can subtract the first from the second to get $y - z = 0$, hence $y = 0$ and $z = 0$, and we are left with $x^2 = 0$, so $x = 0$, albeit with multiplicity 2. Given the two equations

$$x^2 - 1 = 0 \qquad xy - 1 = 0, \tag{8}$$

there might seem to be no row operation available. But in fact we can subtract $x$ times the second equation from $y$ times the first, to get $x - y = 0$. Hence the solutions are $x = \pm 1$, $y = x$. In this case, it's easy enough to verify that these actually are solutions, and are all the solutions, but in general what are we to do? So far this looks rather like the heuristics we saw at the end of Sects. 1.2 and 1.3. It was the genius of Buchberger [11] to translate this into a complete algorithm, and he and many others have produced a complete and effective theory, which is well-treated elsewhere (e.g. [18]).

However, a major problem can be the size of the produced Gröbner bases: it is well known that these can be doubly-exponential in the number of variables [26], even if the ideal is radical (has no multiple components) [13]. An alternative is the method of triangular sets, or more precisely regular chains [22], which, instead of producing a single base, produces a set of regular chains that between them describe all the solutions, and this in only single exponential time [3], though it has to be said that the distinction between $d^{2^n}$ and $d^{5n^3}$ only manifests itself for $n > 14$: currently totally impracticable. It is also not clear how rare the bad cases are for either algorithm.

### 3.2  Equations etc. Over the Reals

Working over the reals is generally described as a problem of *quantifier elimination*: given a statement such as $\exists x : x^2 - y = 0$, produce the equivalent unquantified statement. That this is possible at all is non-trivial [38, 1948, but proved in 1930]. In this case the corresponding statement is $y \geq 0$, which shows that we must allow inequalities in the class of expressions we are talking about. The first practical algorithm was due to Collins [15], and there have been many developments since. Originally independent of the theory in Sect. 3.1, it can now use Gröbner bases [21] or the theory of triangular sets [12].

### 3.3  Integration etc.

The question left over at the end of Sect. 1.3 was that of proving that there was no integral in the allowed class of formulae if we couldn't find one. In fact this theory pre-dates computer algebra, and was basically discovered by Liouville [25], but largely forgotten until computer algebra showed the need for it. For the transcendental elementary functions (i.e. rational functions plus exp and log as defined in Sect. 1.3) the theory is complete, and implemented in many algebra systems. If we also allow algebraic functions, the theory is now complete, but quite complicated and implementations tend to be partial.

There are various extensions to, for example, error functions, but we are lacking a comprehensive extension. Hence the result of the computer algebra system should be three-valued:
$$\begin{cases} \text{I've found } F \\ \text{I have proved there's no } F \\ \text{I can't find } F \text{ (but it might exist)} \end{cases}, \text{ where the}$$
third answer happens when we have an integrand outside the class of functions for which the system has a complete algorithm. Unfortunately the user interfaces of today seem to be unable to distinguish the last two, returning an unevaluated integral in either case. There is surely an argument for some sort of semantic annotation to distinguish the two.

## 4   Modular Methods

These were first invented [9, 10] to solve the greatest common divisor (Sect. 2.1) problem, that we seem to need larger numbers "than are reasonable".

## 4.1   G.c.d. of Univariate Polynomials

If we go back to the example of Sect. 2.1, and compute the g.c.d of $A$ and $B$, but reduce them modulo 5 to $A_5$ and $B_5$, and work modulo 5, we rapidly deduce that $\gcd_5(A_5, B_5) = 1$. Now if $C$ divides $A$ and $B$, $C_5$ divides $A_5$ and $B_5$, so $C_5 = 1$. But the leading coefficient of $C$ can't be a multiple of 5, so in fact $C = 1$.

So far this is a neat trick: to turn it into an algorithm requires (at least) answering these questions.

Q1. Will it always work? The answer in fact is no: consider replacing 5 by 2 above, when $\gcd_2(A_2, B_2) = x - 1$.
Q2. What about leading coefficients in general?
Q3. What if the g.c.d. is non-trivial?
Q4. How big a prime should I take?

The answers to these questions are in fact quite simple.

A1. It works for any prime not dividing $\det(\mathrm{Syl}(A/C, B/C))$ where $C$ is the g.c.d., i.e. all but a finite number of primes. For these bad primes you get too great an answer (so it can't pretend to divide $A$ and $B$ over the integers), never too small an integer.
A2. You shouldn't use a prime that divides *both* leading coefficients—though in fact many implementors avoid primes that divide either.
A3. The problem is that we only know the g.c.d. up to a multiple modulo the prime $p$. For $p = 5$, an answer of $x + 1$ might actually be $2(x + 1) = 2x - 1$ etc. If lc stands for "leading coefficient" we know that $\gcd(A, B) | A$, so $\mathrm{lc}(\gcd(A, B)) | \mathrm{lc}(A)$, so $\mathrm{lc}(\gcd(A, B)) | \gcd(\mathrm{lc}(A), \mathrm{lc}(B))$. Hence we multiply $A$ and $B$ by $\gcd(\mathrm{lc}(A), \mathrm{lc}(B))$, and look for a common factor whose leading coefficient is precisely $\gcd(\mathrm{lc}(A), \mathrm{lc}(B))$, then sort out the integer content (the g.c.d. of all the coefficients) later.
A4. If the true g.c.d. is $x+7$, but I choose the prime 5, I'll think the g.c.d. is $x+2$ not $x+7$. It is tempting to think that the coefficients of $\gcd(A, B)$ can't be larger than those of $A$, $B$, but the following example shows otherwise:

$$
\begin{aligned}
A = \quad & x^5 + 3x^4 + 2x^3 - 2x^2 - 3x - 1 & = (x+1)^4(x-1); \\
B = x^6 + {} & 3x^5 + 3x^4 + 2x^3 + 3x^2 + 3x + 1 & = (x+1)^4(x^2 - x + 1); \\
\gcd(A, B) = \quad & x^4 + 4x^3 + 6x^2 + 4x + 1 & = (x+1)^4.
\end{aligned}
$$

Nevertheless there is a bound [27]. In the case of the example of Sect. 2.1, this Landau–Mignotte bound is 510.2, so we can take any prime greater than twice this (to allow for $\pm$ coefficients), e.g. 1021.

We may still feel that 1021 is too big (and there are reasons later on why we should). But clearly if we use primes smaller than the coefficients in the answer we can't get the right answer *from a single prime*. The solution is to use several primes and the following result.

**Theorem 1 (Chinese Remainder Theorem).** *Two simultaneous congruences $X \equiv a \pmod{M}$ and $X \equiv b \pmod{N}$, where $M$ and $N$ are relatively prime, are precisely equivalent to one congruence $X \equiv c \pmod{MN}$, and $c$ can be efficiently calculated as $a + \lambda M(b - a)$ where $\lambda M + \mu N = 1$.*

Hence we can take several primes $p_i$, whose *product* is greater than twice the Landau–Mignotte bound, and use Theorem 1 to combine the g.c.d.s modulo the $p_i$ into one modulo $\prod p_i$. If the $p_i$ disagree about the degree of the g.c.d., A1 says that we must discard the $p_i$ giving higher degrees. They are definitely wrong: the remainder *may* still be wrong, so we always need to check that the answer does divide *both* of the inputs, and if not, try again with different primes.

If we assume that the coefficients of the answer are in fact no larger than the input coefficients, the complexity is $O(n^3)$, where $n$ is the greater of the degrees, and various tricks can make this $O(n^2)$.

## 4.2   G.c.d. of Bivariate Polynomials

Consider two polynomials $A, B \in K[x, y]$. If after substituting the value $a \in K$ for $y$, we find $\gcd(A_{y=a}, B_{y=a}) = 1$, and if the leading coefficients $\mathrm{lc}_x(A), \mathrm{lc}_y(B)$ do not vanish when we substitute $y = a$, we can deduce that the polynomials are relatively prime w.r.t. $x$, i.e. any g.c.d. is in $K[y]$ only. If, however, the g.c.d is non-trivial, we can use this theorem to deduce the true result from various different evaluations.

**Theorem 2 (Chinese Remainder Theorem in $K[y]$).** *Two simultaneous congruences $X \equiv a \pmod{M}$ and $X \equiv b \pmod{N}$, where $a, b \in K[y]$ and $M$ and $N$ are relatively prime polynomials in $K[y]$, are precisely equivalent to one congruence $X \equiv c \pmod{MN}$, and $c$ can be efficiently calculated as $a + \lambda M(b - a)$ where $\lambda M + \mu N = 1$.*

To turn this into an algorithm requires (at least) answering these questions.

$Q_2 1$. Will it always work? Not always, consider what happens when $A = x - 1$, $B = x - y$, but $y = 1$.

$Q_2 2$. What about leading coefficients in general?

$Q_2 3$. How is a non-trivial g.c.d. computed?

$Q_2 4$. How many evaluations should I take?

The answers to these questions are in fact simpler than Sect. 4.1.

$A_2 1$. It works for any $a$ not a root of $\det(\mathrm{Syl}_x(A/C, B/C))$ where $C$ is the g.c.d., i.e. all but a finite number of values. For these bad values you get too great an answer, in terms of $x$-degree.

$A_2 2$. You shouldn't use a value that nullifies *both* leading coefficients—though in fact many implementors avoid values that nullify either.

$A_2 3$. The problem is that we only know the g.c.d. up to a multiple in $K[y]$. As in A3 above, we multiply $A$ and $B$ by $\gcd(\mathrm{lc}_x(A), \mathrm{lc}_x(B))$, and look for a common factor whose leading coefficient is precisely $\gcd(\mathrm{lc}_x(A), \mathrm{lc}_x(B))$, then sort out the $K[y]$ content later.

$A_2$4. This is much simpler: $\deg_y(\gcd(A,B)) \leq \min(\deg_y(A), \deg_y(B))$, so we take $1 + \min(\deg_y(A), \deg_y(B))$ good evaluations. Again, we discard evaluations that give too great an $x$-degree.

The complexity is again $O(n^2)$ where $n = \max(\deg_x(A), \deg_x(B))$ but the dependence on $\deg_y$ is messier because of the leading coefficients in step $A_2$3.

## 4.3   G.c.d. of Multivariate Polynomials

In principle the same methods apply to multivariate polynomials in $k$ variables. Under reasonable assumptions and suitable implementation improvements, the complexity is $O(n^{k+1})$ [10, (95)]. In practice there are significant challenges for large multivariate polynomials.

1. Even if $A$ and $B$ are relatively prime, $\gcd(\mathrm{lc}_x(A), \mathrm{lc}_x(B))$ may be an expensive computation. We can check for $\gcd(A,B) = 1$ before computing $\gcd(\mathrm{lc}_x(A), \mathrm{lc}_x(B))$, but then there are cases when $\gcd(A,B)$ is small but not 1.
2. if $\gcd(\mathrm{lc}_x(A), \mathrm{lc}_x(B))$ is large, multiplying by it may make $A$ and $B$ much larger.
3. The complexity is not much larger than the *maximal* size of a polynomial in $k$ variables with degrees bounded by $n$. However, most multivariate polynomials of high degree are sparse. This is not just a theoretical objection: the intermediate results in these g.c.d. computations can be much larger than the sparse results, even if the sparse results are of high degree.

## 4.4   Sparsity

In practice nearly all computer algebra systems use *sparse* storage, i.e. only storing the non-zero terms in an expression. However traditional complexity theory deals mostly with dense representation, i.e. counting all the zeros. This isn't just being naïve: sparsity has major challenges [19,32]. These are easy to explain with univariate examples, even though they are most blatant in the multivariate case as we have just seen. Suppose the polynomial $f$ has degree $d_f$ and $t_f$ non-zero terms (so $t_f \leq d_f + 1$). Then the basic operations have the following complexity properties.

$h := f + g$: $d_h \leq \max(d_f, d_g)$, $t_h \leq t_f + t_g$;
$h := f \times g$: $d_h = d_f + d_g$, $t_h \leq t_f \times t_g$;
$h := f/g$: $d_h = d_f - d_g$, $t_h \leq d_f - d_g + 1$—consider $\frac{x^{d_f}-1}{x-1}$;
$h := \gcd(f,g)$: $d_h \leq \min(d_f, d_g)$, $t_f \leq \max(\min(t_f, t_g), \min(d_f, d_g) - 1)$.

The last follows from the neat example of [35]:

$$\gcd(x^{pq} - 1, x^{p+q} - x^p - x^q + 1) = x^{p+q-1} - x^{p+q-2} \pm \cdots - 1.$$

Dense complexity theory measures the cost in terms of $\max(d_f, d_g)$, but we can see that sparse complexity theory has to measure the cost in terms of $\max(t_f, t_g, t_h)$. Even this isn't easy.

**Theorem 3** ([31]). *It is NP-hard (in terms of $t_f, t_g$) to determine whether* $\gcd(f, g) = 1$.

However, as [19] points out, this example involves cyclotomic polynomials (factors of $x^n - 1$), and we may be able to do better by excluding such. Hence, although we seem to have efficient modular algorithms in practice for computing the g.c.d. of sparse polynomials [43], it is difficult to make useful statements about the complexity.

## 4.5   Other Applications

Many linear algebra problems can be solved by modular methods, and this is particularly applicable if we only want the values of a subset of the variables, as only these values need to be reconstructed by the Chinese Remainder Theorem. These values may be fractions or rational functions, but it is possible to reconstruct these by modular methods [40].

Buchberger's algorithm (or any other) for Gröbner Bases (Sect. 3.1) can produce very large intermediate numbers, hence would seem to be crying out for the use of modular methods. However, the answers to key questions are less helpful.

$A_{GB}$1. It works for any primes that do not divide the denominator of any rational that would occur if we did it over the integers, hence there are only a finite number of bad primes. However, there is no simple way of deciding which of two prmes giving differently-shaped results is the bad one, i.e. no equivalent of "too great an answer" as in A1, though [4] gives a partial answer.

$A_{GB}$2. One clearly shouldn't use a prime that divides any denominator, or any leading coefficients of the input polynomials.

$A_{GB}$3. If the Gröbner Base is non-trivial, it can be reconstructed from compatible modular images by Theorem 1.

$A_{GB}$4. But there are no known *a priori* bounds equivalent to the Landau–Mignotte bounds. [4] again gives a partial answer to the related question "how can I tell when I have used enough primes".

## 5   *p*-adic (Hensel) Methods

The classic use of these methods in computer algebra is for the factorisation of polynomials over the integers/rational numbers, for which there is no "obvious" algorithm (see Sect. 1.2). Almost all polynomials are irreducible in the sense that

$$\forall d > 0 \lim_{H \to \infty} \frac{|\{\text{such polynomials that factor}\}|}{|\{\text{polynomials of degree } d, \text{ coefficients } \leq H\}|} = 0, \qquad (9)$$

so we have a constant-time (or linear if we write out the answer) almost sure algorithm that just returns "irreducible".

## 5.1   Univariate Polynomials

There are efficient algorithms for factoring polynomials modulo a prime [6,7], so why don't we apply these and the techniques of Sect. 4? The Landau–Mignotte inequality will still bound the maximum size of any coefficient of any factor, which may well be larger than the coefficients of the input polynomial: see [2] for examples and a survey of various bounds $B(f)$ on the factors of $f$.

The first difficulty is that factoring polynomials modulo a prime $p_1$ returns a *set* of factors, and modulo $p_2$ another *set* of factors, and there is no way in general of deciding which factor modulo $p_1$ corresponds to which factor modulo $p_2$—of course if the factors have different degrees, then we can use the degrees to identify matching factors, but the factors might all have the same degree. This is inevitable: the polynomials over the integers modulo $p_1$ (and modulo $p_2$) and a unique factorisation domain, but the polynomials over the integers modulo $p_1p_2$ are not. Hence using the methods of Sect. 4 will lead to a combinatorial explosion as we try all possibilities of combinations of factors modulo $p_1, p_2, \ldots$ (and we may need many $p_i$).

Another difficulty is that, with modular methods we knew (A1, A$_2$1) that there were only finitely many "bad" primes where the answer modulo $p$ did not correspond to the answer over the integers, but here every prime can be bad: for example $x^4 + 1$ is irreducible over the integers, but factors into two quadratics or more modulo every prime, and this is not an isolated example [37]. However, we should note that squarefreeness of $f$ corresponds to $\gcd(f, f') = 1$, so A1 above means that there are only finitely many bad primes for squarefreeness.

Hence the idea, due to [42] is to factor $f$ over the integers by the following procedure.

1. Ensure $f$ is squarefree by computing $\gcd(f, f')$ etc., factoring each squarefree component separately.
2. Factor $f$ modulo several "good for squarefreeness" primes.
3. Choose the best $p$ (or deduce irreducibility).
4. Lift ("Hensel's Lemma") the factorisation modulo $p$ to one modulo $p^k$ for $p^k > 2B(f)$.
5. Worry about the leading coefficients.
6. If this isn't a factorisation over the integers, combine factors modulo $p^k$ or otherwise deduce the factorisation over the integers

The rationale behind steps 2–3 is that an irreducible polynomial may well factor modulo every prime chosen[2], but those factorisations may be incompatible: for example a quintic (5) might factor as (3,2) modulo one prime, but (4,1) modulo another, and these are not compatible and imply that the polynomial must be irreducible over the integers. So what should "several" be? Traditional wisdom [28] is five, but recent research [30] suggests that the correct answer is seven.

There are various ways of conducting step 4: the simplest method is to proceed $p, p^2, p^3, \ldots, p^k$, but it might be more efficient to proceed $p, p^2, p^4, \ldots, p^{2^l} \geq$

---

[2] The probability of a "random" irreducible polynomial $f$ remaining irreducible modulo $p$ is $1/\deg(f)$.

$2B(f)$, or in a hybrid fashion [1]. Step 5 is an equivalent of A3 for g.c.d.s: we only know the factors up to units modulo $p^k$, and, if we choose the "wrong" versions these may not correspond to factors over the integers. The solution is similar: if there are $\ell$ factors being lifted, we multiply $f$ by $\mathrm{lc}_x(f)^{\ell-1}$ and insist that every factor have leading coefficient the original $\mathrm{lc}_x(f)$.

Step 6 is potentially the most expensive. The original solution was to try each subset of the $\ell$ factors, hence possibly $2^{\ell-1}$ sets (we do not need to try a set and its complement). The famous LLL algorithm was invented [24] to avoid this: they took a factor $g$ of $f$ modulo $p^{k'}$ and produced the irreducible factor of $f$ over the integers which is divisible by $g$ modulo $p^{k'}$. While polynomial time, this was expensive in practice, not least because $k'$ was substantially greater than $k$. There have been many improvements since, which are too technical to go into here.

## 5.2  Bivariate Polynomials

The process for factoring polynomials in $K[x, y]$ is similar to Sect. 5.1, as Sect. 4.2 is to Sect. 4.1: we replace working modulo $p$ by evaluations $y = a$, and we use a variant of Hensel's Lemma to lift the factorisation of $f_{y=a}$ (which is a factorisation of $f$ modulo $y - a$) to one modulo $(y - a)^2$, $(y - a)^3$, etc. (quadratic lifting is also possible, but rarely used). Again, there is no complicated question of bounds: it suffices to work modulo $(y - a)^{\deg_f(f)+1}$.

A slight simplification is that Hilbert's Irreducibility Theorem means that there are infinitely many $a$ such that the evaluation $y = a$ preserves the factorisation of $f$. Nevertheless, there may also be infinitely many $y = a$ that do not, so it is still necessary to follow steps 2–3 and take several evaluations. Implementors tend to use five by analogy with [28], though this has no strong theoretical foundation.

## 5.3  Multivariate Polynomials

Again this generalises in principle to multivariate polynomials. In practice there are two major stumbling blocks. One is sparsity, as with the g.c.d. problem. Though there has been much research in this area, it is probably fair to say that a really good solution eludes us.

The other is the analogy of step 5, where we multiply $f$ by $\mathrm{lc}_x(f)^{\ell-1}$ if there are $\ell$ factors being lifted. This tends to make $f$ both much larger and much denser. There is an ingenious solution in [39], but again it has a price to pay in terms of losing sparsity.

## 5.4  Sparsity

Again, Plaisted's results [31] show that, even for univariates, many factorisation problems for sparse polynomials are NP-hard, and examples such as the factorisation of $x^n - 1$ for highly composite $n$ show that the output can be much larger than a sparse input. There are even examples [16] that show that just the square-free decomposition can be arbitrarily larger than the input.

## 5.5   Other Uses of *p*-adic Methods

For simplicity we describe the univariate version, though the multivariate is similar. It is possible to compute g.c.d.s with Hensel's Lemma: if $g_p = \gcd_p(A_p, B_p)$, then we have that $A_p = g_p h_p$ and we ought to be able to lift this (after imposing leading coefficients as in Sect. 4.1) to $A = gh$ over the integers. Hensel's Lemma only applies when $g_p$ and $h_p$ are relatively prime, though, and this might not be the case. However, for random integers $\lambda, \mu$, we will with high probability have $\lambda A_p + \mu B_p = g_p h_p$ with $g_p$ and $h_p$ relatively prime, so we lift this.

# 6   Methodologies of Reporting Research

So far we have spoken about the methodologies computer algebra uses—improvements of straightforward algorithms (Sect. 2), advanced algorithms still working in the same setting (Sect. 3), working modulo primes and reconstructing (Sect. 4) or working modulo a prime and then its powers (Sect. 5). There are also questions of how research proceeds. There is little doubt that, in the minds of most researchers, the ideal paper consists of a problem statement, a new algorithm, a complexity analysis and a few validating examples. There are many such great papers [10, 14, 24].

However, these complexity results tend to be in the dense setting, while most practical work is done in the sparse setting. In that setting many of these problems are NP-hard (Sects. 4.4, 5.4), or have exponential or doubly-exponential complexity (end of Sect. 3.1), though possibly only on rare cases. The SAT community, and its developments such as SMT, have wrestled with the problem of measuring efficiency and progress when solving NP-hard problems, and their solution is very different: lots of benchmarks and genuine contests. It is at least arguable (and is argued in [8]) that the computer algebra community ought to move further in this direction. However, because of (9), such benchmarks can't be chosen "at random", and there is currently very little community consensus on benchmarks.

**Acknowledgments.** I am grateful to Russell Bradford, Akshar Nair, Zak Tonks and the AISC 2018 organisers for useful comments and corrections. As always, I am grateful to David Carlisle for TeXnical advice.

# References

1. Abbott, J.A.: Factorisation of polynomials over algebraic number fields. Ph.D. thesis, University of Bath (1988)
2. Abbott, J.A.: Bounds on factors in **Z**[*x*]. J. Symb. Comp. **50**, 532–563 (2013)
3. Amzallag, E., Pogudin, G., Sun, M., Vo, N.T.: Complexity of triangular representations of algebraic sets. https://arxiv.org/abs/1609.09824v6 (2018)
4. Arnold, E.A.: Modular algorithms for computing Gröbner bases. J. Symb. Comp. **35**, 403–419 (2003)

5. Bareiss, E.H.: Sylvester's identity and multistep integer-preserving Gaussian elimination. Math. Comp. **22**, 565–578 (1968)
6. Berlekamp, E.R.: Factoring polynomials over finite fields. Bell. Syst. Tech. J. **46**, 1853–1859 (1967)
7. Berlekamp, E.R.: Factoring polynomials over large finite fields. Math. Comp. **24**, 713–735 (1970)
8. Brain, M.N., Davenport, J.H., Griggio, A.: Benchmarking solvers, SAT-style. In: SC$^2$ 2017 Satisfiability Checking and Symbolic Computation CEUR Workshop 1974, no. RP3, pp. 1–15 (2017)
9. Brown, W.S.: On Euclid's algorithm and the computation of polynomial greatest common divisors. In: Proceedings of SYMSAC 1971, pp. 195–211 (1971)
10. Brown, W.S.: On Euclid's algorithm and the computation of polynomial greatest common divisors. J. ACM **18**, 478–504 (1971)
11. Buchberger, B.: Ein Algorithmus zum Auffinden des Basiselemente des Restklassenringes nach einem nulldimensionalen Polynomideal. Ph.D. thesis, Math. Inst. University of Innsbruck (1965)
12. Chen, C., Moreno Maza, M.: Quantifier elimination by cylindrical algebraic decomposition based on regular chains. J. Symb. Comp. **75**, 74–93 (2016)
13. Chistov, A.L.: Double-exponential lower bound for the degree of any system of generators of a polynomial prime ideal. St. Petersb. Math. J. **20**, 983–1001 (2009)
14. Collins, G.E.: Subresultants and reduced polynomial remainder sequences. J. ACM **14**, 128–142 (1967)
15. Collins, G.E.: Quantifier elimination for real closed fields by cylindrical algebraic decompostion. In: Brakhage, H. (ed.) Automata Theory and Formal Languages 2nd GI Conference Kaiserslautern. LNCS, vol. 33, pp. 134–183. Springer, Heidelberg (1975). https://doi.org/10.1007/3-540-07407-4_17
16. Coppersmith, D., Davenport, J.H.: Polynomials whose powers are sparse. Acta Arith. **58**, 79–87 (1991)
17. Cox, D.A.: Why Eisenstein proved the Eisenstein criterion and why Schönemann discovered it first. Am. Math. Monthly **118**, 3–31 (2011)
18. Cox, D.A., Little, J., O'Shea, D.: Ideals, Varieties, and Algorithms. Undergraduate Texts in Mathematics. Springer, Heidelberg (2015). https://doi.org/10.1007/978-0-387-35651-8
19. Davenport, J.H., Carette, J.: The sparsity challenges. In: Watt, S., et al. (eds.) Proceeding of SYNASC 2009, pp. 3–7 (2010)
20. Dodgson, C.L.: Condensation of determinants, being a new and brief method for computing their algebraic value. Proc. R. Soc. Ser. A **15**, 150–155 (1866)
21. England, M., Davenport, J.H.: The complexity of cylindrical algebraic decomposition with respect to polynomial degree. In: Gerdt, V.P., Koepf, W., Seiler, W.M., Vorozhtsov, E.V. (eds.) CASC 2016. LNCS, vol. 9890, pp. 172–192. Springer, Cham (2016). https://doi.org/10.1007/978-3-319-45641-6_12
22. Kalkbrener, M.: A generalized Euclidean algorithm for computing triangular representations of algebraic varieties. J. Symb. Comp. **15**, 143–167 (1993)
23. Kaltofen, E., Li, B., Yang, Z., Zhi, L.: Exact certification of global optimality of approximate factorizations via rationalizing sums-of-squares with floating point scalars. In: Jeffrey, D.J. (ed.) Proceedings of ISSAC 2008, pp. 155–164 (2008)
24. Lenstra, A.K., Lenstra Jun, H.W., Lovász, L.: Factoring polynomials with rational coefficients. Math. Ann. **261**, 515–534 (1982)
25. Liouville, J.: Premier Mémoire sur la Détermination des Intégrales dont la Valeur est Algébrique. J. l'École Polytech. **14**(22), 124–148 (1833)

26. Mayr, E.W., Ritscher, S.: Dimension-dependent bounds for Gröbner bases of polynomial ideals. J. Symb. Comp. **49**, 78–94 (2013)
27. Mignotte, M.: An inequality about factors of polynomials. Math. Comp. **28**, 1153–1157 (1974)
28. Musser, D.R.: On the efficiency of a polynomial irreducibility test. J. ACM **25**, 271–282 (1978)
29. Nguyễn, P.Q., Stehlé, D.: An LLL algorithm with quadratic complexity. SIAM J. Comput. **39**, 874–903 (2009)
30. Pemantle, R., Peres, Y., Rivin, I.: Four random permutations conjugated by an adversary generate $S_n$ with high probability. Random Struct. Algorithms **49**, 409–428 (2015)
31. Plaisted, D.A.: Some polynomial and integer divisibility problems are $NP$-hard. SIAM J. Comp. **7**, 458–464 (1978)
32. Roche, D.S.: What can (and can't) we do with sparse polynomials? In: Proceedings of ISSAC 2018, pp. 25–30 (2018)
33. Sasaki, T., Sasaki, M.: Analysis of accuracy decreasing in polynomial remainder sequence and floating-point number coefficients. J. Inform. Proc. **12**, 394–403 (1989)
34. Sasaki, T., Yamaguchi, S.: An analysis of cancellation error in multivariate Hensel construction with floating-point arithmetic. In: Gloor, O. (ed.) Proceedings of ISSAC 1998, pp. 1–8 (1998)
35. Schinzel, A.: On the greatest common divisor of two univariate polynomials, I. In: A Panorama of Number Theory or the View from Baker's Garden, pp. 337–352. C.U.P. (2003)
36. Shirayanagi, K.: Floating point Gröbner bases. Math. Comput. Simul. **42**, 509–528 (1996)
37. Swinnerton-Dyer, H.P.F.: Letter to E.H. Berlekamp. Mentioned in [7] (1970)
38. Tarski, A.: A decision method for elementary algebra and geometry. In: Caviness, B.F., Johnson, J.R. (eds.) Quantifier Elimination and Cylindrical Algebraic Decomposition, pp. 24–84. Springer, Vienna (1998). https://doi.org/10.1007/978-3-7091-9459-1_3
39. Wang, P.S.: An improved multivariable polynomial factorising algorithm. Math. Comp. **32**, 1215–1231 (1978)
40. Wang, P.S., Guy, M.J.T., Davenport, J.H.: $p$-adic reconstruction of rational numbers. SIGSAM Bull. **16**(2), 2–3 (1982)
41. Wu, W.-T.: Basic principles of mechanical theorem proving in elementary geometries. J. Syst. Sci. and Math. Sci. (Beijing) **4**, 207–235 (1984)
42. Zassenhaus, H.: On Hensel factorization I. J. Number Theor. **1**, 291–311 (1969)
43. Zippel, R.E.: Effective Polynomial Computation. Kluwer Academic Publishers, Boston (1993)

# Artificial Intelligence, Theorem Proving and SAT Solving

# A Formal Proof of the Computation of Hermite Normal Form in a General Setting

Jose Divasón$^{(\boxtimes)}$ and Jesús Aransay

Universidad de La Rioja, Logroño, La Rioja, Spain
{jose.divason,jesus-maria.aransay}@unirioja.es

**Abstract.** In this work, we present a formal proof of an algorithm to compute the Hermite normal form of a matrix based on our existing framework for the formalisation, execution, and refinement of linear algebra algorithms in Isabelle/HOL. The Hermite normal form is a well-known canonical matrix analogue of reduced echelon form of matrices over fields, but involving matrices over more general rings, such as Bézout domains. We prove the correctness of this algorithm and formalise the uniqueness of the Hermite normal form of a matrix. The succinctness and clarity of the formalisation validate the usability of the framework.

**Keywords:** Hermite normal form · Bézout domains
Parametrised algorithms · Linear algebra · HOL

## 1 Introduction

Computer algebra systems are neither perfect nor error-free, and sometimes they return erroneous calculations [18]. Proof assistants, such as Isabelle [36] and Coq [15] allow users to formalise mathematical results, that is, to give a formal proof which is mechanically checked by a computer. Two examples of the success of proof assistants are the formalisation of the four colour theorem by Gonthier [20] and the formal proof of Gödel's incompleteness theorems by Paulson [39]. They are also used in software [33] and hardware verification [28]. Normally, there exists a gap between the performance of a verified program obtained from a proof assistant and a non-verified one. However, research in this area is filling this gap to obtain efficient and verified programs which can be used for real applications and not just restricted to toy examples [6]. Linear algebra algorithms are widely used in mathematics and computer software due to their numerous applications in various fields, such as modern 3D graphics, search engines and modern compression algorithms. In this paper, we present a formalisation of an algorithm to compute the Hermite normal form of a matrix.

This work has been supported by the projects MTM2014-54151-P and MTM2017-88804-P from Ministerio de Economía y Competitividad (Gobierno de España).

© Springer Nature Switzerland AG 2018
J. Fleuriot et al. (Eds.): AISC 2018, LNAI 11110, pp. 37–53, 2018.
https://doi.org/10.1007/978-3-319-99957-9_3

The Hermite normal form is a well-known canonical matrix that plays an important role in different fields. It can be used to solve algorithmic problems in lattices [21], cryptography [45], loop optimisation techniques [40], solution of systems of linear diophantine equations [10], and integer programming [25], among other applications.

The paper is structured as follows. We present a brief introduction to the Isabelle interactive theorem prover in Sect. 2. In Sect. 3 we describe the main features of our existing framework, where algorithms over matrices can be formalised, executed, refined, and coupled with their mathematical meaning as well as we introduce some benchmarks and improvements that we have carried out in this work. As a use case, we present in Sect. 4 the main contribution of this paper, *i.e.*, a formal proof of an algorithm to compute the Hermite normal form of a matrix in a general setting and the uniqueness of Hermite normal forms. A study of related and further work is given in Sect. 5. Finally, we show the conclusions in Sect. 6.

## 2   A Brief Introduction to Isabelle/HOL

Isabelle is a generic interactive proof assistant in which several logics are implemented. The most used of them is HOL (Isabelle/HOL), a version of classical higher-order logic similar to the one of the HOL System [5]. The Isabelle/HOL type system resembles that of functional programming languages such as Haskell [23]. There are base types (such as $bool$), function types representing total functions (*i.e.* $\Rightarrow$), type constructors (such as $list$), and type variables (such as $'a$ and $'b$).

For instance, $f :: 'a \Rightarrow 'b\ set$ indicates that $f$ is a function that maps an element of type $'a$ to a set of elements of type $'b$. Isabelle/HOL also introduces type classes in a Haskell-like manner. A type class is just a group of types with a common interface: all types in that class must provide the functions in the interface. A type class not only provides an interface, but also allows to encode properties of the types. A type $'a$ being in a class $B$ is written $'a :: B$. Since our formalisation is based on Isabelle/HOL, throughout the paper we present the theorems and definitions following its syntax. Isabelle's keywords are written in **bold**. For a complete introduction on this proof assistant we refer the reader to [37]. All our quoted developments are publicly available in the Archive of Formal Proofs (AFP) [1,17], which is a refereed repository of formal proof libraries developed in Isabelle.

## 3   Framework

In the last few years, we have completed several linear algebra developments in Isabelle/HOL [6–8]. They are based on the HOL Analysis (*HA*) library where a vector (type $vec$) is encoded as a function over a finite type (following the seminal work by Harrison [22]) and, consequently, a matrix is represented as a vector of vectors. More concretely, we developed a framework where linear algebra algorithms can be formalised, executed, refined, and coupled with their mathematical meaning. This framework includes, for instance, connection between linear maps and matrices, necessary generalisations of the HA library, formalisation of elementary operations of matrices, formalisation of the fundamental theorem of linear algebra, symbolic execution, a full library of algebraic structures (Bézout domains, principal ideal domains, GCD rings, *etc.*), connection with the Cayley-Hamilton theorem and so on.

As use cases, we implemented the Gauss-Jordan algorithm, the $QR$ decomposition and the echelon form algorithm. All of them are available in the AFP [1]. Thiemann and Yamada used the framework in their formalisation of Jordan normal forms [44] (it is worth noting that they use a different representation of vectors from the one we use, by means of functions over the natural numbers with explicit dimensions associated to them). Also Li and Paulson reused our generalisations of the HA library in their work on real algebraic numbers [32]. Some parts have also been moved to the standard Isabelle/HOL library.

### 3.1   Main Parts

The main parts of this framework are as follows:

1. Formalisation of elementary operations of matrices. We have defined them in Isabelle/HOL using the `vec` representation. For instance, we show here the definition of interchanging two rows of a matrix:

   **definition** `interchange_rows A a b = (χ i j. if i = a then A $ b $ j else if i = b then A $ a $ j else A $ i $ j)`

   In the above definition, $\chi$ denotes the morphism from functions to type `vec` and `$` is the access operator for `vec`.

2. Refinements from `vec` to executable representations (details can be found in [6]). We developed a *natural* refinement, from `vec` to functions over a finite type. We also developed a refinement to immutable arrays, or `iarray` to improve performance. An example of code lemma to transform from `vec` to `iarray` follows:

   **lemma** `[code-unfold]: matrix_to_iarray (interchange_rows A i j)`
   `= interchange_rows_iarray (matrix_to_iarray A) (to_nat i)`
   `(to_nat j)`

3. Serialisations to obtain better performance when generating code to functional programming languages (SML and Haskell). We have used two kinds of serialisations:
   - Immutable arrays (the efficient type used to represent vectors and matrices).
   - $\mathbb{Z}_2$, $\mathbb{Q}$, and $\mathbb{R}$ numbers (the types of the coefficients of the matrices).

   The latter ones are trivial. The first one in SML [4] was a part of the library. Regarding the serialisation of arrays in Haskell [2], we have serialised the iarray Isabelle/HOL datatype to the *Data.Array.IArray.array* (or shorter, *IArray.array*) constructor present in the standard Haskell library. Let us note that arrays are a natural way to represent *dense matrices*, which are the ones we are focusing in (we compute normal forms of matrices by means of elementary row operations). There exist also sparse representations of matrices in Isabelle/HOL by means of lists (see for instance the work by Obua and Nipkow [38]).

In the next subsection, we present some computational experiments that we completed and that justify our choice of immutable arrays for generating code of linear algebra algorithms from Isabelle/HOL specifications.

## 3.2  Performance

There exist different implementations of immutable arrays in Haskell, such as *IArrays* (*Data.Array.IArray.array*) or *UArrays* (*Data.Array.Unboxed.array*). In the case of the code generated from our Isabelle/HOL developments, we have empirically tested that *IArray.array* performs slightly better than *Unboxed.array*. As an example, the computation of the determinant of a $1500 \times 1500$ $\mathbb{Z}_2$ matrix by means of the code generated to Haskell from our verified Gauss-Jordan algorithm took $6.09s$ using *IArray.array* and $6.37s$ using *Unboxed.array*.[1] A more specific Haskell module for immutable arrays is *Data.Array* (where the *Data.Array.array* constructor is involved). As in the case of unboxed immutable arrays, the use of *Data.Array.array* does not imply an empirical advantage in terms of performance in our particular setup with respect to *Data.Array.IArray.array*.

We also perform some benchmarks in order to compare the performance of vec implemented as functions over finite domains, as immutable arrays, and also as lists (using an existing AFP entry about an implementation of matrices as lists of lists [42]). To do that, we define recursive functions (one for each representation: *function over finite domains*, *immutable arrays*, and *lists*) which take a rational matrix $A$ as their input, and in each iteration interchange two rows of $A + A$.

Benchmarks are carried out for $10n \times 10n$ identity matrices, $n$ being the number of iterations. Concretely, we execute the previous functions in two cases: applied to a $50 \times 50$ identity matrix with $n = 5$ and to the $100 \times 100$ identity matrix with $n = 10$. The algorithm is applied to identity matrices to minimize arithmetic time consumption. Table 1 shows the performance obtained when executing them within Isabelle/HOL (by means of the simplifier, with fully symbolic evaluation and highest confidence), and exporting code to SML by means of the command code to obtain better performance (in the second case, part of the code generation process is not verified, and needs to be trusted). Results show that the case $n = 5$ is usable in practice with any of the three representations. However, for bigger matrices, functions over finite domains become too slow.

---

[1] The Isabelle file that serialises iarray to *UArrays* is available from our website [16].

Immutable arrays outperform functions and lists in any case when exporting code, as expected. It is worth noting that inside Isabelle/HOL (but not when code is exported), `iarray` is just a wrapper of the type `list`. Thus, in the quest for performance, immutable arrays yield reasonable performance.

Focusing on our linear algebra algorithms, the performance obtained using functions over finite domains makes algorithms based on this representation unusable in practice. For instance, the computation of the Gauss-Jordan algorithm over $15 \times 15$ matrices is rather slow (several minutes). Using immutable arrays the computation is done immediately.

**Table 1.** Comparative among matrix representations.

|          |      | functions | iarray  | list    |
|----------|------|-----------|---------|---------|
| $n = 5$  | simp | 241.158s  | -       | 20.860s |
|          | code | 0.639s    | 0.159s  | 0.971s  |
| $n = 10$ | code | 827.673s  | 0.881s  | 1.824s  |

The benchmarks and the execution examples presented throughout the paper have been carried out in a laptop with an Intel® Core™ i7-4810MQ processor with 16 GiB of RAM and Ubuntu GNU/Linux 16.04. The code developed to carry out the benchmarks can be obtained online [16] and works for Isabelle 2017.

*Mutable arrays* (and imperative programming) should also be a good choice. Nevertheless, we compared the performance of using immutable arrays and mutable arrays in our formalisation of the Gauss-Jordan algorithm and obtained similar results [6].

## 4    A Formalisation of the Hermite Normal Form of a Matrix

The Hermite normal form is commonly defined for integer matrices, but it also exists for more general matrices. Following a similar approach as the one that we followed in the formalisation of an algorithm to compute the echelon form of a matrix [8], we implemented an algorithm to compute the Hermite normal form for matrices whose elements belong to a Bézout domain. Execution is guaranteed for matrices over any Euclidean domain, since there always exists an executable operation for computing Bézout coefficients. This executable operation over Euclidean domains is already implemented in the Isabelle/HOL standard library by Eberl. One could also execute the algorithm with matrices over a Bézout domain, as long as an executable operation to compute Bézout coefficients is provided.

### 4.1    Definition of Hermite Normal Form

Our formalisation of the Hermite normal form is built from many pieces. Essentially we need matrices and polynomials from the standard library and from our framework:

- Generalisations of the HA library.
- Elementary operations over matrices, executability of the *vec* representation, serialisations to obtain efficient code.
- An algorithm to compute the echelon form of a matrix.
- Ring theory (some fragments were already present in the standard library).

Let us stress that there is no unique definition of Hermite normal form in the literature. For instance, some authors, like Newman [35], restrict their definitions to the case of square nonsingular matrices (that is, invertible matrices). Other authors, like Cohen [13], just work with integer matrices. Furthermore, given a matrix $A$ its Hermite normal form $H$ can be defined to be upper triangular [43] or lower triangular [35]. In addition, the transformation from $A$ to $H$ can be made by means of elementary row operations [35] or elementary column operations [13]. In this formalisation, we work as generally as possible, so we do not impose restrictions in the input matrix (coefficients must belong to a Bézout domain and both square and non-square matrices are accepted).

In our algorithm the transformation to the Hermite normal form is carried out by means of *elementary row operations*, obtaining $H$ as an upper triangular matrix. This design decision will allow us to reuse our previous work. Moreover, any algorithm or theorem using an alternative definition of Hermite normal form (for example, in terms of column operations and/or lower triangularity) can be moulded into the form of Definition 4.

Firstly, we have to define the concepts of *complete set of nonassociates* and *complete set of residues modulo* $\mu$. Let $\mathcal{R}$ be a commutative ring with unit.

**Definition 1 (Complete set of nonassociates).** *An element $a \in \mathcal{R}$ is said to be an* associate *of an element $b \in \mathcal{R}$ if there exists an invertible element $u \in \mathcal{R}$ such that $a = ub$. This is an equivalence relation over $\mathcal{R}$. A set of elements of $\mathcal{R}$, one from each equivalence class, is said to be a* complete set of nonassociates.

**Definition 2 (Complete set of residues).** *Let $\mu$ be any nonzero element of $\mathcal{R}$. Let $a, b \in \mathcal{R}$; $a$ is* congruent *to $b$ modulo $\mu$ if $\mu$ divides $a - b$. This is an equivalence relation over $\mathcal{R}$. A set of elements of $\mathcal{R}$, one from each equivalence class, is said to be a* complete set of residues modulo $\mu$ *(or shorter, a* complete set of residues of $\mu$).

Let us start introducing the Isabelle/HOL implementation of associated (due to Eberl) and congruent elements (this and the following definitions are available from file *Hermite.thy* of our development [17]). In the definitions, x dvd y means that the element x divides the element y:

**definition** associated x y ⟷ x dvd y ∧ y dvd x
**definition** cong a b u = (u dvd (a - b))

We easily connect Eberl's definition of associated elements with Definition 1 and show they are equivalent.

**lemma** associated a b = (∃u∈Units. a = u * b)

Next, we define the corresponding relations of associates and congruence introduced by the definitions. We define the relations by means of sets. Two elements $(a, b)$ belong to the set if they are related. Hence:

**definition** `associated_rel = {(a, b). associated a b}`
**definition** `congruent_rel u = {(a, b). cong a b u}`

We prove both of them to be reflexive, transitive, and symmetric (*i.e.*, they are equivalence relations over `UNIV`, where `UNIV` represents the set of all elements of the ring).

**lemma** `equiv UNIV associated_rel`
**lemma** `equiv UNIV (congruent_rel u)`

From the definitions of associated and congruent elements, we introduce the *complete set of nonassociates* and *complete sets of residues modulo an element*. Authors usually avoid these definitions imposing additional conditions to the Hermite normal form. For instance, in the particular case of integers, the residues $r$ modulo $\mu$ are usually chosen such that $0 \le r < \mu$ (see [13]), but $-\mu < r \le 0$ (see [10]) and $-\frac{\mu}{2} < r \le \frac{\mu}{2}$ (see [3]) are also valid choices. Every possibility fits selecting a complete set of nonassociates and complete sets of residues.

A function $f$ is an *associates function* if for all $a \in \mathcal{R}$, then $a$ and $f(a)$ are associated. In order to obtain a complete set of nonassociates, we impose the elements belonging to the range of $f$ to be pairwise nonassociates. Hence, a set $S$ will be a complete set of nonassociates if there exists an associates function $f$ whose range is $S$.

**definition** `ass_function f = ((∀a. associated a (f a)) ∧`
  `pairwise (λa b. ¬ associated a b) (range f))`
**definition** `Complete_set_non_associates S =`
  `(∃f. ass_function f ∧ range f = S)`

Such definitions satisfy the following properties:

**lemma assumes** `ass_function f`
**shows** `Complete_set_non_associates (range f)`
**lemma assumes** `Complete_set_non_associates S`
**and** `x ∈ S` **and** `y ∈ S` **and** `x ≠ y` **shows** `¬ associated x y`

A function $f$ is a *residues function* if, given $u \in \mathcal{R}$, the following conditions hold:

1. For all $a, b \in \mathcal{R}$, $a$ and $b$ are congruent modulo $u$ if and only if $f\ u\ a = f\ u\ b$.
2. The elements which belong to the range of $f$ are pairwise noncongruent modulo $u$.
3. For all $a \in \mathcal{R}$, $f\ u\ a$ and $a$ are congruent modulo $u$.

**definition** `res_function f =`
`(∀u. (∀a b. cong a b u ⟷ f u a = f u b)`
`∧ pairwise (λa b. ¬ cong a b u) (range (f u))`
`∧ (∀a. cong (f u a) a u))`

Essentially, the residue function picks out an element for each residue class. From the latter condition it follows that the elements (such as $a$) above a leading entry (such as $u$) can be converted to $f\ u\ a$ by elementary operations, since there exists $k \in \mathcal{R}$ such that $f\ u\ a = a + ku$. There exists a complete set of residues for each element $u \in \mathcal{R}$. Thus, $g$ models a complete sets of residues if there exists a residues function $f$ such that each set $g\ u$ is exactly the range of $f\ u$:

**definition** `Complete_set_residues g =`
`(∃f. res_function f ∧ (∀u. g u = range (f u)))`

The function satisfies the expected properties:

**lemma assumes** `f: res_function f`
**shows** `Complete_set_residues (λu. range (f u))`
**lemma assumes** `Complete_set_residues g`
**and** `x ∈ g b` **and** `y ∈ g b` **and** `x ≠ y` **shows** `¬ cong x y b`

We can provide (executable) associates and residues functions involving elements over Euclidean domains:

**definition** `ass_function_euclidean p = normalize p`
**definition** `res_function_euclidean b n= (if b=0 then n else n mod b)`

In the above definitions, `normalize` specifies a canonical representant for each equivalence class in the Euclidean domain. For instance, in the case of the integers, `normalize` corresponds to the absolute value. The functions are proven to be associates and residues functions respectively:

**lemma** `ass_function ass_function_euclidean`
**lemma** `res_function res_function_euclidean`

We could also provide other different instances of associates and residues functions. For instance, the minus absolute value can be used as an associates function for integer elements:

**lemma** `ass_function (λn::int. -abs n)`
**lemma** `range (λn::int. -abs n) = {x. x ≤ 0}`

With the previous ingredients we can now introduce the definition of the Hermite normal form.

**Definition 3 (Echelon form).** *A matrix $H \in M_{m \times n}(\mathcal{R})$ is said to be in echelon form if:*

1. *All rows consisting only of 0's appear at the bottom of the matrix.*
2. *For any two consecutive nonzero rows, the leading entry of the lower row is to the right of the leading entry of the upper row.*

**Definition 4 (Hermite normal form).** *Given a complete set of nonassociates $S$ and complete sets of residues $G$, a matrix $H \in M_{m \times n}(\mathcal{R})$ is said to be in Hermite normal form if:*

1. *$H$ is in echelon form.*
2. *The leading entry of every nonzero row belongs to $S$.*
3. *Let $h$ be the leading entry of a nonzero row. Then each element above $h$ belongs to $G\ h$.*

Our Isabelle/HOL implementation of the definition is parametrised by a matrix $A$ and two functions, `associates` and `residues`, which are demanded to be associates and residues functions respectively. The operator `LEAST n. P n` returns the least element $n$ that satisfies a property `P`, in our case the least index $n$ such that $A\$i\$n \neq 0$.

```
definition Hermite associates residues A =
(Complete_set_non_associates associates
∧ Complete_set_residues residues ∧ echelon_form A
∧ (∀i. ¬ is_zero_row i A ⟶
       A $ i $ (LEAST n. A $ i $ n ≠ 0) ∈ associates)
∧ (∀i. ¬ is_zero_row i A ⟶
       (∀j. j<i ⟶ A$j$(LEAST n. A $ i $ n ≠ 0)
       ∈ residues (A$i$(LEAST n. A$i$n ≠ 0)))))
```

**Definition 5 (Hermite normal form of a matrix).** *A matrix $H \in M_{m \times n}(\mathcal{R})$ is the Hermite normal form of a matrix $A \in M_{m \times n}(\mathcal{R})$ if:*

1. *There exists an invertible matrix $P$ such that $A = PH$.*
2. *$H$ is in Hermite normal form.*

## 4.2    An Algorithm to Compute the Hermite Normal Form of a Matrix

Any matrix over a Bézout domain can be transformed by means of elementary operations to its Hermite normal form. A schema of the computation of the Hermite normal form is presented in Algorithm 1. There exist more efficient (in both computational cost and space consumption) algorithms to compute the Hermite normal form of a matrix. Normally they are restricted to specific domains, such as polynomial matrices [26].

We have implemented the Hermite algorithm in Isabelle/HOL iterating over rows. That is, we have defined an operation that carries out the transformations over one row and then we have defined the Hermite algorithm folding such

---

**Algorithm 1.** An algorithm to compute the Hermite normal form of a matrix $A$

---

**Input**: $A \in M_{m \times n}(\mathcal{B})$ and complete sets of nonassociates and residues.

**Output**: A matrix $H$ such that $\exists P.\ A = PH$, where $P \in M_{m \times m}(\mathcal{B})$ is invertible and $H \in M_{m \times n}(\mathcal{B})$ is in Hermite normal form with respect to the given complete sets of nonassociates and residues.

1 Transform the matrix $A$ to its corresponding echelon form;

2 Transform each row such that its leading entry belongs to the complete set of nonassociates, multiplying each row by an appropriate constant;

3 Transform the elements above each leading entry, *i.e.*, such elements must belong to the corresponding complete set of residues with respect to the leading entry. This is done by adding to each row above the leading entry, the row of the leading entry multiplied by a constant (that is, the transformation is carried out by means of elementary operations).

---

an operation over all rows. Our Hermite algorithm relies on our previous version of the echelon form algorithm [8]. The algorithm is parametrised by three functions:[2]

- A function that computes Bézout's identity of two elements (required for the echelon form).
- An associates function whose range is a complete set of nonassociates.
- A residues function whose range consists of complete sets of residues.

The Hermite algorithm must be parametrised with the functions that satisfy the required properties presented above. The proof of correctness of the algorithm will assume that such functions are really Bézout, associates, and residues functions respectively. These requirements are expressed by means of premises.

The following Isabelle functions reproduce the steps in Algorithm 1. Step 1 corresponds with the function *echelon_form_of* of our previous work. Step 3 is performed by means of *Hermite_reduce_above* starting from the proper index (one of its parameters is the residues function). We use a primitive recursive definition over the representation of the row-indexes as natural numbers.

```
primrec Hermite_reduce_above A 0 i j res  = A
    | Hermite_reduce_above A (Suc n) i j res =
(let i'=((from_nat n)::'rows); Aij = A $ i $ j;  Ai'j = A$i'$j;
     k = (((res Aij (Ai'j))-(Ai'j)) div Aij) in
Hermite_reduce_above (row_add A i' i k) n i j res)
```

This function is reused in *Hermite_of_row_i*, which performs Step 2 (it also has both the associates and the residues functions as parameters).

```
definition Hermite_of_row_i ass res A i =
```

---

[2] Neither records nor locales [9] are used for this task, although they are a valid alternative.

```
(if is_zero_row i A then A else
  let j = (LEAST n. A $ i $ n ≠ 0); Aij= (A $ i $ j);
  A' = mult_row A i ((ass Aij) div Aij)
in Hermite_reduce_above A' (to_nat i) i j res)
```

The function `Hermite_of_upt_row_i` iterates the process up to a row $i$.

**definition** `Hermite_of_upt_row_i A i ass res =`
  `foldl (Hermite_of_row_i ass res) A (map from_nat [0..<i])`

Finally, `Hermite_of` takes echelon form as starting point and applies the function `Hermite_of_upt_row_i` to its rows:

**definition** `Hermite_of A ass res bezout = (let A'= echelon_form_of A bezout`
`in Hermite_of_upt_row_i A' (nrows A) ass res)`

The soundness of the algorithm can be split into four parts:

1. The output matrix is in echelon form.
2. Each leading entry belongs to the complete set of nonassociates.
3. Each element above a leading entry belongs to the corresponding complete set of residues.
4. The algorithm is carried out by means of elementary row operations (therefore, the *output* matrix is the Hermite normal form of the input matrix).

Part 1 takes advantage of our previous proof about echelon forms [8], and requires proving that `Hermite_of_upt_row_i` preserves echelon forms. This property follows from the definition of `Hermite_reduce_above`, since it performs elementary row operations only above the leading coefficients of each row. Thus, it does not alter any of the properties of the echelon form. Part 2 easily follows from the definition `Hermite_of_row_i`. The proof of Part 3 is based on the definition of `Hermite_reduce_above`. The proof is more intricate, since we have to prove the result for one row, and then apply inductively the result to the rest of rows (proving that the previous ones are preserved in each iteration). The crucial lemma for this part states that the property holds when the algorithm is iteratively applied up to the $k - th$ row:

**lemma defines** `n=(LEAST n. A $ i $ n ≠ 0)`
**assumes** ¬ `is_zero_row i A` **and** `echelon_form A` **and** `ass_function ass`
**and** `res_function res` **and** `to_nat i < k` **and** `k ≤ nrows A` **and** `j < i`
**shows** `(Hermite_of_upt_row_i A k ass res) $ j $ n`
  `∈ range (res (Hermite_of_upt_row_i A k ass res $ i $ n))`

Finally, Part 4 is established by proving that the required steps to compute the Hermite normal form can be expressed as invertible matrices (here we also reuse results of our previous developments), and therefore are equivalent to elementary operations. We refer the interested reader to the file *Hermite.thy* of the development for the full-detailed proofs and statements. In a modest 1400 code lines we obtain the final theorem:

```
theorem assumes ass_function ass
and res_function res and is_bezout_ext bezout
shows ∃P. invertible P ∧ (Hermite_of A ass res bezout) = P ** A ∧
Hermite (range ass) (λc. range (res c)) (Hermite_of A ass res bezout)
```

In *ca.* 150 Isabelle/HOL code lines, we refine the algorithm to immutable arrays and generate its SML and Haskell versions (see file *Hermite_IArrays.thy*).

## 4.3  Uniqueness

**Theorem 1** *Fixing a complete set of nonassociates and complete sets of residues, if $A \in M_{n \times n}(\mathcal{R})$ is a nonsingular matrix, then its Hermite normal form is unique.*

Let us note that the Hermite normal form of an invertible matrix is the identity matrix when the standard associates and residues functions over euclidean domains are chosen in the algorithm, but this does not hold in general. In order to prove Theorem 1, we follow the proof by Newman [35, Theorem II.3]. Where Newman considers the Hermite normal form as a lower triangular matrix we consider it upper triangular.

```
lemma assumes A = P ** H and A = Q ** K and invertible A
and invertible P and invertible Q
and Hermite associates residues H
and Hermite associates residues K shows H = K
```

The original proof comprises 28 lines [35, Theorem II.3]. The argument proceeds as follows: let us suppose that, for a given nonsingular matrix $A$, there are two different upper triangular Hermite forms (*wrt* the same sets of associates and residues), $H$ and $K$. Then, there exists a unit matrix $U$ such that $H = UK$. $U$ must also be upper triangular. Its diagonal elements are 1, since $h_{ii} = u_{ii}k_{ii}$ with both $h_{ii}$, $k_{ii}$ in the same set of nonassociates. The remaining elements of the matrix must be equal to 0. Let $s \in \{0 \ldots n - 1\}$ (any valid row). Since the matrix is upper triangular, we consider the element $u_{s,s+j}$ with $j \in \{1 \ldots (\text{ncols } A - s)\}$ (*i.e.*, any element above the diagonal). We apply *total induction* in $j$, and therefore we assume that $u_{s,s+1}, \ldots, u_{s,s+(j-1)}$ are equal to 0. Hence:

$$h_{s,s+j} = \sum_{t=0}^{nrows\, A} u_{st} k_{t(s+j)} \tag{1}$$

$$= \sum_{t=s}^{s+j} u_{st} k_{t(s+j)} \tag{2}$$

$$= u_{ss}k_{s,s+j} + u_{s,s+1}k_{s+1,s+j} + \cdots + u_{s,s+j}k_{s+j,s+j} \tag{3}$$

$$= u_{ss}k_{s,s+j} + u_{s,s+j}k_{s+j,s+j} \tag{4}$$

$$= k_{s,s+j} + u_{s,s+j}k_{s+j,s+j} \tag{5}$$

Step 2 follows from $K$ being upper triangular; step 4 follows from the induction hypothesis; step 5 follows from the first part of the proof ($u_{ii} = 1$). Therefore, $h_{s,s+j} \equiv k_{s,s+j} \bmod k_{s+j,s+j}$, from where it follows that $h_{s,s+j} \equiv k_{s,s+j}$, since both elements belong to the same complete set of residues of $k_{s+j,s+j}$ (and hence $u_{s,s+j} = 0$).

This inductive reasoning, which in the original proof took 18 lines, required 88 lines in our formalisation. The proof itself is not particularly intricate, but it demands a correct manipulation of the indexes. The complete proof took 150 lines, thanks to the strong reuse of previous results already available in the framework. It firmly follows the book proof line by line.

## 4.4    Examples of Execution

We provide two examples of execution of our formalised algorithm. Both use the *standard* associates and residues functions, which are defined for Euclidean domains. Let us choose a rectangular random integer matrix $A$ and a polynomial matrix $B$.

$$A = \begin{bmatrix} 37 & 9 & 10 & 28 & 40 & 23 & 59 & 25 & 73 & 79 \\ 5 & 96 & 93 & 7 & 71 & 44 & 63 & 90 & 27 & 89 \\ 70 & 65 & 36 & 69 & 2 & 81 & 14 & 30 & 92 & 60 \\ 16 & 98 & 100 & 50 & 64 & 21 & 39 & 95 & 80 & 34 \end{bmatrix}, B = \begin{bmatrix} 5x^2 + 4x + 3 & x - 2 \\ 2x^2 - 1 & x^3 + 4x^2 + x \end{bmatrix}$$

Their Hermite normal forms are computed in Isabelle/HOL similarly:

```
value[code] matrix_to_list_of_list (Hermite_of M ass_function_euclidean
res_function_euclidean euclid_ext2)
```

Where M is a matrix in Isabelle/HOL that corresponds to $A$ or $B$, depending on the example we are executing. The function `Hermite_of` has four parameters: the input matrix $M$, the *standard* associates function for Euclidean domains (`ass_function_euclidean`), the *standard* residues function for Euclidean domains (`res_function_euclidean`), and the function which computes Bézout coefficients in Euclidean domains (`euclid_ext2`), which is required by the echelon form algorithm. Let us note that the type inference will determine which version of the associates and residues functions must be executed, depending on the type of the input matrix. The function `matrix_to_list_of_list` eases outputting matrices. The obtained results follow:

$$\begin{bmatrix} 1 & 0 & 0 & 2126849 & -2040340 & -1544323 & -3517370 & -665650 & 1303207 & -5664981 \\ 0 & 1 & 0 & 3330071 & -3194626 & -2417993 & -5507258 & -1042230 & 2040466 & -8869838 \\ 0 & 0 & 1 & 1681610 & -1613209 & -1221033 & -2781035 & -526300 & 1030392 & -4479062 \\ 0 & 0 & 0 & 3802428 & -3647768 & -2760977 & -6288437 & -1190065 & 2329900 & -10127986 \end{bmatrix}$$

$$\begin{bmatrix} 1 & -\dfrac{44}{89} + \dfrac{31}{89}x - \dfrac{68}{89}x^2 + \dfrac{137}{89}x^3 + \dfrac{40}{89}x^4 \\ 0 & -\dfrac{2}{5} + \dfrac{4}{5}x + 4x^2 + \dfrac{22}{5}x^3 + \dfrac{24}{5}x^4 + x^5 \end{bmatrix}$$

The results satisfy the expected properties (for instance, the polynomial matrix has monic polynomials as leading entries in each row). Let us also note the growth on the size of the elements. Both matrices are computed instantly. Such examples can also be executed inside the logic, but they take 30 and 8 minutes respectively.

The performance of the algorithm is highly dependent on several factors. Some of them follow from our design choices, such as the selection of associates and residues functions, and the function to compute the Bézout identity. Some others depend on the system configuration, such as the serialisations employed. Finally, the chosen algorithm to compute the Hermite normal form itself can be extremely space consuming (there exist versions that bound the size of the intermediate entries computed [27]). With our particular version, the time to compute the Hermite normal form of a $20 \times 20$ integer matrix with random entries between 0 and 100 making use of the *standard* associates, residues, and Bézout functions is negligible using the refinement to `iarray`. The resulting matrix has elements with more than 50 digits. The same happens with a $25 \times 25$ integer matrix. Memory issues appear with higher dimensions and greater elements.

# 5    Related and Further Work

Linear algebra has been formalised in many theorem provers: Isabelle/HOL [44], Coq [12], Mizar [41], HOL Light [22], PVS [34], and ACL2 [19] are just a few examples of it. On the contrary, the verification and implementation of linear algebra algorithms have not been so widely explored, especially involving matrices over rings. The most similar works have been carried out in Coq. It is worth citing the CoqEAL development [12], which contains several linear algebra algorithms formalised in Coq, such as the Sasaki-Murao [14] algorithm for computing determinants of matrices over rings. The closest work to ours is the one done by Cano *et al.* [11] also in Coq, which presents a formalisation of the Smith normal form (SNF) of a matrix. Their formalisation is restricted to explicit division rings, such as constructive Bézout domains, whereas in our case we can work with more abstracts structures where the existence of divisions and greatest common divisors are known, but maybe not how to compute them. In any case, the SNF algorithm is distinct from the Hermite normal form. SNF requires both row and column operations and the result is a diagonal matrix. The Hermite normal form sometimes can be view as a previous step, but is not required to compute the SNF. The computation of the echelon form of integer matrices has been recently formalised in ACL2 by Lambán *et al.* [30] as an application of abstract single threaded objects. A formal proof of the SNF would be desirable in Isabelle/HOL. Most of the algorithms to compute the SNF of a matrix are based on submatrices [43]. Unfortunately, submatrices are a delicate issue in the HA library: Since Isabelle does not feature dependent types, we cannot use the size of the matrix in the definition of submatrix. Thiemann and Yamada already faced this problem when formalising Jordan normal forms of matrices [44], a kind of forms whose construction is done by means of block matrices. As a solution, they propose a new matrix representation which is indeed an abstraction of the HA representation, but flexible for dimensions. We aim to formalise the Smith normal form in Isabelle using such a representation, also connecting it to the HA library and our framework by means of the lifting and transfer package [24] and the new addition of local type definitions when necessary [29]. This would also allow us to implement some decision procedures based on linear algebra methods, such as the decision algorithm proposed by Li *et al.* [31].

## 6    Conclusions

We have presented a formalisation of the Hermite normal form of a matrix that reuses our previous developments. The Hermite normal form of a matrix is a well-known canonical matrix over rings. We have not only proved the correctness of the algorithm, but we have also refined it to immutable arrays and we have formalised its uniqueness as well. As far as we know, this is the first formalisation of the Hermite normal form in any theorem prover, even only considering the case of integer matrices.

The formalisation could be seen as a proof pearl because of two reasons: we have formalised a non-trivial and well-known linear algebra algorithm in a modest number of lines (*ca.* 2300, to be compared with more than 10000 that we needed for similar results with the Gauss-Jordan algorithm) thanks to a strong reuse of the infrastructure presented in Sect. 3; and we have focused on obtaining the most general version by means of a parametrised algorithm. The formalisation has been carried out involving matrices over Bézout domains and it is not restricted to the common case of integer matrices. Furthermore, the algorithm has been parametrised by functions so that it can compute every definition of the Hermite normal form in the literature.

## References

1. Archive of Formal Proofs. http://afp.sourceforge.net/
2. Immutable non-strict arrays in Haskell. http://hackage.haskell.org/package/array-0.5.2.0/docs/Data-Array.html
3. Matlab documentation. Definition of Hermite Normal Form. http://es.mathworks.com/help/symbolic/hermiteform.html#butzrp_-5
4. The Vector structure in SML. http://sml-family.org/Basis/vector.html
5. HOL interactive theorem prover (2016). https://hol-theorem-prover.org/
6. Aransay, J., Divasón, J.: Formalisation in higher-order logic and code generation to functional languages of the Gauss-Jordan algorithm. J. Funct. Program. **25**, 22 p. (2015). https://doi.org/10.1017/S0956796815000155
7. Aransay, J., Divasón, J.: A formalisation in HOL of the fundamental theorem of linear algebra and its application to the solution of the least squares problem. J. Autom. Reason. **58**, 509–535 (2017). https://doi.org/10.1007/s10817-016-9379-z
8. Aransay, J., Divasón, J.: Formalisation of the computation of the echelon form of a matrix in Isabelle/HOL. Formal Aspects Comput. **28**(6), 1005–1026 (2016)

9. Ballarin, C.: Locales: a module system for mathematical theories. J. Autom. Reason. **52**(2), 123–153 (2014)
10. Bradley, G.H.: Algorithms for Hermite and Smith normal matrices and linear diophantine equations. Math. Comput. **25**(116), 897–907 (1971)
11. Cano, G., Cohen, C., Dénès, M., Mörtberg, A., Siles, V.: Formalized linear algebra over elementary divisor rings in Coq. Logical Methods Comput. Sci. **12**(2) (2016)
12. Cohen, C., Dénès, M., Mörtberg, A.: Refinements for free!. In: Gonthier, G., Norrish, M. (eds.) CPP 2013. LNCS, vol. 8307, pp. 147–162. Springer, Cham (2013). https://doi.org/10.1007/978-3-319-03545-1_10
13. Cohen, H.: A Course in Computational Algebraic Number Theory. Springer, New York (1993). https://doi.org/10.1007/978-3-662-02945-9
14. Coquand, T., Mörtberg, A., Siles, V.: A formal proof of Sasaki-Murao algorithm. J. Formalized Reason. **5**(1), 27–36 (2012)
15. The Coq development team. The Coq proof assistant reference manual. LogiCal Project (2018). Version 8.8.0. http://coq.inria.fr
16. Divasón, J.: Additional files to the Hermite normal form development (2018). http://www.unirioja.es/cu/jodivaso/Isabelle/Hermite/Hermite.html
17. Divasón, J., Aransay, J.: Hermite normal form. Archive of Formal Proofs, July 2015
18. Durán, A.J., Pérez, M., Varona, J.L.: The Misfortunes of a trio of mathematicians using computer algebra systems. Can we trust in them? Not. AMS **61**(10), 1249–1252 (2014)
19. Gamboa, R., Cowles, J., Baalen, J.V.: Using ACL2 arrays to formalise matrix algebra. In: Fourth International Workshop on the ACL2 Theorem Prover and Its Applications (2003)
20. Gonthier, G.: Formal proof - the four-color theorem. Not. AMS **55**(11), 1382–1393 (2008)
21. Hafner, J.L., McCurley, K.S.: A rigorous subexponential algorithm for computation of class groups. J. Am. Math. Soc. **2**(4), 837–850 (1989)
22. Harrison, J.: The HOL Light theory of euclidean space. J. Autom. Reason. **50**(2), 173–190 (2013)
23. The Haskell Programming Language (2016). http://www.haskell.org/
24. Huffman, B., Kunčar, O.: Lifting and transfer: a modular design for quotients in Isabelle/HOL. In: Gonthier, G., Norrish, M. (eds.) CPP 2013. LNCS, vol. 8307, pp. 131–146. Springer, Cham (2013). https://doi.org/10.1007/978-3-319-03545-1_9
25. Hung, M.S., Rom, W.O.: An application of the Hermite normal form in integer programming. Linear Algebra Appl. **140**, 163–179 (1990)
26. Kaltofen, E., Krishnamoorthy, M.S., Saunders, B.D.: Fast parallel computation of Hermite and Smith forms of polynomial matrices. SIAM J. Algebraic Discrete Methods **8**(4), 683–690 (1987)
27. Kannan, R., Bachem, A.: Polynomial algorithms for computing the Smith and Hermite normal forms of an integer matrix. SIAM J. Comput. **8**(4), 499–507 (1979)
28. Klein , G., et al.: seL4: formal verification of an OS kernel. In: Proceedings of the 22nd ACM Symposium on Operating Systems Principles, SOSP 2009, Big Sky, Montana, USA, pp. 207–220 (2009)
29. Kunčar, O., Popescu, A.: From types to sets by local type definitions in higher-order logic. In: Blanchette, J.C., Merz, S. (eds.) Interactive Theorem Proving. pp, pp. 200–218. Springer International Publishing, Cham (2016)
30. Lambán, L., Martín-Mateos, F.J., Rubio, J., Ruiz-Reina, J.-L.: Using abstract stobjs in ACL2 to compute matrix normal forms. In: Ayala-Rincón, M., Muñoz,

C.A. (eds.) ITP 2017. LNCS, vol. 10499, pp. 354–370. Springer, Cham (2017). https://doi.org/10.1007/978-3-319-66107-0_23

31. Li, L., Li, H., Liu, Y.: A decision algorithm for linear sentences on a PFM. Ann. Pure Appl. Logic **59**, 273–286 (1993)

32. Li, W., Paulson, L.C.: A modular, efficient formalisation of real algebraic numbers. In: Proceedings of the 5th ACM SIGPLAN Conference on Certified Programs and Proofs, Saint Petersburg, FL, USA, 20–22 January 2016, pp. 66–75 (2016)

33. Lochbihler, A.: Verifying a compiler for Java threads. In: Gordon, A.D. (ed.) ESOP 2010. LNCS, vol. 6012, pp. 427–447. Springer, Heidelberg (2010). https://doi.org/10.1007/978-3-642-11957-6_23

34. Narkawicz, A., Muoz, C., Dutle, A.: Formally-verified decision procedures for univariate polynomial computation based on Sturm's and Tarski's theorems. J. Autom. Reason. **54**(4), 285–326 (2015)

35. Newman, M.: Integral Matrices. Pure and Applied Mathematics. Elsevier Science, New York (1972)

36. Nipkow, T., Wenzel, M., Paulson, L.C. (eds.): Isabelle/HOL. A Proof Assistant for Higher-Order Logic. LNCS, vol. 2283. Springer, Heidelberg (2002). https://doi.org/10.1007/3-540-45949-9

37. Nipkow, T., Paulson, L.C., Wenzel, M.: Isabelle/HOL - A Proof Assistant for Higher-Order Logic (2018). Updated version of the book with the same title and authors

38. Obua, S., Nipkow, T.: Flyspeck II: the basic linear programs. Ann. Math. Artif. Intell. **56**(3–4), 245–272 (2009)

39. Paulson, L.C.: A mechanised proof of Gödel's incompleteness theorems using nominal Isabelle. J. Autom. Reason. **55**(1), 1–37 (2015)

40. Ramanujam, J.: Beyond unimodular transformations. J. Supercomput. **9**(4), 365–389 (1995)

41. Rudnicki, P., Schwarzweller, C., Trybulec, A.: Commutative algebra in the Mizar system. J. Symbolic Comput. **32**(1/2), 143–169 (2001)

42. Sternagel, C., Thiemann, R.: Executable matrix operations on matrices of arbitrary dimensions. Archive of Formal Proofs, June 2010

43. Storjohann, A.: Algorithms for matrix canonical forms. Ph.D. thesis, Swiss Federal Institute of Technology, Zurich (2000)

44. Thiemann, R., Yamada, A.: Formalizing Jordan normal forms in Isabelle/HOL. In: Proceedings of the 5th ACM SIGPLAN Conference on Certified Programs and Proofs, Saint Petersburg, FL, USA, 20–22 January 2016, pp. 88–99 (2016)

45. Tourloupis, V.E.: Hermite normal forms and its cryptographic applications. Master's thesis, University of Wollongong (2013)

# Formalizing Some "Small" Finite Models of Projective Geometry in Coq

David Braun, Nicolas Magaud[(⊠)], and Pascal Schreck

Icube UMR 7357 CNRS - Université de Strasbourg (IGG), Illkirch Cedex, France
{david.braun,magaud,schreck}@unistra.fr

**Abstract.** We study two different descriptions of incidence projective geometry: a synthetic, mathematics-oriented one and a more practical, computation-oriented one, based on the combinatorial concept of rank of a set of points. Using both axiom systems, we prove that some specific finite planes (resp. spaces) verify the axioms of projective plane (resp. space) geometry and Desargues' property. It requires using repeated case analysis on all variables of some finite inductive data-types and leads to numerous (sub-)goals in the Coq proof assistant. We thus investigate to what extend Coq can deal with such a combinatorial explosion in the number of cases to handle. We propose some easy-to-implement but relevant proof optimizations which, combined together, lead to an efficient way to deal with such large proofs.

**Keywords:** Coq · Proof automation · Combinatorial explosion
Finite inductive types · Projective geometry · Finite geometry
Desargues' property

## 1 Introduction

Incidence projective geometry is one of the simplest and most expressive frameworks used to describe some aspects of geometry. It is a good candidate for formalization: few axioms are needed and some key geometric properties such as Desargues' one can be formally stated and proved correct under some specific assumptions (see [11,12]).

The notion of incidence projective plane is mainly defined by two axioms: two distinct points define a single line and two lines concur in a single point. A third axiom is usually used to catch precisely the dimension of geometry. For higher dimensions, the second axiom is a bit more complicated and defined as the two following statements: (1) two lines concur in at most one point and (2) *Pasch*'s axiom: given four different points A, B, C and D, if lines AB and CD concur, so do lines AC and BD. Moreover, other axioms can be added to avoid degenerate cases.

Proving properties in projective geometry or proving that some planes or spaces are actual models of projective geometry is usually based on analyzing a few general configurations as well as numerous degenerate cases. Using a proof

© Springer Nature Switzerland AG 2018
J. Fleuriot et al. (Eds.): AISC 2018, LNAI 11110, pp. 54–69, 2018.
https://doi.org/10.1007/978-3-319-99957-9_4

assistant such as Coq [3,6] makes it easier for the user to write a correct and comprehensive proof. Indeed, Coq forces the programmer to handle each possible case in the proof. In addition, all details of the proof must be provided, which improves the confidence in it and allows the system to verify the proofs (by type-checking). The drawback is that it represents a tremendous amount of work for the proof developer. Thankfully, the Coq proof assistant and its tactic language Ltac allow to build *ad-hoc* tactics to automate large parts of the proofs efficiently.

We use two equivalent formal descriptions of projective geometry: a synthetic one and an alternative one using a matroid structure operating on points [4]. We check to what extent each of them allows to perform tractable, readable, easy-to-write and easy-to-process proofs. To achieve this goal, we work with some finite models of projective geometry: $pg(2,2)$, also known as Fano plane, $pg(2,3)$ and $pg(2,5)$; as well as the smallest finite projective space $pg(3,2)$ (see Subsect. 2.3). As models grow bigger, we need smarter proof techniques to cope with the inherent complexity and to keep memory usage, proof search and compile time under control.

**Related Work.** Finite geometry has been studied since the late 19th century and is intrinsically linked to the development of algebraic structures like division rings, near fields or ternary rings. There has been a renewed interest with its application to computational domains like cryptography or planning (see [2,5] for a comprehensive state of the art). The theoretical aspects are out of the scope of this paper. Rather we are interested in efficiently automating proofs with numerous cases within the Coq proof assistant. The use of ranks to carry out proofs in projective geometry was first introduced by Michelucci and Schreck [14]. Our work reuses some ideas of the mathematical components library about finite types [13] but we choose to refactor parts of it to suit our own needs.

**Outline.** This article is organized as follows. In Sect. 2, we present two different ways of specifying projective geometry, directly or by using rank theory. We also introduce some common properties (e.g. Desargues' property) and describe some finite models of projective geometry. In Sect. 3, we study the inherent complexity of the finite models and describe some techniques to handle these complexity issues properly in Coq. In Sect. 4, we present some more practical tools to help the user to write formal proofs easily via proof structuring and automation. Finally, in Sect. 5, we summarize our contributions and present some suitable perspectives.

**Notation.** We name axioms $AXYN$. $A$ stands for axiom, $X$ is the axiom number, $Y$ may take two values ($P$ = projective, $R$ = rank) and $N$ denotes the dimension.

## 2  Formal Specification of Projective Geometry, Rank Theory and Finite Fields

We define two equivalent axiom systems for incidence projective geometry: one based on the usual synthetic description, and another one based on the combinatorial notion of rank provided by the matroid structure of incidence projective geometry. Then, we prove, using these two specifications, that some finite planes/spaces are models of incidence projective geometry and we study Desargues' theorem.

### 2.1  Axiom Systems for Incidence Projective Geometry

Incidence Geometry is a simple view of geometry, where only points and lines, together with the incidence relation linking them are kept. Projective geometry is obtained by assuming that two coplanar lines always meet. Incidence projective geometry can be easily described as a small set of axioms, as shown in Coxeter's book [7].

**Plane.** The axiom system for projective plane geometry consists of five axioms presented in Fig. 1. Axioms (A1P2) and (A2P2) deal with construction of points and lines. Axiom (A3P2) concerns uniqueness of points and lines. Finally, axiom (A4P2) states that each line contains at least three points; and axiom (A5P2) expresses that there always exists two distinct lines.

(A1P2) **Line-Existence** : $\forall$ A B : Point, $\exists$ l : Line, A $\in$ l $\wedge$ B $\in$ l

(A2P2) **Point-Existence** : $\forall$ l m : Line, $\exists$ A : Point, A $\in$ l $\wedge$ A $\in$ m

(A3P2) **Uniqueness** : $\forall$ A B : Point, $\forall$ l m : Line, A $\in$ l $\wedge$ B $\in$ l $\wedge$
                    A $\in$ m $\wedge$ B $\in$ m $\Rightarrow$ A = B $\vee$ l = m

(A4P2) **Three-Points** : $\forall$ l : Line, $\exists$ A B C : Point,
                    A $\neq$ B $\wedge$ B $\neq$ C $\wedge$ A $\neq$ C $\wedge$ A $\in$ l $\wedge$ B $\in$ l $\wedge$ C $\in$ l

(A5P2) **Lower-Dimension** : $\exists$ l m: Line, $l \neq m$

**Fig. 1.** Axiom system for projective plane geometry

**Space and Higher Dimensions.** Similarly, we define an axiom system to capture projective space geometry in Fig. 2 by extending the previous one. The system still contains five axioms with three of them remaining unchanged (A1P3, A3P3, A4P3). *Pasch*'s axiom replaces (A2P2) and assumes that two coplanar lines always meet. Furthermore, we modify the axiom *Lower-Dimension* to capture projective geometry for spaces of dimension greater or equal than 3. It is possible to limit this to spatial geometry by adding the optional axiom (A6P3) to constrain the dimension to be exactly 3.

**(A1P3) Line-Existence** : $\forall$ A B : Point, $\exists$ l : Line, A $\in$ l $\wedge$ B $\in$ l

**(A2P3) Pasch** : $\forall$ A B C D : Point, $\forall$ $l_{AB}$ $l_{CD}$ $l_{AC}$ $l_{BD}$ : Line,
$$A \neq B \wedge A \neq C \wedge A \neq D \wedge B \neq C \wedge B \neq D \wedge C \neq D \wedge$$
$$A \in l_{AB} \wedge B \in l_{AB} \wedge C \in l_{CD} \wedge D \in l_{CD} \wedge$$
$$A \in l_{AC} \wedge C \in l_{AC} \wedge B \in l_{BD} \wedge D \in l_{BD} \wedge$$
$$(\exists \text{ I : Point, I} \in l_{AB} \wedge \text{I} \in l_{CD}) \Rightarrow$$
$$(\exists \text{ J : Point, J} \in l_{AC} \wedge \text{J} \in l_{BD})$$

**(A3P3) Uniqueness** : $\forall$ A B : Point, $\forall$ l m : Line,
$$A \in l \wedge B \in l \wedge A \in m \wedge B \in m \Rightarrow A = B \vee l = m$$

**(A4P3) Three-Points** : $\forall$ l : Line, $\exists$ A B C : Point,
$$A \neq B \wedge B \neq C \wedge A \neq C \wedge A \in l \wedge B \in l \wedge C \in l$$

**(A5P3) Lower-Dimension** : $\exists$ l m : Line, $\forall$ p : Point, p $\notin$ l $\vee$ p $\notin$ m

**(A6P3) Upper-Dimension** : $\forall$ l1 l2 l3 : Line, l1 $\neq$ l2 $\wedge$ l1 $\neq$ l3 $\wedge$ l2 $\neq$ l3 $\Rightarrow$
$$\exists \text{ l4 : Line}, \exists \text{ P1 P2 P3 : Point}, \text{P1} \in \text{l1} \wedge$$
$$\text{P1} \in \text{l4} \wedge \text{P2} \in \text{l2} \wedge \text{P2} \in \text{l4} \wedge \text{P3} \in \text{l3} \wedge \text{P3} \in \text{l4}$$

**Fig. 2.** Axiom system for projective space geometry

## 2.2 A Rank-Based Axiom Systems

Ranks are based on matroids [16] and they allow a combinatorial approach to theorem proving in projective geometry. Matroid theory allows us to capture and generalize the main set of properties of linear dependence in vector spaces. When combined with a finite set of points, it captures incidence (collinearity, coplanarity, ...) between these points without handling directly lines or planes. It makes the computational content of projective geometry more accessible, the price to pay being less readable statements and proofs. It is quite similar to analytic geometry which also favors computability at the expense of readability.

A rank function is an integer-valued function on a finite set of objects E that can be associated to a matroid if and only if the following conditions of Fig. 3 are satisfied. To illustrate rank function, we give an intuitive interpretation of how the synthetic and rank-based descriptions correspond (see Table 1).

**(A1R2-R3) nonnegative and subcardinal** : $\forall$ X $\subseteq$ E, $0 \leq rk(X) \leq |X|$

**(A2R2-R3) nondecreasing** : $\forall$ X $\subseteq$ Y, $rk(X) \leq rk(Y)$

**(A3R2-R3) submodular** : $\forall$ X,Y $\subseteq$ E, $rk(X \cup Y) + rk(X \cap Y) \leq rk(X) + rk(Y)$

**Fig. 3.** Matroid properties for the rank function

**Table 1.** Some rank statements and their geometric interpretations

| | |
|---|---|
| rk{A, B} = 1 | A = B |
| rk{A, B} = 2 | A ≠ B |
| rk{A, B, C} = 2 | A, B, C are collinear with at least two of them distinct |
| rk{A, B, C} ≤ 2 | A, B, C are collinear |
| rk{A, B, C} = 3 | A, B, C are not collinear |
| rk{A, B, C, D} = 3 | A, B, C, D are coplanar, not all collinear |
| rk{A, B, C, D} = 4 | A, B, C, D are not coplanar |

**Plane.** To capture projective geometry entirely, we need to add some more *geometry-oriented* axioms. These five additional axioms are presented in Fig. 4. The first two ones establish the non-degeneracy of the rank function. The other ones are more or less direct translations of the axioms of projective geometry.

**(A4R2) Rk-Singleton** : $\forall$ P : Point, rk{P} $\geq$ 1

**(A5R2) Rk-Couple** : $\forall$ P Q: Point, P $\neq$ Q $\Rightarrow$ rk{P, Q} $\geq$ 2

**(A6R2) Rk-Inter** : $\forall$ A B C D, $\exists$ J, rk{A, B, J} = rk{C, D, J} = 2

**(A7R2) Rk-Three-Points** : $\forall$ A B, $\exists$ C, rk{A, B, C} = rk{B, C} = rk{A, C} = 2

**(A8R2) Rk-Lower-Dimension** : $\exists$ A B C, rk{A, B, C} $\geq$ 3

**Fig. 4.** Rank-based axiom system for projective plane geometry

**Space.** Finally, we define a rank-based axiom system to describe projective space in Fig. 5. Again, only the axioms *Pasch* and *Lower-Dimension* are modified. To restrict the dimension to 3, we add the optional axiom (A9R3).

**Equivalence Proof.** We recently proved [4] that the two descriptions of incidence geometry presented above are equivalent:

**Theorem.** *The axiom system based on incidence projective geometry and the rank-based axiom system are equivalent respectively in 2D, $\geq$3D and 3D.*

This equivalence gives us the possibility to choose the most adequate theory to prove a lemma. Indeed, statements can be bilaterally translated. This important fact allows us both to compare proofs carried out with two different approaches but also to complete some demonstrations when one of the two theories is not conducive to a tractable proof.

**(A4R3) Rk-Singleton** : $\forall$ P : Point, rk$\{$P$\} \geq 1$

**(A5R3) Rk-Couple** : $\forall$ P Q: Point, P $\neq$ Q $\Rightarrow$ rk$\{$P, Q$\} \geq 2$

**(A6R3) Rk-Pasch** : $\forall$ A B C D, rk$\{$A, B, C, D$\} \leq 3 \Rightarrow \exists$ J,
  rk$\{$A, B, J$\} = $ rk$\{$C, D, J$\} = 2$

**(A7R3) Rk-Three-Points** : $\forall$ A B, $\exists$ C, rk$\{$A, B, C$\} = $ rk$\{$B, C$\} = $ rk$\{$A, C$\} = 2$

**(A8R3) Rk-Lower-Dimension** : $\exists$ A B C D, rk$\{$A, B, C, D$\} \geq 4$

**(A9R3) Rk-Upper-Dimension** : $\forall$ A B C D E, rk$\{$A, B, C, D, E$\} \leq 4$

**Fig. 5.** Rank-based axiom system for projective space geometry

## 2.3   Finite Models

The first examples of incidence geometries are built with fields. For instance, affine planes often arise from $F^2$, where $F$ is a field, via a coordinate system and projective planes from $F^3$ via a homogeneous coordinate system. Considering finite fields leads to classical examples of finite geometries. For instance, Fano spaces come from field $\mathbb{Z}/2\mathbb{Z}$.

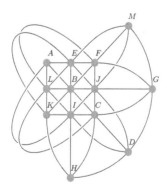

**Fig. 6.** A configuration of $pg(2,3)$: 13 points and 13 lines (e.g. AEFG, CELM, DILF).

Finite fields of cardinality $n$ denoted by $GF(n)$ are called Galois fields as they are isomorphic to the field $\mathbb{Z}_p[X]/f(X)$ where $p$ is a prime number, $\mathbb{Z}_p$ stands for $\mathbb{Z}/p\mathbb{Z}$ and $f$ is an irreducible polynomial over $\mathbb{Z}_p[X]$. It follows that, $k$ being the degree of $f$, such a finite field has cardinality $n = p^k$ and each line of a corresponding affine space (resp. projective space) has cardinality $n$ (resp. $n+1$). Finite projective spaces arising from $GF(n)$ are then denoted by $pg(d, n)$ where $d$ is the dimension of the space and $n$ the order of the underlying field. Table 2 summarizes cardinalities and Fig. 6 represents $pg(2,3)$.

**Table 2.** Description of several finite projective plane/space.

|          | Point(s) | Line(s) | Plane(s) |
|----------|----------|---------|----------|
| pg(2, 2) | 7        | 7       | 1        |
| pg(2, 3) | 13       | 13      | 1        |
| pg(2, 4) | 21       | 21      | 1        |
| pg(2, 5) | 31       | 31      | 1        |
| pg(3, 2) | 15       | 35      | 15       |

Forgetting the way that such spaces are built, $pg(d, n)$ spaces offer a convenient benchmark to test our strategies for mechanizing proofs in Coq. Although we work in the context of $pg(d, n)$, we only take into account its geometric characteristics. This means that while keeping in mind the theoretical results, we do not use coordinates in our formalization.

## 2.4    Desargues' Property

It is well known that Desargues' property (see Fig. 7) holds in any projective space of dimension higher or equal to 3. This was formally proven in [12]. However, when considering projective planes, Desargues' property is independent of the axiom system of Fig. 1. This means that there exists Desarguesian and

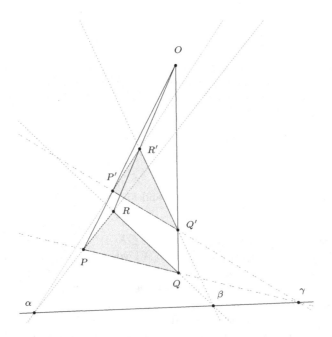

**Fig. 7.** A configuration of Desargues' property.

non-Desarguesian planes. For instance, Moulton's plane (see [11,15] for details) or Hall's [9] planes of order 9 are non-Desarguesian planes. Desargues' theorem states:

**Theorem.** *If the three lines joining the corresponding vertices of two triangles PQR and P'Q'R' all meet in a point O called the perspector[1], then the three intersections of pairs of corresponding triangle sides lie on a line $\alpha\beta\gamma$. Equivalently, if two triangles are perspective from a point, then they are perspective from a line.*

Now that the geometric framework is depicted, we shall investigate possibilities of automation within proofs. Throughout this paper, we aim at proving that some finite structures described using only points, lines and an incidence relation are models of these axiom systems. When dealing with plane projective geometry, we also analyze whether Desargues' property actually holds.

# 3 Dealing with Complexity in Building Some Finite Models of Incidence Projective Geometry

## 3.1 Plane

We use finite projective models to study the large-scale automation of proofs of geometric properties. One can prove fairly easily that the axioms of projective plane geometry hold for $pg(2,2)$, $pg(2,3)$ and $pg(2,5)$. In the same way, we show that the axioms of rank theory hold for $pg(2,2)$ and $pg(2,3)$. We use these examples to show how to manage the proof complexity in Coq.

We identify several criteria (e.g. the geometric context, the formulation of the statements) which can strongly influence the complexity of the proofs. As an example, we compare some proofs which have been mechanized in Coq using both incidence projective geometry and rank theory.

**Finite Model.** First of all, we work on a finite domain. In this context, to carry out geometric reasoning, it is necessary to know all the points and lines (and planes) that describe our finite projective plane (resp. projective space). For instance the description of $pg(2,3)$ contains 13 points and 13 lines (see Fig. 6) in incidence projective geometry and looks like:

```
Inductive ind_Point : Set := A | B | C | ... | K | L | M.

Inductive ind_line : Set := ABCD | AEFG | AIJM | AHKL | BEHI | BGJL
| BFKM | CELM | CFHJ | CGIK | DEJK | DGHM | DFIL.

Definition Incid_bool (P:Point) (l:Line) : bool := match P with
  | A => match l with
```

---

[1] The perspector is the point at which the three lines connecting the vertices of two perspective triangles concur.

```
                    | ABCD | AEFG | AIJM | AHKL => true
                    | _ => false
             end
      [...]
end.
```

The description of finite models can be easily generated algorithmically by only specifying all points and lines. In this way, the relation of incidence linking these two objects is thus automatically created. The size of the specification of $pg(2,n)$ increases quickly as $n$ grows bigger, indeed $pg(2,n)$ has $n^2+n+1$ points and as many lines.

**Case Analysis.** In such a finite model, to prove a geometric statement requires to check all the possible configurations of this theorem, i.e. to perform case analysis on both points and lines. Most often a brute-force approach leads to too many cases, which makes the proof not tractable in Coq. Let us illustrate this case analysis issue on one of the axioms of the incidence projective geometry in the finite projective plane $pg(2,3)$. For instance, the (A3P2) *Uniqueness* axiom:

```
Lemma uniqueness : forall A B :Point, forall l m : Line,
   Incid A l -> Incid B l -> Incid A m -> Incid B m -> A=B \/ l=m.
```

As $pg(2,3)$ contains 13 points and 13 lines, basic case analysis leads to $13^4 = 28\ 051$ cases to be dealt with. This situation is not yet critical, such a proof is still easily performed. It becomes more tedious when dealing with $pg(2,5)$ and its 31 points and 31 lines, where 923 521 cases must be studied. For a given $n$, the projective plane $pg(2,n)$ has $(n^2 + n + 1)^4$ possible combinations to be investigated, so such proofs are tractable only for some small $n$.

**Formulation and Choice of Theory.** A second factor strongly influencing complexity is the formulation of statements. This question is well known and studied in the theory of complexity especially in the problem SAT [1,17]. Criteria such as the size of the clauses, number of propositions and the order of propositions have a significant impact on the resolution time of a proof. For example, let us consider the two definitions of intersection existence in $pg(2,3)$ first using incidence geometry, and second using ranks.

```
Lemma point_existence : forall (l1 l2 :Line),
   exists A : Point, Incid A l1 /\ Incid A l2.
```

```
Lemma rk_inter : forall A B C D : Point,
   exists J, rk(triple A B J) = 2 /\ rk(triple C D J) = 2.
```

Case analysis in the first description generates $13^2 = 169$ cases before providing a witness to the existential quantifier whereas in the second statement we again face $13^4 = 28\ 051$ cases. It would be necessary to create a method of resolution of the existential formula one hundred times faster in rank theory to obtain the same execution time as in incidence projective geometry. So choosing an appropriate description of a formula is utterly relevant to make the proofs doable in practise. The best way to deal properly with the combinatorial explosion caused by successive case analysis is to manage the pruning of the proof tree as early as possible.

**Proof Tree Pruning.** Let us consider again the axiom of *Uniqueness* (A3P2) and its proof in the finite projective plane $pg(2,3)$:

```
Lemma uniqueness : forall A B : Point, forall l m : Line,
Incid A l -> Incid B l -> Incid A m -> Incid B m -> A=B \/ l=m.
Proof.
induction A;induction B;induction l;induction m;
try discriminate;try (left;reflexivity);try (right;reflexivity).
Qed.
```

Basic case analysis without pruning and quantifier management gives rise to 28 051 cases. A brute-force execution takes on a standard machine[2] about 40 s in this situation. More clever strategies are required to ensure that the proofs remain tractable. The variable $A$ is linked to $l$ in the hypothesis `Incid A l`. It is thus possible to prune the proof tree after the induction on line $l$ when the point $A$ is not incident to the line $l$. Another improvement consists in solving directly the left hand side of the goals right after the induction on $B$ when the equality $A = B$ (i.e. the left side of the disjonction) holds. It is not necessary to carry on and perform case analysis on the next variable $m$ if the goal can be discarded or is already verified. These two adjustments allow the proof to be built in less than 1 s.

```
Lemma uniqueness : forall A B : Point, forall l m : Line,
Incid A l -> Incid B l -> Incid A m -> Incid B m -> A=B \/ l=m.
Proof.
induction A;induction l;try discriminate;
induction B;try discriminate;try (left;reflexivity);
induction m;try discriminate;try (right;reflexivity).
Qed.
```

**Constraining Hypothesis.** Scheduling quantifiers based on assumptions can have a strong impact on proof tree pruning. In other words, the order in which the case analysis is performed is important. Furthermore, it is important to consider the pruning power of each hypothesis. The idea is to use first the most restrictive assumptions to prune as much as possible and as soon as possible to limit the width of the proof tree. Let us consider the assumptions $A \neq B$ and `Incid A l` in $pg(2,3)$. After performing induction on all variables, the first

---

[2] Computer setup: Intel(R) Core(TM) i5-4460 CPU @ 3.20 GHz with 16G of memory.

assumption allows to eliminate 7 cases out of 49 while the second one removes 28 cases out of 49. It is therefore more interesting to take the incidence hypothesis into account to quickly eliminate goals.

**Pseudo Depth-First Search.** In highly-branching proofs, when the previous optimizations are not sufficient (because memory consumption is too big), we adapt the classical breadth-first search of Coq (`tac1;tac2;tac3`). By taking advantage of the right assiociativity, we carry out pseudo depth-first search in order to limit number of cases at each level of the demonstration (`tac1;(tac2;tac3)`). Finally, we work with the `abstract` [6] tactic to prove a sub-goal as a separate lemma to structure huge proof terms and to facilitate type checking.

These optimizations are independent of each others and allow to prove more lemmas, even when the combinatorial is huge.

## 3.2 Space

The above-mentioned techniques are even more relevant when dealing with the smallest projective space $pg(3,2)^3$. It features 15 points and 35 lines. In the same way as in the plane, we can prove that the axioms of projective space geometry hold for $pg(3,2)$. However, it is a little more challenging to prove this. Indeed, while writing and feeding Coq with the proofs, we face strong limitations related to memory usage. Tactics have to be carefully designed and decomposition should be smart enough to avoid facing thousands of millions of sub-goals at the same level. Consider for instance the statement of *Pasch*'s axiom in $pg(3,2)$:

```
Lemma pasch : forall A B C D : Point, forall lAB lCD lAC lBD : Line,
              all_distinct A B C D ->
              Incid A lAB /\ Incid B lAB ->
              Incid C lCD /\ Incid D lCD ->
              Incid A lAC /\ Incid C lAC ->
              Incid B lBD /\ Incid D lBD ->
              (exists I : Point, (Incid I lAB /\ Incid I lCD)) ->
              exists J : Point, (Incid J lAC /\ Incid J lBD).
```

As finite space $pg(3,2)$ contains 15 points and 35 lines, case analysis leads to $15^4 \times 35^4 = 75\,969\,140\,625$ cases to be dealt with. It is thus essential to limit the size of proof tree by eliminating as many cases as soon as possible. The order in which we perform inductions is no longer sufficient to maintain a tractable proof.

Proof parts usually proved using Ltac sophisticated tactics without user interaction need to be factorized into relevant lemmas and a careful decomposition into several intermediate lemmas is mandatory to complete the proof. In the

---

[3] An interactive representation of $pg(3,2)$ can be viewed on wolfram web site: http://demonstrations.wolfram.com/15PointProjectiveSpace/.

proof of *Pasch*'s property, we state the following intermediate lemma which provides the actual line which carries two given (distinct) points $T$ and $Z$. The function l_from_points computes a line which goes throught the two points $T$ and $Z$ (this line is unique when we have $T \neq Z$).

Here, the proofs-as-programs paradigm is fully exploited. Indeed, this function can be written as a simplified (non-dependent) version of the property (A1P3) *Line-existence* which can be directly used as a program[4]. It allows us to perform case analysis on lines without adding further cases (only one case is correct at each step).

Similarly, a program which retrieves the points which belongs to a given line $l$ can easily be extracted from theorem (A4P3) *Three-Points*.

```
Lemma points_line : forall T Z : Point, forall x : Line,
Incid T x -> Incid Z x -> T<>Z -> x=(l_from_points(T,Z)).
```

In this way, we reduce the overall number of cases to check to $15^4 = 50625$ cases, before performing the elimination of the existential hypothesis in *Pasch*'s axiom: exists I :Point, (Incid I lAB /\ Incid I lCD).

So far we made proofs manageable by the system, but we still need to help the user to write proofs. That is what we shall study in the next section.

## 4  Automating Proofs of Desargues's Property

All the techniques presented above in order to prove that some small planes or projective spaces are models of the projective incidence geometry can be reused to carry out the proof of Desargues' theorem in each of these models.

```
Lemma Desargues : forall O P Q R P' Q' R' X Y Z X' Y' Z'
X'' Y'' Z'' alpha beta gamma,
all_distinct O X Y Z X' Y' Z' X'' Y'' Z'' ->
rk(O,X,Y,Z)=2 -> rk(O,X',Y',Z')=2 -> rk(O,X'',Y'',Z'')=2 ->
rk(P,Q,gamma)=2 -> rk(P',Q',gamma)=2 -> rk(P,R,beta)=2 ->
rk(P',R',beta)=2 -> rk(Q,R,alpha)=2 -> rk(Q',R',alpha)=2 ->
rk(P,O,X,Y,Z)=2 -> rk(P',O,X,Y,Z)=2 ->
rk(Q,O,X',Y',Z')=2 -> rk(Q',O,X',Y',Z')=2 ->
rk(R,O,X'',Y'',Z'')=2 -> rk(R',O,X'',Y'',Z'')=2 ->
rk(O,P,P')=2 -> rk(O,Q,Q')=2 -> rk(O,R,R')=2 -> rk(O,P,Q)=3 ->
rk(O,P,R)=3 -> rk(O,Q,R)=3 -> rk(P,Q,R)=3 -> rk(P',Q',R')=3 ->
( rk(P,P')=2 \/ rk(Q,Q')=2 \/ rk(R,R')=2 ) ->
rk(alpha,beta,gamma)=2.
```

It is well-known that the projective planes $pg(2, n)$ are Desarguesian. We formally prove these results in Coq for $pg(2,2)$ and $pg(2,3)$. As in the previous proofs, using a naive approach leads to intractable proofs. The property of

---

[4] Fully-specified functions can be automatically defined using the proof search capabilities of Coq.

Desargues is expressed using 10 points. The last three ones can be automatically calculated from the first seven ones. In $pg(2,3)$, induction on the first 7 points yields several billion cases to be treated without pruning.

In this case, ranks provide a much more efficient approach to handle the numerous configurations that we need to check. It is tractable if we prune the proof tree as much as possible during inductions on the ten points of the property. Automating this proof relies on some geometric aspects of Desargues' property and on the data structure of ranks.

### 4.1   Automation Through Geometry

First of all, we take advantage of some symmetries in Desargues' property. In the first place, we use the symmetry of the problem w.r.t. the center of perspective. By fixing this center as one of the points of the model $pg(2,x)$ and proving that the permutation of the points in a finite model remains a finite model, it is possible to prove that the property of Desargues holds whatever the center of perspective selected. Intuitively, this symmetry allows us to avoid induction on the perspector point.

The second symmetry that we use to decompose the problem follows from the permutation of the concurrent lines at the center of perspective. Let $A$ be the perspector, it is possible to fix the straight lines containing $A$ to form the two triangles. Subsequently, we show that every permutation of these lines always satisfies the property.

Finally, we take advantage of the conditions of non-degeneracy to quickly eliminate the degenerate cases of Desargues' theorem and thus limit the combinatorial explosion. For example, it is possible to consider a more general theorem where the two triangles can share at most two points in common. This theorem leads to a contradiction in the specification of the line $\alpha\beta\gamma$ (some lines are confused). By restricting the theorem to the case where triangles can have only one point in common, we eliminate approximately 33% of the goals at all levels of the demonstration.

### 4.2   Automation Thanks to Proof Engineering

Thanks to the rank structure, we can represent homogeneously all incidences of our geometric context by dealing only with points. Intuitively this means that we can avoid performing case analysis on lines without increasing the number of cases on the points. For instance considering Desargues' theorem, six case analyses on lines can be removed. It becomes even more meaningful in the higher dimensions when manipulating planes, etc.

In addition, when writing tactics with Ltac to perform simplifications (e.g. rewriting, elimination of contradiction, attempt to solve), there is no need to consider objects of several types or multiple predicates. We simply match the result returned by the function of rank, as all propositions are of the form $rk(E) = n$ where $E$ is a set of points and $n$ is a natural number representing intuitively the dimension of the set.

Finally, it is better to avoid generic tactics such as `auto`, `intuition` or `omega`, and to use specific lemmas which solve the goal instead. Proofs of statements of the form $rk(E) = n$ usually proceed by first proving separately that $rk(E) \leq n$ and $rk(E) \geq n$, and then use `omega` to deduce the equality from the two inequalities. Of course, if such a proof scheme is heavily used, running `omega` becomes a bottleneck. We can instead write a simple lemma ($\forall n : \mathtt{nat}, rk(E) \leq n \rightarrow rk(E) \geq n \rightarrow rk(E) = n$) and apply it to conclude the proof. An single application of `apply` is always significantly cheaper than calling `omega`. However, the drawback is that we have a more specific proof, which may be less robust to changes in the specification. Finding such bottlenecks can be easily achieved using the Ltac profiler [18].

## 5   Conclusion

We verify that some finite planes (resp. spaces) are actually models of projective plane (resp.space) geometry. We achieve that by using two distinct approaches, a mathematics-oriented one and a computer-science-oriented one featuring ranks. Overall it represents 5000 lines of specification and 2500 lines of proofs. All the results are summarized in Table 3. For each formalization, it presents three key figures: the number of lines of specification, the number of lines of proof as well as the time required to compile it.

**Table 3.** Benchmarks for various proofs using Coq on an Intel(R) Core(TM) i5-4460 CPU @3.20 GHz with 16G of memory. CE means combinatorial explosion.

| | Formalization of projective geometry | | | | | |
| | Using the synthetic description | | | Using ranks | | |
| | Spec. | Proof | Compile time | Spec. | Proof | Compile time |
|---|---|---|---|---|---|---|
| $pg(2,2)$ is a model | 216 | 71 | 2 s | 127 | 42 | 16 s |
| Desargues holds in $pg(2,2)$ | 188 | 205 | 37 s | 297 | 162 | 26 s |
| $pg(2,3)$ | 149 | 46 | 7 s | 309 | 77 | 2055 s |
| Desargues *in* $pg(2,3)$ | 191 | 225 | CE | 2089 | 386 | 10700 s |
| $pg(2,5)$ | 74 | 28 | 90 s | CE | | |
| Desargues *in* $pg(2,5)$ | CE | | | CE | | |
| $pg(3,2)$ | 267 | 67 | 4309 s | CE | | |
| Desargues *in* $pg(3,2)$ | Overall proof in 3D thanks to [4,12] | | | | | |

This provides a good stress test for Coq. Indeed, it is a small theory, but proving that the axioms hold requires performing huge proofs with numerous cases. Our experiments shed light on some regression in the efficiency of Coq to perform proofs and type-check them, starting from version 8.5. This issue is currently being addressed by the coqdev team.

The optimizations that we propose allow to go further in the order of magnitude of the planes/spaces that we can handle. Eventually, an interesting goal would be to tackle some of Hall's planes which feature 91 points and 91 lines. Currently, we are working on a more comprehensive benchmark featuring more projective planes/spaces and using various provers using the TPTP framework [17]. Using brute-force, only 3 provers find a proof of Desargues' theorem in a suitable time of 300 s for $pg(2, 2)$ (iprover, Vampire [10] and Z3 [8]). The Vampire SAT seems very promising with solutions 10 times faster. However, provers do not provide a formal checkable proof.

The Coq development is available at https://github.com/Projective Geometry/.

# References

1. Armand, M., Grégoire, G., Keller, B., Théry, L., Werner, B.: Verifying SAT and SMT in Coq for a fully automated decision procedure. In: International Workshop on Proof-Search in Axiomatic Theories and Type Theories (PSATTT 2011) (2011)
2. Batten, L.M.: Combinatorics of Finite Geometries. Cambridge University Press, Cambridge (1997)
3. Bertot, Y., Castéran, P.: Interactive Theorem Proving and Program Development, Coq'Art: The Calculus of Inductive Constructions. Springer, Heidelberg (2004). https://doi.org/10.1007/978-3-662-07964-5
4. Braun, D., Magaud, N., Schreck, P.: An equivalence proof between rank theory and incidence projective geometry. In: Automated Deduction in Geometry (ADG 2016), pp. 62–77 (2016)
5. Buekenhout, F. (ed.): Handbook of Incidence Geometry. North Holland, Amsterdam (1995)
6. Coq Development Team: The Coq Proof Assistant Reference Manual, Version 8.6. LogiCal Project (2017)
7. Coxeter, H.S.M.: Projective Geometry. Springer, New York (2003)
8. de Moura, L., Bjørner, N.: Z3: an efficient SMT solver. In: Ramakrishnan, C.R., Rehof, J. (eds.) TACAS 2008. LNCS, vol. 4963, pp. 337–340. Springer, Heidelberg (2008). https://doi.org/10.1007/978-3-540-78800-3_24
9. Hall, M.: Projective planes. Trans. Am. Math. Soc. **54**(2), 229–277 (1943)
10. Kovács, L., Voronkov, A.: First-order theorem proving and VAMPIRE. In: Sharygina, N., Veith, H. (eds.) CAV 2013. LNCS, vol. 8044, pp. 1–35. Springer, Heidelberg (2013). https://doi.org/10.1007/978-3-642-39799-8_1
11. Magaud, N., Narboux, J., Schreck, P.: Formalizing projective plane geometry in Coq. In: Sturm, T., Zengler, C. (eds.) ADG 2008. LNCS (LNAI), vol. 6301, pp. 141–162. Springer, Heidelberg (2011). https://doi.org/10.1007/978-3-642-21046-4_7
12. Magaud, N., Narboux, J., Schreck, P.: A case study in formalizing projective geometry in Coq: Desargues theorem. Comput. Geom.: Theor. Appl. **45**(8), 406–424 (2012)
13. Mahboubi, A., Tassi, E.: Mathematical Components. Draft (2016)
14. Michelucci, D., Schreck, P.: Incidence constraints: a combinatorial approach. Int. J. Comput. Geom. Appl. **16**(5), 443–460 (2006)
15. Moulton, F.R.: A simple non-desarguesian plane geometry. Trans. Am. Math. Soc. **3**(2), 192–195 (1902)

16. Oxley, J.G.: Matroid Theory, vol. 3. Oxford University Press, Oxford (2006)
17. Sutcliffe, G.: The TPTP problem library and associated infrastructure. J. Autom. Reason. **43**(4), 337 (2009)
18. Tebbi, T., Gross, J.: A profiler for Ltac. In: Coq PL Workshop 2015 (2015)

# Into the Infinite - Theory Exploration for Coinduction

Sólrún Halla Einarsdóttir[1]([⊠]), Moa Johansson[1]([⊠]),
and Johannes Åman Pohjola[2]([⊠])

[1] Chalmers University of Technology, Gothenburg, Sweden
{slrn,moa.johansson}@chalmers.se
[2] Data61/CSIRO, Sydney, Australia
johannes.amanpohjola@data61.csiro.au

**Abstract.** Theory exploration is a technique for automating the discovery of lemmas in formalizations of mathematical theories, using testing and automated proof techniques. Automated theory exploration has previously been successfully applied to discover lemmas for inductive theories, about recursive datatypes and functions. We present an extension of theory exploration to coinductive theories, allowing us to explore the dual notions of corecursive datatypes and functions. This required development of new methods for testing infinite values, and for proof automation. Our work has been implemented in the Hipster system, a theory exploration tool for the proof assistant Isabelle/HOL.

## 1 Introduction

Coinduction and corecursion are dual notions to induction and recursion that admit the specification of potentially infinite structures and functions that operate on them. Their many applications in theoretical computer science include, to name a few: defining and verifying behavioral equivalence of processes [21], Hoare logic for non-terminating programs [23], total functional programming in the presence of non-termination [29], and accounting for lazy data in functional languages like Haskell. Recently, support for coinduction in proof assistants has matured significantly, with powerful definitional packages and reasoning tools [1,5,6].

In this paper, we extend a technique, called *theory exploration* [7], and present a tool that automatically discovers and proves equational properties about corecursive functions in the proof assistant Isabelle/HOL [24], a widely used interactive theorem proving system featuring both automated and interactive proof techniques. The purpose of theory exploration is to automate the discovery of basic lemmas when, for instance, developing a new theory. The human user can then focus on inventing and proving more complex conjectures, using the automatically generated background lemmas. As an appetizer, consider this simple example of an Isabelle theory:

```
codatatype (sset: 'a) Stream = SCons (shd: 'a) (stl: "'a Stream")
```

© Springer Nature Switzerland AG 2018
J. Fleuriot et al. (Eds.): AISC 2018, LNAI 11110, pp. 70–86, 2018.
https://doi.org/10.1007/978-3-319-99957-9_5

**primcorec** `smap :: "(’a ⇒ ’b) ⇒ ’a Stream ⇒ ’b Stream"` **where**
   `"smap f xs = SCons (f (shd xs)) (smap f (stl xs))"`

**primcorec** `siterate :: "(’a ⇒ ’a) ⇒ ’a ⇒ ’a Stream"` **where**
   `"siterate f a = SCons a (siterate f (f a))"`

**cohipster** `smap siterate` — tell Hipster to explore these functions

The theory above defines the codatatype `Stream` of infinite sequences, the function `smap` that maps a function onto every element of a stream, and the function `siterate` that given a function $f$ and an initial element $x$ generates the sequence $f(x), f(f(x)), f(f(f(x))), \ldots$. The verbatim output of our tool, Hipster, is as follows:

lemma `lemma_a [thy_expl]:` `"smap y (siterate y z) = siterate y (y z)"`
   **by** `(coinduction arbitrary: y z rule: Stream.coinduct_strong)`
      `auto`

lemma `lemma_aa [thy_expl]:` `"SCons (y z) (smap y x2) = smap y (SCons`
`z x2)"`
   **by** `(coinduction arbitrary: x2 y z rule: Stream.coinduct_strong)`
      `simp`

lemma `lemma_ab [thy_expl]:` `"smap z (SCons y (siterate z x2)) = SCons`
`(z y) (siterate z (z x2))"`
   **by** `(coinduction arbitrary: x2 y z rule: Stream.coinduct_strong)`
      `(simp add: lemma_a)`

This Isabelle snippet, when pasted into the theory (simply by a mouse-click), proves the discovered laws about `smap` and `siterate` by coinduction. The first lemma, `lemma_a`, may appear familiar as it describes the *map-iterate property* [3]. The whole process of generation and proof took Hipster less than 10 s on a regular laptop computer. Moreover, the generated proofs are formal proofs, machine-checked down to the axioms of higher-order logic.

   Note that at no point did the user need to supply the conjectures or proofs. Hipster uses a specialized conjecture discovery subsystem, called QuickSpec [28], which heuristically generates type-correct terms and uses automated testing to invent interesting candidate lemmas. We give a brief introduction to QuickSpec in Sect. 2, along with a lightweight introduction to coinduction.

   Earlier versions of Hipster [14, 16] supported only induction and recursive datatypes. The main difference when we also treat codatatypes is in the testing phase, when conjectures are generated. Naively testing and evaluating terms for equivalence cannot be done in the same way as for regular datatypes, since instances of a codatatype like `Stream` are infinite, so testing would not terminate. Our solution to this conundrum is that for testing purposes, we generate step-indexed *observer functions* for the codatatypes under consideration. These operate on a copy of the codatatype with an extra nullary constructor, that we return when the step-index reaches 0. The step-indexing guarantees that testing will terminate. Section 3 describes this in more detail, along with our approach to coinductive proof exploration.

We evaluate our tool by testing it on several examples of codatatypes and corecursive functions in Sect. 5. Results are encouraging: we can discover and prove many well-known and useful properties. Similar theory exploration systems can be found in the literature [9,15,20,22], but ours is the first system capable of discovering and proving properties of coinductive types and corecursive functions. We integrate inductive and coinductive reasoning, so that in a theory featuring both recursion and corecursion, both inductive and coinductive proofs can be discovered even when one depends on the other. The source code and examples are available online.[1]

## 2    Background

We give a brief introduction of coinduction for readers unfamiliar with the concept, followed by an introduction to the proof assistant Isabelle/HOL and the Hipster theory exploration system.

*Coinduction.* Coinduction is the mathematical dual of structural induction, relying on deconstructing structures top-down instead of constructing them bottom-up as induction does. Consider lists with elements of type a, defined by: `List a = Nil | Cons a (List a)`.
The inductive reading of this declaration is that it specifies everything that can be constructed from the empty list `Nil` in a finite number of steps, by using the `Cons` constructor to add elements. The coinductive reading is that it specifies everything that is either `Nil` or can be decomposed ("destructed") into a head and a tail, where the tail is either `Nil` or something that can be destructed into another head and tail, and so on. The latter reading encompasses not only `Nil`-terminated lists, but also infinite lists built from `Cons` only. We say that the first reading defines a *datatype* while the second defines a *codatatype*.

Since codata need not bottom out in a base case, proof by induction does not apply; instead we resort to the dual notion of coinduction, which allows us to prove equalities between elements $x, y$ of a codatatype by exhibiting a *candidate relation* $R$ such that $x \ R \ y$ and $R$ is closed under destruction. For example, here is the coinduction principle for the `Stream` type introduced in Sect. 1:

$$\frac{\forall s_1, s_2 \quad \dfrac{R \ s_1 \ s_2}{shd \ s_1 = shd \ s_2 \land R \ (stl \ s_1) \ (stl \ s_2)}}{s = s'} R \ s \ s'$$

In words: to show that $s = s'$, we must prove that for all pairs $s_1, s_2$ related by $R$, $s_1$ and $s_2$ have the same heads and $R$-related tails. Interested readers can find a more detailed introduction to coinduction in [27] or [13].

---

[1] https://github.com/moajohansson/IsaHipster.

*Isabelle/HOL.* Isabelle/HOL is an interactive proof assistant for higher-order logic [24]. Users write definitions and proofs in *theory files*, which are checked by running them through Isabelle's small trusted logical kernel to ensure each step in a proof is correct. More complex proof techniques, called *tactics*, can be built up using combinations of basic inference rules from the trusted kernel. Isabelle is an *interactive system*, meaning that there are both automated and semi-automated tactics available. An example of the former is the *simplifier*, which performs equational reasoning automatically. An example of the latter is Isabelle's (co)induction tactics, which applies a (maybe user given) induction rule to a subgoal while leaving it to the user how to prove the resulting subgoals. Sledgehammer is a useful tool in Isabelle which allows outsourcing proofs to fully automated external first-order (FO) or SMT-solvers [25]. When the external provers report back, the proof is reconstructed inside Isabelle's trusted kernel. In our work on Hipster, we combine Isabelle's interactive tactics with Sledgehammer to provide automation for (co)inductive proofs.

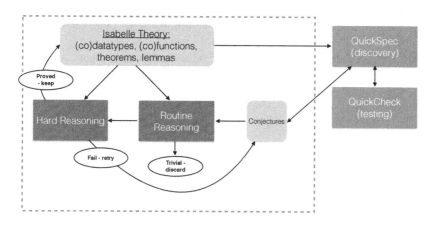

**Fig. 1.** The architecture of the Hipster system.

*Hipster.* The architecture of the Hipster system is shown in Fig. 1. Hipster outsources conjecture generation to the external tool QuickSpec. QuickSpec generates type-correct terms in order of size, up to a given limit. At each step, it evaluates the terms on randomly generated test data, using the property-based testing tool QuickCheck [8]. Based on the results of testing, terms are divided into equivalence classes from which equational conjectures are extracted. For a full description of QuickSpec's conjecture generation algorithm and its heuristics we refer the reader to [28]. The conjectures produced by QuickSpec are then read back into the Isabelle/HOL environment for proof. The conjectures have been thoroughly tested at this point, so we have quite good reasons to believe they may actually be true. However, not all of them might be considered interesting by a human. In particular, statements that have trivial proofs are rarely exciting.

Hipster therefore takes two reasoning strategies as parameters: *routine reasoning* (often just rewriting), and *hard reasoning* (for instance coinduction). Depending on the exact configuration of the routine and hard reasoning strategies, we can tweak Hipster to produce slightly different output: the conjectures that follow from using only routine reasoning are discarded, while those proved by the hard reasoning strategy are reported back to the user. Whenever Hipster proves a lemma, it may use it in subsequent proofs. This means that during exploration, its automated proof strategies become more powerful as more lemmas are found. Should some conjecture fail to be proved by either of the proof strategies, it is also presented to the user, who can try a manual proof.

## 3    Testing Infinite Structures

Recall from Sect. 2 that Hipster's conjecture generation subsystem, QuickSpec, relies on being able to test terms on randomly generated values. When a codatatype has no finite instances, as in the case of streams, QuickSpec cannot directly check the equality of any of the generated terms, since that would take an infinite amount of time due to their infinite size. Thus testing will not work.

When an Isabelle user invokes Hipster on a coinductive theory, an *observer type* and *observer function* are generated for every type under consideration. These types and functions ensure that QuickSpec only tests a (randomly chosen) finite prefix of any infinite values, using support for *observational equivalence*. This allows Hipster to discover lemmas about codatatypes without finite instances.

*Observational Equivalence in QuickSpec.* When used interactively through its Haskell interface, QuickSpec supports observational equivalence to deal with types that for instance have no finite instances, and thus cannot be directly compared [28]. Note that in this case, the user must define a function for observing such a type and state that two values of the type are equivalent if all such observations make them equal. We have extended this functionality by developing a method to *automatically generate* observer functions for the codatatypes being explored and added it to the interface between Hipster and QuickSpec.

More specifically, observer functions are used as follows: For any type $T$, QuickSpec can be given an observer function of type $Obs \rightarrow T \rightarrow Res$, where $Obs$ can be any type that QuickSpec can generate random data for, and $Res$ any type that can be compared for equality. QuickSpec will then include a random value of type $Obs$ as part of each test case, and will compare values of type $T$ by applying this observer function using the random value of type $Obs$ and comparing the resulting values of type $Res$. For instance, we can define an observer function for streams:

$$obsStream :: Int \rightarrow Stream \rightarrow List,$$

where $obsStream\ n\ s$ returns a list containing the first $n$ elements of the stream $s$. If we supply this observer function to QuickSpec it will generate a random integer $n$ for each test case where streams are to be observed, and assume that two streams are equal if their first $n$ elements are equal in every case.

*Generating Observer Functions.* For Hipster, we want to relieve the user of having to define the observer function by hand, and instead generate it automatically. Our method of generating observer functions is inspired by the *Approximation Lemma* [2,12]. Here, a so called *approximation function*, *approx*, is defined in the same way as the recursive identity function for a given type, except that it has an additional numeric argument which is decremented at each recursive call. The lemma states that $a = b$ if *approx* $n$ $a$ = *approx* $n$ $b$ for all values of $n$. For the Stream type introduced in Sect. 1, the approximation function is defined as:

$$approx\ (n+1)\ xs = SCons\ (shd\ xs)\ (approx\ n\ (stl\ xs))$$

The function is undefined for $n = 0$ and therefore returns a partial structure, for instance, if *zeroes* is a stream of zeroes then *approx* 1 *zeroes* evaluates to the partial stream $SCons\ 0\ \bot$, where $\bot$ represents an undefined value.

To make our solution practical we, instead of using the undefined value $\bot$, generate a new type that has the same structure as the type being observed, but with an additional nullary constructor. For example, the generated observation type for a stream is:

$$OStream\ a = OSCons\ a\ (OStream\ a)\ \mid\ NullConsStream$$

We then generate an observer function for a given type $T$ with an observer type $ObsT$ in the following manner:

$$obsFunT :: Nat \rightarrow T \rightarrow ObsT$$
$$obsFunT\ 0\ \_\ =\ NullConsT$$
$$obsFunT\ n\ t\ =\ approx'\ n\ t$$

where $approx'$ is like the recursive identity function for $T$ except that it replaces each constructor occurring in $t$ with the equivalent constructor for $ObsT$, and the *fuel* parameter $n$ is decremented at every recursive call, ensuring we will only attempt to observe a finite prefix. As an example, an observer function for streams using the observer type from above is shown below:

$$obsFunStream :: Nat \rightarrow Stream\ a \rightarrow OStream\ a$$
$$obsFunStream\ 0\ \_\ =\ NullConsStream$$
$$obsFunStream\ n\ (SCons\ x\ xs)\ =\ OSCons\ x\ (obsFunStream\ (n-1)\ xs)$$

Some care needs to be taken when decrementing the numeric fuel argument which determines how much more of the structure should be observed, as using $n - 1$ in every step results in testing being too slow for structures with larger branching factors, such as trees. For now, we use a heuristic measure which decrements $n$ to $n/\#constructors - 1$ in each recursive call. For *OStream*, this is simply $(n-1)$, while for e.g. binary trees, defined:

$$Tree\ a = TNode\ a\ (Tree\ a)\ (Tree\ a)$$

with an observer type defined:

$$OTree\ a = OTNode\ a\ (OTree\ a)\ (OTree\ a)\ |\ NullConsTree$$

the fuel counter is decremented to $n/2-1$ for each branch, as seen in the observer function definition below:

$$obsFunTree :: Nat \to Tree\ a \to OTree\ a$$
$$obsFunTree\ 0\ \_\ =\ NullConsTree$$
$$obsFunStream\ n\ (TNode\ x\ l\ r)\ =$$
$$OTNode\ x\ (obsFunTree\ (n/2 - 1)\ l)\ (obsFunTree\ (n/2 - 1)\ r)$$

# 4    Automating Proofs of Coinductive Lemmas

Isabelle/HOL features a built-in `coinduction` tactic that applies a coinduction principle to a goal, with the candidate relation instantiated to be the singleton relation containing the equation in the conclusion. After applying this tactic the user must decide how to finish the proof after the coinductive step. However, the ability to automatically prove lemmas without user involvement is crucial in lemma discovery by automated theory exploration. Therefore we have extended Hipster with an automated tactic for proving coinductive lemmas. In order to do this, we must automatically determine the parameters for our call to Isabelle/HOL's coinduction tactic, and then automate the subgoal proofs.

*Automatically Determining Parameters.* Isabelle/HOL's coinduction tactic has parameters to set which variables are *arbitrary*, meaning that they appear universally quantified in the candidate relation (and hence existentially quantified in the conclusion of the resulting subgoal). It also has an optional parameter to specify which coinduction rule to use.

Our default setting is to set all free variables in the current goal as arbitrary. This yields weaker proof obligations, at the expense of introducing existential quantifiers in the goal, which is sometimes less automation-friendly since it may require guessing an instantiation to discharge the goal. Our experience is that setting at least some variables to arbitrary is necessary for all but the most trivial of proofs; for the rest, the goal statements are simple enough that the extra existentials do not cause any difficulty in practice.

The built-in `coinduction` tactic also has an optional parameter to specify what coinduction rule should be used for the proof. We must again make a tradeoff between one that can be applied to prove as wide a range of lemmas as possible, such as coinduction up-to the codatatype's companion function [26];

and one that yields simple and automation-friendly subgoals, such as the (weak) coinduction principle associated with the datatype.

For reasoning about functions defined with primitive corecursion, we find that the strong coinduction principle generated by the datatype package works well in practice. It allows one to close the proof by proving either equality or membership in the candidate relation. For example, here is the strong coinduction principle for the $\mathtt{Stream}$ type defined in Sect. 1:

$$R\ s\ s' \quad \forall s_1, s_2 \ \frac{R\ s_1\ s_2}{shd\ s_1 = shd\ s_2 \wedge (R\ (stl\ s_1)\ (stl\ s_2) \vee stl\ s_1 = stl\ s_2)))}{s = s'}$$

Note that the (weak) coinduction principle shown in Sect. 2 differs by omitting the right-hand side $stl\ s1 = stl\ s2$ of the disjunction. The extra disjunction is lightweight enough not to confuse the simplifier, and the equality has very important consequences: it allows equations that have previously been proven by coinduction to be re-used in the proof, without having to include them in the candidate relation. This allows us to automatically prove, e.g., the associativity of append on lazy lists as seen in Sect. 5.1.

The recent AmiCo definitional package by Blanchette et al. [4] allows a form of non-primitive corecursion where corecursive calls may be guarded by *friends* in addition to constructors. A friend is a function that consumes at most one constructor to produce a constructor. For functions with friend-guarded corecursive calls, the strong coinduction rule often results in an unsuccessful proof attempt: terms on the shape required by the candidate relation tend to occur as arguments to friends rather than at top-level. Fortunately, the AmiCo package generates a coinduction principle up-to friendly contexts covering precisely this use case. Hence we prioritize such coinduction principles over the strong coinduction principle whenever they are relevant, i.e., whenever the goal state mentions a function symbol defined using non-primitive corecursion.

*Proving Subgoals.* After applying coinduction, Hipster's $\mathtt{simp\_}$ or sledgehammer tactic is applied to the current proof state in an attempt to prove the remaining subgoals and conclude the proof of the lemma. This tactic first attempts to complete the proof using Isabelle's automatic simplification procedure $\mathtt{simp}$. If this does not suffice it uses Isabelle's automated proof construction tool Sledgehammer [25] to attempt to construct a proof. Since Sledgehammer is quite powerful, this tactic is sufficient to conclude the proofs of a wide range of lemmas.

*Mixed Induction and Coinduction.* In practice, theories are neither purely inductive nor purely coinductive—coinductive definitions of datatypes and functions may use auxiliary inductive definitions, and vice versa. In order to cope with such theories, it is important that we integrate Hipster's inductive and coinductive functionality. For conjecture discovery, this integration comes for free since Isabelle's code generator maps both data and codata to identical Haskell code.

For proof search, we must decide whether to tackle our conjectures using induction, coinduction or both. For this, we use a simple heuristic that appears to work well in practice: if the conjecture contains a free variable whose type has an induction principle, we invoke the inductive proof search procedure; if the left- and right-hand sides of the conjecture are of a type that has a coinduction principle, we invoke the coinductive proof search; if both, we try both and keep the first successful proof attempt. This architecture allows us to find proofs of inductive lemmas that require coinductive auxiliary lemmas, such as the fact that *append* distributes over the *to_llist* function on finite lists (see Sect. 5).

## 5   Evaluation and Results

We apply Hipster to several theories of common codatatypes found in the literature: lazy lists, extended natural numbers, streams, and two kinds of infinite trees. Our goal is to demonstrate how a user can invoke Hipster to discover useful lemmas in their coinductive theory development, showing that our method for testing infinite structures, as described in Sect. 3, is effective in discovering coinductive properties and that our automated coinduction tactic, described in Sect. 4, is effective in proving those properties.

We restrict each Hipster call to a small number of functions, to explore how those functions relate to each other, rather than exploring all the functions in a theory at once. This is how we envision typical users will interact with the tool, since in practice it tends to yield quicker and more relevant results.

The evaluation was performed with Isabelle 2017 using Isabelle/jEdit, on a ThinkPad X260 laptop with a 2.5 GHz Intel i7-6500U processor and 16 GB of RAM running 64-bit Linux. The Isabelle theory files used to attain these results are available online[2].

### 5.1   Case Study: Lazy Lists and Extended Natural Numbers

In this section we demonstrate the results attained when using Hipster to explore a theory of lazy lists (lists of potentially infinite length). We define some common functions for this type: *lappend* to append two lazy lists, a map function *lmap*, *iterates* which generates a lazy list by iteratively applying a function to an element, *llist_of* which maps a standard Isabelle/HOL list to a lazy list, *llength* which returns the length, and *ltake* which takes a given number of elements. We also define a codatatype *ENat* for extended natural numbers (natural numbers of potentially infinite size) and an addition function *eplus* on *ENats*.

We check which of the lemmas we discover are stated and proved in the Coinductive library [18] in the archive of formal proofs[3], which is a collection of formalizations about coinductive types and functions. For the extended naturals we refer to the *Extended_Nat* theory from the Isabelle/HOL library[4]. Since

---

[2] https://github.com/moajohansson/IsaHipster/tree/master/benchmark/AISC18.

[3] https://www.isa-afp.org/.

[4] http://isabelle.in.tum.de/library/HOL/HOL-Library/Extended_Nat.html.

the lemmas in these libraries have been collected and hand-proved by Isabelle experts, we conclude that they must be interesting and/or useful for Isabelle theory development.

Table 1 shows the results of exploration on this theory. The column args shows the names of the functions explored in the particular Hipster call, Expl is the amount of time (in seconds) spent in exploration and testing, Expl+Proof is the amount of time (in seconds) spent in exploration, testing, and proving, # properties shows the number of properties Hipster discovers, # library lemmas shows how many of those properties are lemmas stated and proved in the libraries mentioned above. For these experiments, Hipster's routine tactic was configured to only do simplification, and the hard tactic was our automated coinduction and induction tactic as described in Sect. 4.

In our 13 calls to Hipster, we discover 33 coinductive or inductive properties. Of these 33 properties, 13 are stated and proved as lemmas in Isabelle libraries, leading us to believe that they are of interest to Isabelle users. Of the other 20, most are rather trivial consequences of function definitions and/or other discovered lemmas, which our routine tactic does not suffice to prove. Some of the discovered properties may however be interesting to users despite not appearing in the libraries, for instance that $llength(lappend\ xs\ ys) = llength(lappend\ ys\ xs)$.

The discovered properties include the associativity of append, *lappend* $(lappend\ x\ y)\ z = lappend\ x\ (lappend\ y\ z)$, and that mapping preserves length, $llength\ (lmap\ f\ x) = llength\ x$. The exploration involving $llist\_of$, which maps a standard list to a lazy list, results in lemmas showing the correspondence between our lazy list functions and Isabelle/HOL's built-in list functions, for example $lmap\ f\ (llist\_of x) = llist\_of\ (map\ f x)$. The previous lemma is proved by induction, demonstrating Hipster's capabilities in exploring mixed inductive and coinductive theories.

All of the discovered properties are proved by our automated proof tactic, except for the commutativity of *eplus*. This was due to our rather short timeout for Sledgehammer, which was just set to 10 s. in this experiment. If we allow a 30 s. timeout (which is the standard when Sledgehammer is used interactively), a proof is found. As can be seen from Table 1, the time it takes for Hipster to discover and prove properties varies between 2–90 s. As all calls took less than 90 s to complete, and most took less than a minute, we can see that the user does not have to wait very long for Hipster to come up with lemmas for their functions. We believe that for most Isabelle users, making a call to Hipster would be much faster than writing down and proving the same lemmas manually, not to mention coming up with them. In Table 1 we also compare the runtime of the calls: most of the time is spent trying to prove properties (we give each call to Sledgehammer a timeout limit of 10 s), while the time to discover and test the properties is just a few seconds. There is however a configuration option in Hipster for very impatient users to only do exploration, leaving the proofs to the user altogether.

**Table 1.** An overview of the results of exploring our lazy list theory.

| cohipster args | Expl | Expl+Proof | # properties | # library lemmas |
|---|---|---|---|---|
| *lappend* | 2.5 s | 25 s | 4 | 2 |
| *lmap* | 3.2 s | 7 s | 3 | 0 |
| *lappend lmap* | 4.1 s | 17 s | 1 | 1 |
| *llist_of lappend append* | 4.9 s | 28 s | 1 | 1 |
| *llist_of lmap map* | 4.9 s | 21 s | 1 | 1 |
| *llength* | 2.1 s | 2 s | 1 | 0 |
| *llength lmap* | 4.0 s | 11 s | 1 | 1 |
| *eplus* | 2.9 s | 39 s | 4 | 3 |
| *llength lappend eplus* | 5.2 s | 87 s | 5 | 1 |
| *ltake* | 4.1 s | 76 s | 7 | 0 |
| *ltake lmap* | 5.7 s | 23 s | 2 | 1 |
| *lmap iterates* | 4.2 s | 18 s | 2 | 1 |
| *lappend iterates* | 4.6 s | 15 s | 1 | 1 |

## 5.2   Case Study: Stream Laws

We already saw in Sect. 1 that Hipster can discover and prove the *map-iterate* property for streams. In this section, our aim is to quantify the degree to which Hipster discovers stream equations that a human would find interesting. That is of course subjective, but for the purposes of this section we operationalize "interesting" as being any of the 18 laws of Hinze's Stream Calculus [11], which according the author *"provides an account of the most important properties of streams"*. Of the 18 laws given by Hinze, three are beyond the scope of Hipster's current capabilities: lambda-expressions are not supported, nor are conditional statements with term depth >1 in the antecedent. The remaining 15 are all equational statements. With respect to these 15 laws, we analyze Hipster's precision (percentage of the lemmas we find that are among Hinze's laws) and recall (percentage of Hinze's laws that we find).

First, we will briefly recapitulate the relevant notation. *pure x* denotes a stream where every element is $x$. $\diamond$ is lifted function application, defined by the observations $hd(f \diamond x) = (hd\ f)\ (hd\ x)$ and $tl(f \diamond x) = (tl\ f) \diamond (tl\ x)$. The interleaving of two streams $x, y$ is written $x \curlyvee y$. Tabulation, written *tabulate f*, is the stream whose $n$:th element is $f(n)$. Lookup, written *lookup s n*, is the $n$:th element of stream $s$. *zip x y* merges two streams into a stream of pairs. *recurse* is defined by the observations $hd(recurse\ f\ a) = a$ and $tl(recurse\ f\ a) = map\ f\ (recurse\ f\ a)$. Unfolding satisfies $hd(unfold\ g\ f\ a) = g\ a$ and $tl(unfold\ g\ f\ a) = unfold\ g\ f\ (f\ a)$.

The results are shown in Table 2. The lemmas' precision, recall and time have been explored together by invoking Hipster with every function mentioned in each lemma; e.g., to search for laws 7–9 we invoke **cohipster** `map zip fst snd`.

**Table 2.** An overview of the stream properties discovered and proved by Hipster. Lemmas in gray are not in scope.

| | Property | Found | Precision | Recall | Time |
|---|---|---|---|---|---|
| 1 | $pure\ id \diamond u = u$ | X | 22% | 67% | 44 s |
| 2 | $pure(\circ) \diamond u \diamond v \diamond w \diamond u = u$ | – | | | |
| 3 | $pure\ f \diamond pure\ x = pure\ (f\ x)$ | X | | | |
| 4 | $u \diamond pure\ x = pure\ (\lambda f.\ f\ x) \diamond u$ | | | | |
| 5 | $map\ id\ x = x$ | X | 50% | 100% | 29 s |
| 6 | $map\ (f \circ g)\ x = map\ f\ (map\ g\ x)$ | X | | | |
| 7 | $map\ fst\ (zip\ s\ t) = s$ | – | 0% | 0% | 255 s |
| 8 | $map\ snd\ (zip\ s\ t) = t$ | – | | | |
| 9 | $zip\ (map\ fst\ p)\ (map\ snd\ p) = p$ | – | | | |
| 10 | $pure\ a \curlyvee pure\ a = pure\ a$ | X | 25% | 50% | 18 s |
| 11 | $(s_1 \diamond s_2) \curlyvee (t_1 \diamond t_2) = (s_1 \curlyvee t_1) \diamond (s_2 \curlyvee t_2)$ | – | | | |
| 12 | $map\ f\ (tabulate\ g) = tabulate\ (f \circ g)$ | X | 100% | 100% | 87s |
| 13 | $f(lookup\ t\ x) = lookup\ (map\ f\ t)\ x$ | X | 33% | 100% | 57s |
| 14 | $recurse\ f\ a = iterate\ f\ a$ | X | 33% | 100% | 73s |
| 15 | $map\ h \circ iterate\ f_1 = iterate\ f_2 \circ h \impliedby h \circ f_1 = f_2 \circ h$ | | | | |
| 16 | $unfold\ hd\ tl\ x = x$ | – | 0% | 0% | 21s |
| 17 | $unfold\ g\ f \circ h = unfold\ g'\ f' \impliedby g \circ h = g' \wedge f \circ h = h \circ f'$ | | | | |
| 18 | $map\ h\ (unfold\ g\ f\ x) = unfold\ (h \circ g)\ f\ x$ | X | 50% | 100% | 18s |
| | | | **21%** | **60%** | **602 s** |

We also report total precision and recall over all such invocations at the bottom. For these experiments, Hipster has been configured to use a Sledgehammer timeout of 10 s, a routine tactic that does only simplification, and a hard tactic that tries coinduction and induction, in each case followed by simplification or sledgehammer, as described in Sect. 4.

We see that in total, Hipster discovers 9 out of the 15 properties in scope, i.e. 60% recall. Note in particular property 13, where Hipster discovers a proof by induction, and property 14, where Hipster discovers a proof by coinduction up-to friendly contexts. The 21% overall precision can be improved by using a more powerful routine tactic, such as simplification interleaved with stream expansion.

The properties that are in scope, but not discovered, are all attributable to QuickSpec's heuristics for restricting the search space. Properties involving variables denoting streams of functions such as Property 2 cannot be tested, and instantiation of type variables is restricted in ways that rule out, e.g., conjectures where $fst$ occurs as an argument to $map$. It seems difficult to lift these restrictions in ways that do not make the search space intractable—this would be an interesting direction for future work.

## 5.3   Case Study: Infinite Trees

We have experimented with two different kinds of corecursive trees: A codatatype representing an infinitely deep binary tree, and another representing an infinitely deep *rose tree*, with arbitrary branching at each node. The purpose here is to demonstrate Hipster on a different kind of codatatype than the previous case-studies. Hipster was configured to use simplification as the *routine tactic*, and as the *hard tactic*, either just Sledgehammer or coinduction followed by Sledgehammer.

*Infinite Binary Trees.* We define an infinite depth binary tree as follows:

```
codatatype 'a Tree = Node (lt: "'a Tree") (lab: 'a) (rt: "'a Tree")
```

We defined three functions over this codatatype: *mirror* (which switches the left and right branches of each node), *tmap* which applies a function to each label in the tree and *tsum* which sums the labels of a tree of natural numbers. A summary of the results is given in Table 3. Hipster discovers the expected properties about the given functions (associativity, distributivity etc.) as well as a few additional properties which perhaps are of less interest. We note that these are presented to the user as Isabelle's simplifier is a rather weak tactic in this context, while another choice for the routine tactic would have pruned out more properties.

**Table 3.** Overview of properties discovered about infinite depth binary trees. Due to space restrictions mainly properties proved by coinduction are listed, full results are available online.

| **cohipster** args | Expl | Expl+Proof | Properties discovered | Proved |
|---|---|---|---|---|
| *mirror* | 3.4 s | 39 s | $mirror\ (mirror\ y) = y$ <br> + 3 more proved by Sledgehammer | coinduction+simp |
| *mirror tmap* | 4.3 s | 35 s | $tmap\ z\ (mirror\ x) = mirror$ <br> $(tmap\ z\ x)$ | coinduction+smt |
| *mirror tsum* | 6.1 s | 112 s | $tsum\ y\ x = tsum\ x\ y$ <br> $tsum\ (tsum\ x\ y)\ z = tsum\ x$ <br> $(tsum\ y\ z)$ <br> $mirror\ (tsum\ y\ (mirror\ x)) =$ <br> $tsum\ x\ (mirror\ y)$ <br> $tsum\ (mirror\ x)\ (mirror\ y) =$ <br> $mirror\ (tsum\ x\ y)$ <br> + 2 more proved by Sledgehammer | coinduction+smt <br> coinduction+smt <br> coinduction+smt <br> Sledgehammer <br> (using above <br> lemmas) |

*Rose Trees: a Nested Codatatype.* We also conducted an experiment with a
nested codatatype representing arbitrarily branching *rose trees*:

```
codatatype 'a RoseTree = Node (lab: 'a) (sub: "'a RoseTree list")
```

We defined functions *mirror* (reversing the list of subtrees), *tmap* (mapping a
function over the labels of each node) and *tsum* (summing the labels of a tree
of natural numbers). Note that unlike for the infinite binary trees, *mirror* and
*tmap* are not corecursive.

For this theory, we noticed that the runtimes varied a great deal from run
to run of the same command. For example, in a series of runs of Hipster on
the function mirror only, the runtime varied from as little as 21 s to as much as
125 s. This is due to how our observer function interacts with the random length
lists being generated for the branches at each node. It decreases its fuel linearly
in this case, so if the list is long observing each child tree recursively is time-
consuming. Implementing smarter observer functions, for instance taking length
of the list of a node's child trees into account to only observe an appropriately
small subtree of each child, is future work.

**Table 4.** Overview of properties discovered about rose trees. Note that timings here
are from one sample run, and can vary quite a lot due to randomness in testing.

| **cohipster** args | Expl+Proof | Properties discovered | Proved |
|---|---|---|---|
| *mirror* | 29 s | *mirror (mirror y) = y* | Sledgehammer |
| *mirror tmap* | 102 s | *tmap z (mirror x) = mirror (tmap z x)* | Sledgehammer |
| *mirror tsum* | 597 s | *tsum (mirror x) (mirror x) = mirror (tsum x x)* <br> *tsum y x = tsum x y* <br> *tsum (tsum x y) z = tsum x (tsum y z)* <br> + 4 more unproved about tsum/mirror | Sledgehammer <br> no <br> no |

As can be seen in Table 4, only a few properties are proved automatically (by
Sledgehammer, no coinduction needed). This is because our automated coinduction tactic is not flexible enough to deal with nested datatypes. We believe a
customized tactic, also able to perform some form of nested induction over the
list of branches, would do a better job, but such domain specific tactics are left
as further work at this stage.

## 6 Related Work

There is substantial recent work on making Isabelle/HOL more expressive for
working with codatatypes and corecursive functions [4,5]. Our extension to Hipster can help Isabelle/HOL users who want to program with these new methods
discover and prove new properties about their theories.

There has been prior work on automating coinductive proofs and reasoning. In [17] Leino and Moskal present a method for automated reasoning about coinductive properties in the Dafny verifier. CIRC [19] is a tool for automated inductive and coinductive theorem proving which uses circular coinductive reasoning. It has been successfully used to prove many properties of infinite structures such as streams and infinite binary trees. However, none of the other systems has the theory exploration capabilities of Hipster.

In the setting of resolution for Horn clause logic with coinductive entailment, Fu et al. [10] present a method for automatically generating appropriate candidate lemmas for proving such entailments. The application is to devise a method for e.g. type class resolution in Haskell that is stronger than cycle detection. Whereas Hipster uses testing to generate candidate lemmas, Fu et al. uses the structure of partial proof attempts. Given a partially unfolded resolution tree, the candidate lemma that gets generated states that the root of the tree is entailed by the conjunction of all leaves that mention fewer symbols than the root. This is also unlike Hipster in that Hipster strives for lemmas that will be generally useful for any further theory development using the types and functions under consideration, whereas Fu et al. are interested in finding which lemmas, were they true, could be used to prove a particular sequent.

IsaCoSy [15] and IsaScheme [22] are other theory exploration systems for Isabelle/HOL, both of which focus on the discovery and proof of inductive properties. MATHsAiD [20] is a tool for automated theorem discovery, aimed at aiding mathematicians in exploring mathematical theories. It can discover and prove theorems whose proofs consist of logical and transitive reasoning as well as induction. Hipster is the first theory exploration system capable of discovering and proving coinductive properties. Furthermore, it is considerably faster than IsaCoSy and IsaScheme thanks to using QuickSpec as a backend [9].

# 7   Conclusion

We have extended the theory exploration system Hipster with the capabilities to discover and prove not only inductive lemmas, but also lemmas in coinductive theories involving potentially infinite types such as streams, lazy lists and trees. We have shown that the system can discover and prove many standard lemmas about these codatatypes. This goes beyond the capabilities of previous theory exploration systems, that do not consider coinduction at all.

In the long term, we envision that invoking a theory exploration system such as Hipster will be a natural first step for the working proof engineer when developing a new theory. This nicely complements tools like Isabelle's Sledgehammer. In a new theory, Sledgehammer is unlikely to be of much help until we have proven at least some basic lemmas, which is exactly what theory exploration can automate.

There are many interesting directions for further work. As seen in the case study on rose trees, we would benefit from specialized observation functions and proof methods for nested (co-)datatypes. The case studies in this paper are mostly in the domain of lazy data in the style of functional programming. It would be interesting to explore if we can extend our work to other uses of coinduction. For example, discovering algebraic laws about coinductively defined behavioral equivalences, or discovering Hoare triples about non-terminating programs. This would require developing a technique to test *relations* as opposed to functions.

**Acknowledgments.** The authors would like to thank Nicholas Smallbone for technical assistance with QuickSpec. The first author was partially supported by the GRACeFUL project, grant agreement No. 640954, which has received funding from the European Union's Horizon 2020 research and innovation program.

# References

1. Abel, A., Pientka, B.: Well-founded recursion with copatterns and sized types. J. Funct. Program. **26**, e2 (2016)
2. Bird, R.: Introduction to Functional Programming, 2nd edn. Pearson Education, London (1998)
3. Bird, R., Wadler, P.: An Introduction to Functional Programming. Prentice Hall International (UK) Ltd., Hertfordshire (1988)
4. Blanchette, J.C., Bouzy, A., Lochbihler, A., Popescu, A., Traytel, D.: Friends with benefits. In: Yang, H. (ed.) ESOP 2017. LNCS, vol. 10201, pp. 111–140. Springer, Heidelberg (2017). https://doi.org/10.1007/978-3-662-54434-1_5
5. Blanchette, J.C., Hölzl, J., Lochbihler, A., Panny, L., Popescu, A., Traytel, D.: Truly modular (Co)datatypes for Isabelle/HOL. In: Klein, G., Gamboa, R. (eds.) ITP 2014. LNCS, vol. 8558, pp. 93–110. Springer, Cham (2014). https://doi.org/10.1007/978-3-319-08970-6_7
6. Blanchette, J.C., Meier, F., Popescu, A., Traytel, D.: Foundational nonuniform (co)datatypes for higher-order logic. In: 2017 32nd Annual ACM/IEEE Symposium on Logic in Computer Science (LICS), pp. 1–12, June 2017
7. Buchberger, B.: Theory exploration with Theorema. Analele Universitatii Din Timisoara, ser. Matematica-Informatica **38**(2), 9–32 (2000)
8. Claessen, K., Hughes, J.: QuickCheck: a lightweight tool for random testing of Haskell programs. In: Proceedings of ICFP, pp. 268–279 (2000)
9. Claessen, K., Johansson, M., Rosén, D., Smallbone, N.: Automating inductive proofs using theory exploration. In: Bonacina, M.P. (ed.) CADE 2013. LNCS (LNAI), vol. 7898, pp. 392–406. Springer, Heidelberg (2013). https://doi.org/10.1007/978-3-642-38574-2_27
10. Fu, P., Komendantskaya, E., Schrijvers, T., Pond, A.: Proof relevant corecursive resolution. In: Kiselyov, O., King, A. (eds.) FLOPS 2016. LNCS, vol. 9613, pp. 126–143. Springer, Cham (2016). https://doi.org/10.1007/978-3-319-29604-3_9
11. Hinze, R.: Concrete stream calculus: an extended study. J. Funct. Program. **20**(5–6), 463–535 (2010)
12. Hutton, G., Gibbons, J.: The generic approximation lemma. Inf. Proces. Lett. **79**, 2001 (2001)
13. Jacobs, B., Rutten, J.: A tutorial on (co)algebras and (co)induction. EATCS Bull. **62**, 222–259 (1997)

14. Johansson, M.: Automated theory exploration for interactive theorem proving. In: Ayala-Rincón, M., Muñoz, C.A. (eds.) ITP 2017. LNCS, vol. 10499, pp. 1–11. Springer, Cham (2017). https://doi.org/10.1007/978-3-319-66107-0_1
15. Johansson, M., Dixon, L., Bundy, A.: Conjecture synthesis for inductive theories. J. Autom. Reason. **47**(3), 251–289 (2011)
16. Johansson, M., Rosén, D., Smallbone, N., Claessen, K.: Hipster: integrating theory exploration in a proof assistant. In: Watt, S.M., Davenport, J.H., Sexton, A.P., Sojka, P., Urban, J. (eds.) CICM 2014. LNCS (LNAI), vol. 8543, pp. 108–122. Springer, Cham (2014). https://doi.org/10.1007/978-3-319-08434-3_9
17. Leino, R., Moskal, M.: Co-induction simply: automatic co-inductive proofs in a program verifier. Technical report, Microsoft Research, July 2013
18. Lochbihler, A.: Coinductive. Archive of Formal Proofs, February 2010. http://isa-afp.org/entries/Coinductive.html. Formal proof development
19. Lucanu, D., Goriac, E.-I., Caltais, G., Roşu, G.: CIRC: a behavioral verification tool based on circular coinduction. In: Kurz, A., Lenisa, M., Tarlecki, A. (eds.) CALCO 2009. LNCS, vol. 5728, pp. 433–442. Springer, Heidelberg (2009). https://doi.org/10.1007/978-3-642-03741-2_30
20. McCasland, R.L., Bundy, A., Smith, P.F.: MATHsAiD: automated mathematical theory exploration. Appl. Intell. **47**, 585–606 (2017)
21. Milner, R.: Communication and Concurrency. Prentice-Hall Inc., Upper Saddle River (1989)
22. Montano-Rivas, O., McCasland, R., Dixon, L., Bundy, A.: Scheme-based theorem discovery and concept invention. Expert Syst. Appl. **39**(2), 1637–1646 (2012)
23. Nakata, K., Uustalu, T.: A Hoare logic for the coinductive trace-based big-step semantics of while. In: Gordon, A.D. (ed.) ESOP 2010. LNCS, vol. 6012, pp. 488–506. Springer, Heidelberg (2010). https://doi.org/10.1007/978-3-642-11957-6_26
24. Nipkow, T., Paulson, L.C., Wenzel, M.: Isabelle/HOL. Springer, Heidelberg (2002). https://doi.org/10.1007/3-540-45949-9. http://isabelle.in.tum.de/dist/Isabelle2017/doc/tutorial.pdf
25. Paulson, L.C., Blanchette, J.C.: Three years of experience with Sledgehammer, a practical link between automatic and interactive theorem provers. In: Proceedings of IWIL-2010 (2010)
26. Pous, D.: Coinduction all the way up. In: Proceedings of LICS, pp. 307–316. ACM, New York (2016)
27. Sangiorgi, D.: Introduction to Bisimulation and Coinduction. Cambridge University Press, New York (2011)
28. Smallbone, N., Johansson, M., Claessen, K., Algehed, M.: Quick specifications for the busy programmer. J. Funct. Program. **27**, e18 (2017)
29. Turner, D.A.: Total functional programming. J. UCS **10**(7), 751–768 (2004)

# Machine Learning for Inductive Theorem Proving

Yaqing Jiang$^{(\boxtimes)}$, Petros Papapanagiotou, and Jacques Fleuriot

School of Informatics, University of Edinburgh,
10 Crichton Street, Edinburgh EH8 9AB, UK
{YQ.Jiang,pe.p,jdf}@ed.ac.uk

**Abstract.** Over the past few years, machine learning has been successfully combined with automated theorem provers to prove conjectures from proof assistants. However, such approaches do not usually focus on inductive proofs. In this work, we explore a combination of machine learning, a simple Boyer-Moore model and ATPs as a means of improving the automation of inductive proofs in the proof assistant HOL Light. We evaluate the framework using a number of inductive proof corpora. In each case, our approach achieves a higher success rate than running ATPs or the Boyer-Moore tool individually.

**Keywords:** Induction · Lemma selection · Theorem proving
Machine learning

## 1 Introduction

Over the past few years, large libraries of formalised theories have been built in interactive theorem provers (ITPs) like Isabelle [26], HOL Light [15] and Coq [1]. Automated, first-order theorem provers (ATPs) like Vampire [23] and E [29], and satisfiability modulo theories (SMT) solvers like Z3 [10] are increasingly being used to facilitate the development of such libraries in large proof corpora.

In order to use such external tools effectively, machine learning (ML) infrastructures have been developed within several proof assistants to automatically select hundreds of potentially relevant lemmas whenever the user tries to prove a goal automatically. Sledgehammer [28] in Isabelle and HOL(y)Hammer [21] in HOL Light are examples of two such ML systems.

Although recursively-defined data types such as lists are widely used in ITPs, the ATPs and SMT solvers do not usually perform well on goals that require inductive theorem proving [9]. However, automated methods for inductive theorem proving do exist. ACL2 [22], for instance, is a system that evolved from the so-called *Boyer-Moore* approach (which we use in our current work) and is successfully being used for the formalization of industrial problems. Inductive theorem proving often requires the manual provision of suitable lemmas to help with the inductive proof (for example as *hints* in ACL2). Identifying such lemmas is a major challenge and the system relies on human expertise and understanding of the problem and its context.

Lemma discovery techniques, which try to automatically speculate relevant lemmas, have been investigated as a solution [8,11]. These include, for example, generalization, which was incorporated in the original Boyer-Moore prover but has had relatively limited success.

In the current work, we investigate the potential use of machine learning to select lemmas from big corpora in support of automated inductive theorem proving. We aim to select a relatively small number of suitable lemmas that can then be used within a Boyer-Moore based inductive theorem prover to make progress with otherwise blocked proofs.

We incorporate proof strategies that make use of machine learning techniques and ATPs within a Boyer-Moore style model and run these in parallel, in a new environment we call a *multi-waterfall*. Our paper is organised as follows: we introduce the Boyer-Moore model and lemma selection approaches in Sect. 2. We present the multi-waterfall model in Sect. 3, together with the application of lemma selection, and other changes to our Boyer-Moore implementation. We evaluate the different strategies on corpora of inductive proofs in Sect. 4, and discuss the results in Sect. 5.

## 2   Background

### 2.1   Recursively-Defined Data Types and Induction

Recursively-defined data types are usually used in inductive theorem proving. For instance, a natural number is either the constant 0, or obtained by applying the *successor* function $s$ to another natural number. Inductive inference involves the use of particular logical rules to prove properties of recursive datatypes that are not otherwise provable [6]. The induction for natural numbers is:

$$\frac{P(0), \forall n.\ P(n) \implies P(s(n))}{\forall x.\ P(x)} \tag{1}$$

Applying this rule allows us to break a subgoal about a particular property $P$ of natural numbers into two new subgoals: the base case $P(0)$ and the step case $P(s(n))$, assuming $P(n)$ for any $n$.

### 2.2   The Boyer-Moore Model

The Boyer-Moore approach [4] covers the key components of an automated theorem prover for inductive proofs. It revolves around the notion of a *waterfall model*, as shown in Fig. 1. In this, conjectures (or proof goals) are poured from the top and through a series of procedures, called *heuristics*. Each heuristic in the waterfall tries to either prove or simplify the goal. It may also determine that the goal is unprovable or, if neither of these is applicable, the heuristic *fails*.

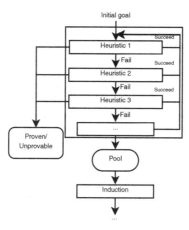

**Fig. 1.** Diagram of the waterfall model

Induction is applied automatically when all heuristics have failed (the goals trickle down to the pool at the bottom of the waterfall). The generated sub-goals (base and step cases) are poured over new waterfalls again. This process is repeated recursively, until all subgoals are proven, in which case a proof of the original goal is reconstructed, or a subgoal is determined to be unprovable.

Examples of heuristics that are relevant to this paper are the following:

- **The Clausal Form heuristic:** This transforms the goals to Clausal Normal Form (CNF), which other Boyer-Moore heuristics take advantage of.
- **The Simplify heuristic:** This applies rewriting to the goal in order to simplify or prove it using function definitions and rewrite rules. Note that termination is not guaranteed and depends on the selection of rewrite rules.
- **The Generalize heuristic:** A lemma speculation process which tries to generalize a subterm in the goal.
- Automated proof procedures in HOL Light such as the model elimination procedure MESON [16] and a simple tautology checker TAUT can also be used as heuristics within Boyer-Moore.

Boyer-Moore uses an additional heuristic at the pool of the waterfall to choose the appropriate induction variable based on the definitions of recursive functions [4]. Note that in our implementation, this heuristic only supports primitively recursive definitions [3].

Based on the above, the system configuration can be tailored to deal with different problems. The most common customizations are the following [27]:

- The rewrite rules for the Simplify heuristic can be elaborately chosen by the user to improve its effectiveness towards proving the subgoals.
- The order and combination of heuristics can also be adjusted for different situations. For instance, some heuristics are unsafe and may render the goal more complicated, or result in an infinite loop.

For our implementation we use the Boyer-Moore system implemented in HOL-Light [27]. An important advantage of both this particular system and HOL Light, particularly in comparison with more sophisticated evolutions of the Boyer-Moore approach such as ACL2, is that they are lightweight implementations with simple structures and thus allow easy, direct access to the inner workings. This makes it easier to manipulate and adjust the Boyer-Moore waterfalls and heuristics, and analyze the effects of machine learning more thoroughly.

### 2.3    Theorem Proving Hammers

As we alluded to in the introduction, ITPs now incorporate so-called *hammers* that act as intermediates between powerful, external ATPs and their built-in proof procedures. With the help of machine learning, these allow users to reconstruct complex formal proofs within ITPs with just one click. Hammers usually consists of four parts [2]:

- A *lemma selection* module to filter relevant lemmas that can be used by ATPs (see Sect. 2.4).
- A *translation module* that translates ITP problems to a first order syntax acceptable to ATPs.
- Links to external ATPs that search for and output proofs.
- A *proof reconstruction* module that reconstructs the output of ATPs to corresponding ITP proofs.

*Sledgehammer* is the original tool that started the whole effort: it is integrated into the Isabelle proof assistant that carries out lemma selection using a combination of relevance filtering *MePo* [25] and Bayesian learning [24].

*HOL(y)Hammer*, the corresponding tool for HOL Light, also uses machine learning for lemma selection. In our work, we incorporate elements from its latest released version[1], such as its *feature extraction* algorithm (see Sect. 2.4).

### 2.4    Machine Learning for Interactive Theorem Proving

*Lemma selection* is an important component of hammers as they provide the external ATPs with the pre-proved results that may lead to a proof. This usually involves training an ML model that can predict the relevance of proven lemmas to new goals and then select those that look more promising. The model is typically trained using existing proofs that have been produced interactively. More specifically, a *dependency tracking* module usually records the definitions and lemmas that have been used during interactive proofs.

In our case, we have developed our own dependency tracking tool that improves upon the one in HOL(y)Hammer by recording additional information such as whether a tracked theorem is a definition and the file that contains it.

Both Sledgehammer and HOL(y)Hammer use ML algorithms, such as naive Bayes, that estimate the relevance of lemmas from the proof library based on

---

[1] http://cl-informatik.uibk.ac.at/software/hh/hh-0.13.tgz.

features generated from the statements of lemmas and the goal at hand. Such features usually consist of strings generated from the constants, subterms, operators, and other parts of the statement [20].

For example, given the HOL-Light theorem $\forall n.\ EVEN\ n \lor ODD\ n$, the following features are extracted:

$$"num",\ "fun",\ "bool",\ "ODD",\ "EVEN",$$
$$"Anum",\ "EVEN\ Anum",\ "ODD\ Anum" \tag{2}$$

In HOL(y)Hammer this is achieved in several ways. For instance, given the term $EVEN\ n : num$ (where "$n : num$" means the type of $n$ is the natural numbers), the default option normalizes the identifier of variable $n$ to an identifier "$A$" followed by the type $num$, i.e. "$Anum$". Moreover, structural information is kept as additional features with entire subterms, e.g. "$EVEN\ Anum$" above, which provides more information for learning [24].

# 3 Methodology

As mentioned in Sect. 2, our work is based on an implementation of the Boyer-Moore model in HOL Light. We followed an experiment-led methodology, using the setup described in Sect. 4. The results of repeated experiments empirically guided our decision making in order to improve and configure the system and expand it with machine learning techniques inspired by hammers. In this section, we summarize the key changes made to the original Boyer-Moore system.

## 3.1 Initial Improvements

Initial experiments were done to form a baseline against which to compare the results of changes and additions. During these experiments we noticed and fixed a number of issues, the most important of which are described next.

**Removing CNF Heuristic.** During our initial experiments, some goals became unprovable by Boyer-Moore after the CNF heuristic was applied. For instance, the heuristic splits the goal $\neg EVEN\ x \iff ODD\ x$ into 2 clauses: $EVEN\ x \lor ODD\ x$ and $\neg EVEN\ x \lor \neg ODD\ x$. In the original formalization, the untransformed goal is proven independently and used as a lemma to be able to prove these 2 clauses. This is an indication that the CNF heuristic does not always make progress in the right direction towards a proof.

Moreover, the CNF heuristic breaks goals that contain logical equivalences (iffs) into subgoals containing implication, leading to the generation of a number of subgoals that is exponential to the number of equivalences encountered in the original goal. Therefore, removing it can significantly reduce the total amount of subgoals.

It is worth noting that removing the CNF heuristic directly affects some of the Boyer-Moore heuristics that follow, which rely on CNF. Despite this side-effect, our experiments showed a significant overall improvement in the performance of Boyer-Moore without the CNF heuristic.

**Generalising Variables.** When applying induction to a formula with more than one universally quantified variable, only one is typically selected for induction, and the others are not affected [6]. For example, applying induction on variable $n$ in the formula $\forall n\ m.\ Q(n, m)$ yields the following step case:

$$\forall n'.\ (\forall m.\ Q(n', m)) \implies (\forall m.\ Q(s(n'), m)) \qquad (3)$$

However, in Boyer-Moore the input formula is always *quantifier-free*, so the step case generated is the following instead:

$$Q(n, m) \implies Q(s(n), m) \qquad (4)$$

This stronger subgoal may be unprovable in certain cases compared to its weaker counterpart (3). Our solution is to generalise all variables other than the one for induction as follows:

$$(\forall m.\ Q(n, m)) \implies Q(s(n), m') \qquad (5)$$

Applying induction then yields the same subgoal (3), though we then remove the quantifiers again to fit to the *quantifier-free* environment of Boyer-Moore.

**HOL Light's Automated Procedures.** During early experiments, we identified (sub)goals that could be proven by HOL Light's automated model elimination procedure MESON. Therefore, MESON was added as a heuristic to the waterfall.

**Forced Induction.** As mentioned previously, the induction heuristic in Boyer-Moore can only handle primitively recursive function definitions. This means Boyer-Moore failed to perform induction in terms containing any non-primitively recursive functions as it was unable to choose an appropriate variable. We address this problem by forcing Boyer-Moore to pick the first free variable with a recursive type for induction if no other suitable selection is found by the original heuristic. For the future, we are considering the use of machine learning techniques as a more sophisticated mechanism for the selection of induction variables.

### 3.2 The Multi-waterfall Model

The original setup of the waterfall works in a serial, monolithic way. Each heuristic is tried sequentially in a static order. However, certain proofs may require different configurations or strategies for different subgoals. Moreover, some of the Boyer-Moore heuristics may naturally get stuck during a proof. For example, certain combinations of rewrite rules may cause the Simplify heuristic to loop endlessly. This is particularly important in the context of automated lemma selection where we have less control over looping rewrite rule sets. Using a different configuration might help unlock and make progress with the proof.

In order to achieve a more flexible implementation that does not rely on a single configuration, we introduce a *Multi-waterfall model*. In this, we run multiple waterfalls with different configurations in parallel and with a preset timeout. We then have the following possible outcomes:

1. One of the waterfalls succeeds and the corresponding (sub)goal is proven. The proof of the (sub)goal is reconstructed and propagated upwards (as in the standard waterfall model), ensuring soundness of the overall proof.
2. One of the waterfalls completes having generated new subgoals that reached their pools. In this case, we apply induction to all unproven goals as in the standard waterfall model (see Sect. 2.2). We then apply the same set of multiple waterfalls to each of the new sugboals generated by induction.
3. All the waterfalls determine the goal is unprovable, or the timeout is reached. In this case, the whole branch of proof search fails and is discarded.

The timeout applied to each waterfall ensures that any waterfalls that take too long are assumed to have failed and are forcibly stopped and their corresponding branches abandoned. This allows the other waterfalls running in parallel to still potentially make progress towards the proof.

An example search tree with 2 waterfalls is shown in Fig. 2. The waterfalls are run in parallel on the same goal. When a waterfall finishes, we apply induction to any unproven subgoals in its pool, constructing new subgoals indicated by the dashed arrows. We then start new waterfalls for each generated subgoal until all subgoals are proven or deemed unprovable.

A full proof can be reconstructed by tracking all successful waterfalls in a branch. This means a proof may be found by a chain of different waterfalls. In Fig. 2, for example, the proof is reconstructed by the waterfalls enclosed in the marked area. Notice that both types of waterfalls were used to make progress on or prove different subgoals.

In our implementation, we spawn the waterfalls for a particular goal using threaded concurrency. If a waterfall fully proves a goal (such as Waterfall 1" in Fig. 2), the other waterfalls working on the same goal (such as Waterfall 2") and their children are forcibly stopped in order to release system resources. Waterfalls could be tried sequentially instead, but this would dramatically increase the time taken for a proof to complete, e.g. because the user would need to wait for different waterfalls to timeout for each and every subgoal.

### 3.3  Lemma Selection for Boyer-Moore

A straightforward way to apply lemma selection in the Boyer-Moore model is to pick rewrite rules for the Simplify heuristic or, more generally, any heuristic that requires relevant lemmas. For this purpose, we train a classifier on the proofs that are encountered up to the current goal (see Sect. 2.4). We then use that to select relevant lemmas for each subgoal encountered in the waterfall.

The main issue with lemma selection in this context is that the number of selected lemmas must be bounded. The larger the rewrite rule set, the more

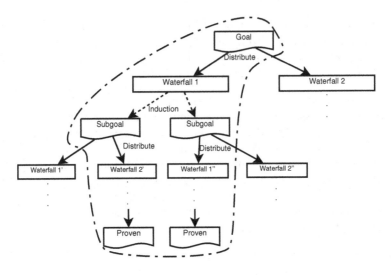

**Fig. 2.** Proof search with multi-waterfall

likely it is that the Simplify heuristic will loop. Selecting fewer lemmas means that key lemmas may be classified as 'not relevant enough' and not be selected.

Replacing the conditional simplifying function SIMP_CONV in the original Boyer-Moore implementation with the simpler rewrite function REWRITE_CONV helped improve our results, but only slightly. The same problem was observed with MESON as it could not handle large sets of lemmas, and timed out. For that reason MESON is currently used on its own without lemmas.

In contrast, ATPs are good at handling large numbers of lemmas in more ways than just simplification (see Sect. 2.3). We take advantage of this by adding a modified version of HOL(y)Hammer (see Sect. 4.4) as a heuristic that can directly prove a (sub)goal. We call this heuristic the *ATP heuristic*.

### 3.4   Direct Induction

It is quite common in manual inductive proofs for the reasoning to begin with induction before any simplification or other proof steps. In Boyer-Moore such proofs may get stuck waiting for the ATP or Simplify heuristics and eventually timing out and failing, whereas applying induction directly could help unlock the proof. Moreover, some goals in our initial experiments were being rewritten to a form that caused Boyer-Moore to either choose a wrong variable for induction or have more complicated subgoals after induction (for example because complex definitions were expanded unnecessarily) and fail.

For these reasons, we constructed a new configuration of the waterfall with no heuristics, but instead induction is applied directly. Including this in our multi-waterfall model (see Sect. 3.2) enables proofs where this waterfall is used first, so that induction is applied directly, then another waterfall uses heuristics to prove the subgoals, thus mimicking the manual proofs mentioned above.

# 4   Experiment Set-Up

## 4.1   Datasets

In order to evaluate our work, we use proven theorems about recursively-defined data types. We note here that the *IsaPlanner benchmark* [19], which has been used by some to test automated inductive theorem provers [8,9], is unsuitable in our case for the following reasons:

1. Many of the definitions use case-expressions, which are not currently supported by HOL Light.
2. The available version[2] contains many theorems that are part of the recursive definitions of the corresponding functions, and so can be proven trivially.
3. In the evaluation of HipSpec, 67 theorems were proven without using any auxiliary lemmas, and more than 10 were proven using only rewriting. Therefore, lemma selection would not have any impact in these examples.

Instead, we chose the following corpora for testing[3]:

1. *The core list library in HOL Light*, which we refer to as *List(core)*.
2. *An additional list library* used in the formalization of Hilbert's Foundations of Geometry [30]. We refer to this as *List(hilbert)*.
3. *A polynomials library in HOL Light* with properties about real polynomials represented as lists of coefficients. We refer to this as *Poly*.

The size of the test data is shown in Table 1. Note that conjunctions have been split, meaning that a theorem (or definition) $P \wedge Q$ is automatically split into $P$ and $Q$ as separate goals (or definitions).

**Table 1.** Size of the testing data

|              | Definitions | Theorems | Inductive     |
|--------------|-------------|----------|---------------|
| *List(core)*   | 44          | 97       | 73 (75.26%)   |
| *List(hilbert)* | 22          | 115      | 80 (69.57%)   |
| *Poly*         | 20          | 123      | 67 (54.47%)   |

Note that the number of inductive proofs is a lower bound, obtained by tallying the proofs containing the string "INDUCT". In our current datasets we did not observe any inductive proofs that were not captured in this way, but this is not necessarily true for other libraries. Since induction can be applied in various ways in HOL Light (e.g. by matching different induction rules), it is somewhat difficult to automatically determine the exact number of inductive proofs.

---

[2] https://github.com/tip-org/benchmarks/tree/master/original/isaplanner.
[3] https://github.com/zidongtuili/BM_test.

## 4.2  Experiments

In order to show that the Boyer-Moore model is a good starting point for inductive theorem proving, a comparison between Boyer-Moore and a simple "induction then rewriting" proof strategy was made. Such a strategy is commonly used in manual proofs for a large number of (relatively simple) inductive theorems. We will refer to it as *Ind simp.*

We then performed the following experiments using the methods described in Sect. 3:

1. *Original:* Running the original Boyer-Moore implementation as a baseline.
2. *Initial:* Running Boyer-Moore with the changes from Sect. 3.1.
3. *Multi-waterfall:* Running the multi-waterfall model described in Sect. 3.2, using the three waterfalls shown in Table 2. More specifically, we used a waterfall with the ATP heuristic, a standard waterfall with the Simplify and MESON heuristics, and a waterfall with direct induction (see Sect. 3.4).
4. *ATP:* The combination of lemma selection with the ATP heuristic *outside* Boyer-Moore, i.e. without induction, so that we evaluate and compare the performance of ATPs on inductive proofs independently.

**Table 2.** Heuristic settings for three waterfalls

| Heuristic | Waterfall 1 | Waterfall 2 | Waterfall 3 |
|---|---|---|---|
| Simplify | | × | |
| MESON | | × | |
| Other Heuristics | × | × | |
| HOL(y)Hammer | × | | |
| Induction | × | × | × |

Note that in the experiments without lemma selection (*Original* and *Initial*), the built-in rewrite rules and definitions in Boyer-Moore are used.

## 4.3  Metrics

For each experiment we evaluate the *total* success rate as $n/m$ where $n$ is the total number of theorems proven and $m$ is the total number of tested theorems. We also consider the *inductive* success rate in the same way for the subset of inductive theorems tested.

## 4.4  Environment

Two ATPs were used in our experiments: Vampire 4.1[4] and Epar (a wrapper of E included in HOL(y)Hammer) [31]. *Sparse Naïve Bayes*, as the only ML algo-

---

[4] http://www.cs.miami.edu/tptp/CASC/J8/.

rithm included in the source code of HOL(y)Hammer, was used as the learning algorithm. We ported their optimised implementation from *Mash*[5] [24].

We set the timeout for each waterfall to 30 s, which is a reasonable time that a user would wait for the system as well as the default timeout of Sledgehammer and HOL(y)Hammer [20,24]. For lemma selection we select the top 256 most relevant lemmas, which is the value at which the success rate of Vampire and Epar is known to drop significantly [20]. Such parameters cannot be optimized globally as each goal may require different values (the user could tinker with the values in an interactive setting). We believe that the current settings are reasonable for the automated evaluation of our implementation, and further optimisations can be tested in future experiments.

In order to run multiple waterfalls in parallel, a multi-core machine was used with 2 *Intel(R) Xeon(R) CPU E5-2690 v2 @ 3.00* GHz (40 threads in total) with 64 GB RAM. Note that the actual CPU load varies for different problems and is relatively low in most cases.

## 5    Evaluation

### 5.1    Results

The comparison between the original implementation of Boyer-Moore and *Ind simp* is shown in Table 3. *Ind simp* is weaker overall than Boyer-Moore. Boyer-Moore only failed on 2 theorems proven by *Ind simp* mainly due to the issue with the CNF heuristic mentioned in Sect. 3.1.

**Table 3.** Success rate of *Ind simp* and the original Boyer-Moore.

|          | *List(core)* | *List(hilbert)* | *Poly* |
|----------|--------------|-----------------|--------|
| Ind simp | 24.74%       | 13.04%          | 8.94%  |
| Original | 41.24%       | 14.78%          | 13.01% |

The results of the rest of the experiments are shown below in Table 4.

**Table 4.** Success rates of the different configurations

|                | Total        |                 |        | Induction    |                 |         |
|----------------|--------------|-----------------|--------|--------------|-----------------|---------|
|                | *List(core)* | *List(hilbert)* | *Poly* | *List(core)* | *List(hilbert)* | *Poly*  |
| Original       | 41.24%       | 14.78%          | 13.01% | 36.99%       | 8.75%           | 11.94%  |
| Initial        | 52.58%       | 20.00%          | 14.63% | 45.21%       | 17.50%          | 13.43%  |
| Multi-waterfall| 57.73%       | 63.48%          | 40.65% | 46.58%       | 62.50%          | 37.31%  |
| ATP            | 25.77%       | 36.52%          | 24.39% | 5.48%        | 30.00%          | 10.45%  |

---

[5] https://github.com/seL4/isabelle/blob/master/src/HOL/Tools/Sledgehammer/ sledgehammer_mash.ML.

*Initial* generally outperformed *Original*, which was still able to prove some theorems that *Initial* failed on though, due to the failure of some heuristics that rely on CNF.

Performance was increased in *Multi-waterfall* compared to *Initial* at a different scale for each of the 3 sets, as shown in Table 4. This indicates that the original Boyer-Moore's built-in lemmas are enough to prove theorems in *List(core)*, while lemma selection is more effective for corpora that contain more difficult theorems and a larger variety of useful lemmas.

ATPs performed relatively poorly on inductive theorems (which significantly affected their total success rate as well). However, ATPs had a high success rate in *List(hilbert)*. This shows that with appropriate lemma selection, ATPs can indeed be powerful enough to prove inductive problems.

Figure 3 shows a Venn diagram representation of the theorems proven by *Initial*, *Multi-waterfall*, and *ATP*, demonstrating the percentage of theorems that could only be proven by some of the methods, but not the others. *Multi-waterfall* could prove many theorems that none of the other methods could. This reveals the enhanced potential of combining lemma selection and Boyer-Moore.

In *List(core)* and *List(hilbert)*, *Multi-waterfall* failed to prove some theorems that were proven by *Initial*. This is mainly due to the lack of conditional rewriting (see Sect. 3.3). Moreover, some theorems were proven by *ATP* but not *Multi-waterfall*, because *Multi-waterfall* requires *quantifier-free* goals as input. This affects how the goals are translated to the ATP format, particularly for higher order (i.e. function) variables, and thus impacts the performance of ATPs.

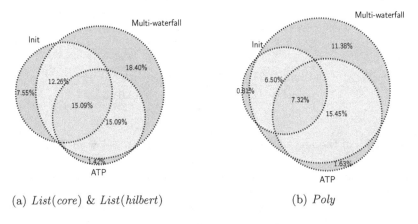

(a) *List(core)* & *List(hilbert)*          (b) *Poly*

**Fig. 3.** Coverage of proven theorems by different methods in Table 4

Examining failed proofs in *Multi-waterfall*, we discovered that in many cases the wrong variable was chosen for induction, particularly when 2 or more induction steps are used in a proof (at least 25% of the time in each data set). Other failed proofs can be attributed to missing key lemmas during lemma selection.

## 5.2   Examples

An example of an inductive theorem is DROP_DROP from *List(hilbert)* shown in
Fig. 4. It is worth comparing the manual proof to the one generated by Boyer-
Moore. With the push of a button, a theorem with a complex manual proof
containing 3 induction steps can be proven by *Multi-waterfall* automatically in
only 2 induction steps. The corresponding proof script for the new proof is automatically generated and verified in HOL Light. Also note that HOL(y)Hammer
was unable to find a proof on its own, neither when supplied with the same
lemmas used in *Multi-waterfall* nor with its own selection of 256 lemmas.

DROP_DROP: $\forall n, m, xs$ : DROP $(n + m)$=DROP $n$ (DROP $m$ $xs$)
Manual proof:

```
INDUCT_TAC THEN REWRITE_TAC [ADD_CLAUSES;DROP]
   THEN INDUCT_TAC THEN ASM_REWRITE_TAC [LENGTH;ADD_CLAUSES;DROP]
   THEN LIST_INDUCT_TAC THEN ASM_REWRITE_TAC [LENGTH;ADD_CLAUSES;DROP]
   THEN REWRITE_TAC [GSYM ADD] THEN ASM_REWRITE_TAC [DROP;ADD_CLAUSES]
```

Proof generated by Boyer-Moore *Multi-waterfall*:

```
REPEAT GEN_TAC THEN REWRITE_TAC[conj 0 ADD_AC] THEN
IND_MP_TAC ['xs:(a)list'] list_INDUCT THEN CONJ_TAC THEN
CONV_TAC (REPEATC (DEPTH_FORALL_CONV RIGHT_IMP_FORALL_CONV)) THEN
(REPEAT GEN_TAC) THENL [REWRITE_TAC[conj 1 DROP];
IND_MP_TAC ['m:num'] num_INDUCTION THEN CONJ_TAC THEN
CONV_TAC (REPEATC (DEPTH_FORALL_CONV RIGHT_IMP_FORALL_CONV)) THEN
(REPEAT GEN_TAC) THENL [REWRITE_TAC [conj 0 DROP;conj 0 ADD];
SIMP_TAC[conj 1 ADD;conj 2 DROP];];]
```

**Fig. 4.** User and Boyer-Moore proofs for DROP_DROP

An example of a failed proof is the LENGTH_REVERSE theorem shown in Fig. 5.
It has a short manual proof with only one induction step and was proven by
*Initial*, but not by *Multi-waterfall*. Further investigation showed that when trying
to prove a particular subgoal, although lemma selection included the 6 lemmas
that were sufficient for the proof, ATPs still failed to find it (even after being
allowed to run for 60 s, i.e. double the time). In our later experiments, a list
of 13 theorems (including definitions and common rewrite rules for lists) can
easily prove many subgoals when used on their own, but not as part of a large
selection. This shows that a small group of carefully picked lemmas can be more
effective than a large number of automatically selected lemmas. This explains
why *Multi-waterfall* failed to prove some theorems that *Initial* proved.

LENGTH_REVERSE: $\forall xs.$ LENGTH (REVERSE $xs$) = LENGTH $xs$
Manual proof:

```
LIST_INDUCT_TAC THEN ASM_REWRITE_TAC
  [LENGTH;REVERSE;LENGTH_APPEND] THEN ARITH_TAC
```

**Fig. 5.** User proof for LENGTH_REVERSE

## 6  Related Work

There is a number of other systems for the automation of inductive proofs. *Isaplanner* [11] is a generic framework for proof planning in Isabelle with lemma speculation techniques [12] that try to derive and prove useful lemmas from a goal. *HipSpec* [7] uses a bottom-up approach to generate lemmas that can be used to prove inductive properties of Haskell programs. *Cruanes* [9] is another system which supports structural induction with an extension to superposition-based provers. *TacticToe* [13,14] is a very recent effort that attempts to learn from human (manual) proofs and uses a Monte Carlo Tree Search [5] as it attempts to construct a proof. Based on a timeout of 60 s, a (very high) success rate of 79.5% is reported when it comes to reproving the theorems in the HOL4 list library. It should be instructive to compare the performance of our approach on the same corpus.

We should also note that there has been some work on combining machine learning techniques with inductive theorem proving in ACL2 [17,18]. The approaches are different from ours in the following ways:

- We apply lemma selection at each subgoal independently, while ACL2(ml) generally applies its search only at the beginning on the whole goal. Our fine-grained approach is possible thanks to the simplicity and accessibility our HOL Light test bed (in contrast to the complicated structure of ACL2).
- Unsupervised learning (clustering), which focuses on the similarity between goals and theorems, was used in ACL2(ml).
- The features used in ACL2(ml) are based on the structure of the formulae, which makes them suitable for selecting lemmas with a desired structure and then mutating them into a simple form of analogical reasoning.

## 7  Conclusion

Experiments with three corpora containing a large number of inductive proofs have demonstrated that the integration of machine learning in a Boyer-Moore model can greatly improve its ability to prove complex inductive theorems. The combination of powerful ATPs with lemma selection techniques and the Boyer-Moore strategies and heuristics for inductive proofs have allowed us to automatically prove a large number of theorems that neither system could prove independently.

This effective combination was enabled by a new multi-waterfall model that allows multiple proof strategies to be used in parallel to prove different subgoals. This model is configurable with respect to the time out and number of selected lemmas, which can be changed to improve its effectiveness, particularly in an interactive setting. However, improvements in the user interaction and feedback provided by the Boyer-Moore tool (perhaps with ideas from ACL2) seem paramount in order to achieve even higher proof success.

The model can also be extended with more than the currently suggested three waterfalls, so as to incorporate additional strategies and techniques in the future. For example, we could add more types or combinations of heuristics, incorporate case splitting, and include better support for non-recursive types.

Our future work will also focus on further uses of machine learning in this setting, for example as a mechanism to select an appropriate induction variable.

We believe our approach is a generic solution for the use of machine learning within proof strategies for automated inductive theorem proving. Using our *Multi-waterfall* model as a skeleton to develop such inductive proof strategies has the potential to greatly enhance the current capabilities of existing systems without sacrificing their individual effectiveness.

**Acknowledgements.** This research was supported by EPSRC grants: ProofPeer: Collaborative Theorem Proving EP/L011794/1 and The Integration and Interaction of Multiple Mathematical Reasoning Processes EP/N014758/1. It was also supported by the China Scholarship Council (CSC).

# References

1. Bertot, Y., Castéran, P.: Interactive Theorem Proving and Program Development: Coq'Art: The Calculus of Inductive Constructions. Springer, Heidelberg (2013). https://doi.org/10.1007/978-3-662-07964-5
2. Blanchette, J.C., Kaliszyk, C., Paulson, L.C., Urban, J.: Hammering towards QED. J. Formaliz. Reason. **9**(1), 101–148 (2016)
3. Boulton, R.: Boyer-Moore automation for the HOL system. In: Higher Order Logic Theorem Proving and Its Applications, pp. 133–142. Elsevier (1993)
4. Boyer, R., Moore, J.: A Computational Logic. ACM Monograph Series. Academic Press, Cambridge (1979)
5. Browne, C., et al.: A survey of Monte Carlo tree search methods. IEEE Trans. Comput. Intell. AI Games **4**(1), 1–43 (2012)
6. Bundy, A.: The automation of proof by mathematical induction. Technical report (1999)
7. Claessen, K., Johansson, M., Rosén, D., Smallbone, N.: HipSpec: automating inductive proofs of program properties. In: ATx/WInG@ IJCAR, pp. 16–25 (2012)
8. Claessen, K., Johansson, M., Rosén, D., Smallbone, N.: Automating inductive proofs using theory exploration. In: Bonacina, M.P. (ed.) CADE 2013. LNCS (LNAI), vol. 7898, pp. 392–406. Springer, Heidelberg (2013). https://doi.org/10.1007/978-3-642-38574-2_27
9. Cruanes, S.: Superposition with structural induction. In: Dixon, C., Finger, M. (eds.) FroCoS 2017. LNCS (LNAI), vol. 10483, pp. 172–188. Springer, Cham (2017). https://doi.org/10.1007/978-3-319-66167-4_10

10. de Moura, L., Bjørner, N.: Z3: an efficient SMT solver. In: Ramakrishnan, C.R., Rehof, J. (eds.) TACAS 2008. LNCS, vol. 4963, pp. 337–340. Springer, Heidelberg (2008). https://doi.org/10.1007/978-3-540-78800-3_24

11. Dixon, L., Fleuriot, J.: IsaPlanner: a prototype proof planner in Isabelle. In: Baader, F. (ed.) CADE 2003. LNCS (LNAI), vol. 2741, pp. 279–283. Springer, Heidelberg (2003). https://doi.org/10.1007/978-3-540-45085-6_22

12. Dixon, L., Johansson, M.: IsaPlanner 2: a proof planner in Isabelle. DReaM Technical report (System description) (2007)

13. Gauthier, T., Kaliszyk, C., Urban, J.: TacticToe: learning to reason with HOL4 tactics. In: 21st International Conference on Logic for Programming, Artificial Intelligence and Reasoning, LPAR-21, vol. 46, pp. 125–143 (2017)

14. Gauthier, T., Kaliszyk, C., Urban, J., Kumar, R., Norrish, M.: Learning to prove with tactics. CoRR abs/1804.00596 (2018). http://arxiv.org/abs/1804.00596

15. Harrison, J.: HOL light: a tutorial introduction. In: Srivas, M., Camilleri, A. (eds.) FMCAD 1996. LNCS, vol. 1166, pp. 265–269. Springer, Heidelberg (1996). https://doi.org/10.1007/BFb0031814

16. Harrison, J.: Optimizing proof search in model elimination. In: McRobbie, M.A., Slaney, J.K. (eds.) CADE 1996. LNCS, vol. 1104, pp. 313–327. Springer, Heidelberg (1996). https://doi.org/10.1007/3-540-61511-3_97

17. Heras, J., Komendantskaya, E.: ACL2(ml): machine-learning for ACL2. arXiv preprint arXiv:1404.3034 (2014)

18. Heras, J., Komendantskaya, E., Johansson, M., Maclean, E.: Proof-pattern recognition and lemma discovery in ACL2. In: McMillan, K., Middeldorp, A., Voronkov, A. (eds.) LPAR 2013. LNCS, vol. 8312, pp. 389–406. Springer, Heidelberg (2013). https://doi.org/10.1007/978-3-642-45221-5_27

19. Johansson, M., Dixon, L., Bundy, A.: Conjecture synthesis for inductive theories. J. Autom. Reason. **47**(3), 251–289 (2011)

20. Kaliszyk, C., Urban, J.: Learning-assisted automated reasoning with flyspeck. J. Autom. Reason. **53**, 1–41 (2014)

21. Kaliszyk, C., Urban, J.: HoL(y)Hammer: online ATP service for HOL Light. Math. Comput. Sci. **9**(1), 5–22 (2015)

22. Kaufmann, M., Manolios, P., Moore, J.: Computer-Aided Reasoning: An Approach. Advances in Formal Methods. Springer, Heidelberg (2000). https://doi.org/10.1007/978-1-4757-3188-0

23. Kovács, L., Voronkov, A.: First-order theorem proving and VAMPIRE. In: Sharygina, N., Veith, H. (eds.) CAV 2013. LNCS, vol. 8044, pp. 1–35. Springer, Heidelberg (2013). https://doi.org/10.1007/978-3-642-39799-8_1

24. Kühlwein, D., Blanchette, J.C., Kaliszyk, C., Urban, J.: MaSh: machine learning for Sledgehammer. In: Blazy, S., Paulin-Mohring, C., Pichardie, D. (eds.) ITP 2013. LNCS, vol. 7998, pp. 35–50. Springer, Heidelberg (2013). https://doi.org/10.1007/978-3-642-39634-2_6

25. Meng, J., Paulson, L.C.: Lightweight relevance filtering for machine-generated resolution problems. J. Appl. Log. **7**(1), 41–57 (2009)

26. Nipkow, T., Paulson, L.C., Wenzel, M.: Isabelle/HOL: A Proof Assistant for High-erorder Logic, vol. 2283. Springer, Heidelberg (2002). https://doi.org/10.1007/3-540-45949-9

27. Papapanagiotou, P., Fleuriot, J.: The Boyer-Moore waterfall model revisited (2011)

28. Paulson, L.C., Blanchette, J.C.: Three years of experience with Sledgehammer, a practical link between automatic and interactive theorem provers. In: IWIL-2010, vol. 1 (2010)

29. Schulz, S.: E-A Brainiac theorem prover. AI Commun. **15**(2, 3), 111–126 (2002)
30. Scott, P.: Ordered geometry in Hilbert's Grundlagen der Geometrie. Ph.D. thesis, The University of Edinburgh (2015)
31. Urban, J.: BliStr: the blind strategymaker. arXiv preprint arXiv:1301.2683 (2013)

# FMUS2: An Efficient Algorithm to Compute Minimal Unsatisfiable Subsets

Shaofan Liu and Jie Luo[✉]

State Key Laboratory of Software Development Environment,
School of Computer Science and Engineering, Beihang University,
Beijing 100191, China
{shaofanliu,luojie}@nlsde.buaa.edu.cn

**Abstract.** In the past few years, much attention has been given to the problem of finding Minimal Unsatisfiable Subsets (MUSes), not only for its theoretical importance but also for its wide range of practical applications, including software testing, hardware verification and knowledge-based validation. In this paper, we propose an algorithm for extracting all MUSes for formulas in the field of propositional logic and the function-free and equality-free fragment of first-order logic. This algorithm extends earlier work, but some changes have been made and a number of optimization strategies have been proposed to improve its efficiency. Experimental results show that our algorithm performs well on many industrial and generated instances, and the strategies adopted can indeed improve the efficiency of our algorithm.

**Keywords:** Minimal unsatisfiable subsets · Heuristic algorithm
Optimization strategy · SAT

## 1 Introduction

Given an unsatisfiable formula in Conjunctive Normal Form (CNF), a minimal unsatisfiable subset (MUS) is a subset of clauses which is (1) unsatisfiable, and (2) minimal, which means removing any one of its elements will make the remaining set satisfiable. Different classes of algorithms have been proposed to efficiently enumerate all or partial MUSes [1, 16, 19]. Early algorithms are based on subset enumeration [3, 8]. In these algorithms, the power set of the input is enumerated in a tree structure and every subset is checked for satisfiability. A MUS can be easily identified by definition. Another class of algorithms [2, 12, 17] relies on the *hitting set duality*. First, all Minimal Correction Subsets (MCSes) are computed. Then, all MUSes are obtained by computing minimal hitting sets of these MCSes. CAMUS [12] is one of the state-of-the-art algorithms for computing all MUSes in this class. Recently, algorithms (e.g. eMUS/MARCO [11, 14]) for partial MUS enumeration were proposed. These algorithms are able to produce the first MUS quickly and early, and the following MUSes are generally produced incrementally.

© Springer Nature Switzerland AG 2018
J. Fleuriot et al. (Eds.): AISC 2018, LNAI 11110, pp. 104–118, 2018.
https://doi.org/10.1007/978-3-319-99957-9_7

Most of the current algorithms rely on a SAT solver for checking the satisfiability of clause sets. The advantage is that they can utilize the power of highly optimized SAT solvers. But they also unavoidably introduce many duplicated computations. For example, if clause set $\{1, 2, 3\}$ is checked unsatisfiable, they should check the satisfiability of $\{1, 2\}, \{1, 3\}, \{2, 3\}$ for determining whether $\{1, 2, 3\}$ is indeed a MUS or not. Although many optimizations (e.g. using the hitting set duality) for these algorithms are proposed to reduce the number of SAT solver calls, there are still many duplicated computations. And when there are a larger number of MUSes in the input, the number of SAT solver calls will be enormous and the time used for duplicated computations will also be obviously large, which will cause a decrease in efficiency.

For the consideration of the shortcoming described above for those algorithms which are based on SAT solvers, we have adopted another approach for enumerating MUSes. This paper extends our earlier work [20] on computing MUSes for a decidable fragment of First-Order Formulas (FOL), and its main contributions can be summarized as follows. First, in contrast to most approaches which make use of variable assignments or an external SAT-solver to check satisfiability, this paper proposes a "decompose-merge" algorithm inspired by the process of logical deduction in belief revision [10,13]. It first decomposes clauses of the given formula into literals to easily identify all inconsistent relations between them, and then assembles all literals back to the original clauses to reveal the minimal inconsistent relations among them. Second, the proposed algorithm uses unification to accomplish "general instantiation". In other words, instead of instantiating all variables by all feasible values, a most general inconsistent subset is used to represent a class of instances which are equivalent under the *more general* relation, which can avoid generating of excessive instances and reduce the searching space. Another contribution of the paper is the optimization strategies used to improve the efficiency of our algorithm. Experimental results show that our algorithm is competitive and has the potential to be even better.

## 2   Preliminaries

This paper focuses on the function-free and equality-free fragment of first-order logic (FEF for short). Satisfiability of formulae from the FEF fragment is decidable, because it is a special case of effectively propositional logic (EPR), also known as the Bernays-Schönfinkel class [15] which is proved to be decidable. Hence it is feasible to design an algorithm to compute all MUSes in the FEF fragment.

Formulas in FEF are represented in CNF. That is, a CNF formula is a conjunction (AND, $\wedge$) of one or more clauses, and each clause is a disjunction (OR, $\vee$) of one or more literals. A literal is an atomic formula or its negation (NOT, $\neg$). The syntax is shown below.

$$F ::= C^1 \wedge \cdots \wedge C^n$$
$$C^i ::= L^1 \vee \cdots \vee L^{m_i}$$
$$L^j ::= A \mid \neg A$$

Following the convention of many other papers (e.g. [4,7]), a CNF formula is treated as a (finite) set of clauses.

Here is an example formula in the FEF fragment.

*Example 1.* The uppercase letter $X$ denotes a variable, while the lowercase letter $a$ and $b$ denote constants.

$$F = (\overset{C^1}{A(a)}) \wedge (\overset{C^2}{\neg A(X) \vee B(X)}) \wedge (\overset{C^3}{\neg B(b)}) \wedge (\overset{C^4}{\neg B(a)})$$

## 3   Algorithm for Computing All MUSes

In this section, we will give an overview of the proposed FMUS2 algorithm for computing all MUSes for formulas in the FEF fragment, which is an improved version of our previous FMUS algorithm [20].

Both FMUS2 and FMUS adopt a constructive "decompose-merge" approach to compute MUSes. First, the clauses of the given formula are decomposed into literals and inconsistent pairs of decomposed literals are all computed, this is the "decompose" procedure. Thus, the initial intermediate results are created, which are sets of literals and indicate the contradictory relations among literals of all clauses. Then, by iteratively merging these intermediate results into larger sets, which are still unsatisfiable during the whole process, the original clauses are restored one by one, this is the "merge" procedure. The merging operation processes literal by literal and clause by clause. After all the literals are merged into original clauses, the final results will contain all MUSes of the clauses in the input formula.

FMUS2 and FMUS both use unification for instantiating clauses with the *most general unifier*, but through different approaches.

**Definition 1 (Most general unifier).** *A substitution $\sigma$ is a most general unifier (MGU) of two literals $L_1$ and $L_2$ if $\sigma$ unifies them, i.e. $(L_1, \sigma) = (L_2, \sigma)$, and for any unifier $\sigma'$ of these two formulas, there exists a substitution $\omega$ such that $\sigma' = \omega \circ \sigma$.*

For FMUS, MGUs are kept along with the whole procedure. There will be a MGU for each intermediate result, which indicates how this intermediate result is unsatisfiable. For example, $I = \{A(a), \neg A(X)[a/X]\}$ is unsatisfiable if we substitute the constant $a$ for the variable $X$. In other words, FMUS uses MGUs instead of explicit instantiation. The implicit way can cause difficulty for identifying whether two substitutions, which look different, are in fact equivalent sometimes. For example, if there are $I_1 = \{A(Y), \neg A(X)[Y/X]\}$ and $I_2 = \{A(Z), \neg A(X)[Z/X]\}$ among all intermediate results, they are equivalent when the variables $Y$ and $Z$ are substituted by the same constant $a$, but they are different when $Y$ is substituted by $a$ and $Z$ is substituted by $b$. When the input formula is complex, the identification will be difficult. Some redundant branches will arise also because of its implicity.

So for FMUS2, we have tried to adopt a new way to solve this problem. We choose to explicitly instantiate the original clauses with ground term (i.e. terms without variables). Before the instantiation, MGUs of decomposed literals are computed, which will be used to confine the scope of instantiation and reduce the number of instantiations. For example, if the variable $X$ from $\neg A(X)$ can be substituted by constants $a, b, c$ but only substitute the constant $a$ for the variable $X$ can lead to contradiction, there is no need for replacing $X$ with $b$ or $c$. Thus the scope of instantiation is confined.

Based on the discussion above, the basic steps of FMUS2 are listed below.

1. **Preprocess.** For the given CNF formula, FMUS2 first parses and decomposes clauses of the CNF formula into literals with labels to indicate their origin, meanwhile overlapping bound variables are renamed to eliminate name ambiguity.
2. **Find initial contradictions.** For each decomposed literal it is checked whether there is another literal which is contradictory to it. This process is accomplished by unification to obtain a MGU.
3. **Instantiation.** If there are variables in the given CNF formula, literals will be instantiated and the MGUs already found will be used to confine the scope of instantiation. After instantiation, the previous step of finding initial contradictions will be processed again for these ground literals in newly instantiated clauses. If there is no variable in the given CNF formula, which means the formula is a ground formula, there is no need for instantiation.
4. **Merge.** After all steps above, the core process of FMUS2—the merge process begins. Literal instances of the same clause are merged to reconstruct instances of their original clause according to certain order, which will be further discussed in Sect. 4.1. The principle for deciding whether two intermediate results can be merged will be discussed in Sect. 4.1 too.
5. **Map back.** If the original CNF formula is a ground formula, then the result is all MUSes of the input. But if the original CNF formula contains variables, one original clause may have many corresponding clause instances. Thus after all steps above, the instance sets need to be mapped back into unsatisfiable subsets of the original clause set. Then all MUSes of the input can be obtained by extracting the minimal ones from the set of all those unsatisfiable subsets.

The pseudo-code for FMUS2 is shown in Algorithm 1.

FMUS2 takes a set of clauses (a CNF formula) $F$ as input, and outputs all MUSes of $F$. If $F$ is satisfiable, the output will be $\emptyset$. Lines 1–4 demonstrate the process of decomposing clauses into literals. Every clause $C^i$ in $F$ is $L_1^i \vee \cdots \vee L_{m_i}^i$ where $m_i$ stands for the number of literals in $C^i$. Lines 5 enumerates all inconsistent pairs among decomposed literals to construct $M_0$. Lines 6–9 show, if $F$ is in the field of first-order logic, all literals will be instantiated and the initial contradictions set $M_0$ will be computed again for $L'$.

The loop in lines 10–16 of FMUS2 is the most interesting but bewildering part. In this loop, we iteratively merge clauses that contain multiple literals. In each iteration, literals from a certain clause are merged to the original form and the unsatisfiable subsets that contain these literals are merged to larger

---

**Algorithm 1.** FMUS2($F$)

---

**Input:** $F$ as a set of clauses $\{C^1, \ldots, C^n\}$
**Output:** The set of all MUSes of $F$
1: **for** $i = 1$ **to** $n$ **do**
2:     Decompose $C^i$ to $\{L_1^i, \ldots, L_{m_i}^i\}$
3: **end for**
4: $L := \bigcup\limits_{i=1}^{n} \{L_1^i, \ldots, L_{m_i}^i\}$
5: Find the initial inconsistent set $M_0$ of $L$
6: **if** there are variables in $L$ **then**
7:     Instantiate $L' := L$
8:     Find the inconsistent set $M_0$ of $L'$
9: **end if**
10: **for** $i = 1$ **to** $n$ **do**
11:     **if** $m_i > 1$ **then**
12:         $M_i := \text{Merge}(i, M_{i-1})$
13:     **else**
14:         $M_i := M_{i-1}$
15:     **end if**
16: **end for**
17: Map instances in $M_n$ back to their corresponding clauses and obtain $M_n'$
18: **return** $M_n'$

---

unsatisfiable sets. Each round of iteration is based on the result of the previous iteration. To give a clearer explanation, let us suppose that the $i$th clause (i.e. $C^i$) is going to be merged and $M_{i-1}$ is the result of the last iteration. So clauses $C^1$ to $C^{i-1}$ have already been merged, and clauses $C^i$ to $C^n$ still appear in the form of literals. The process of merging the $i$th clause is shown as Algorithm 2. Note that when merging, all literals are in propositional logic, which means all substitutions $\sigma$ are empty now. So in Algorithm 2, we do not use the symbol $\sigma$.

In the Merge process, $N_i$ is generated by extracting elements from $M_{i-1}$ which have no intersection with literals in $C^i$ (Line 2). Conversely, $S_i$ is a set of $m_i$-tuples that represent all merging options with respect to $C^i$ (Line 3). The $j$th item in each tuple $(\Phi_1^i, \ldots, \Phi_{m_i}^i)$ is supposed to be an element of $M_{i-1}$ that contains literal $L_j^i$. Then $M_i'$ is constructed through merging all alternative $\Phi_1^i, \ldots, \Phi_{m_i}^i$ (Line 6). As a result, $N_i$ consists of unsatisfiable subsets without $C^i$, while $M_i'$ is formed of unsatisfiable subsets which contain $C^i$. The operation of MS() is to obtain those minimal elements under set inclusion. That is, if $\Theta = \{\Theta_1, \ldots, \Theta_n\}$, where $\Theta_1, \ldots, \Theta_n$ are different sets, then $\text{MS}(\Theta) = \{\Theta' \mid \Theta' \in \Theta$ and there is no $\Theta'' \in \Theta$ such that $\Theta'' \subset \Theta'\}$. After all formulas are merged, we get $M_n$, the set that contains all MUSes of instances of original clauses. Finally, by processing the **Map Back** step that is Algorithm 1 Line 17, all MUSes are extracted.

Since the input set consists of finite clauses, and the number of intermediate results generated during the procedure of FMUS2 is also finite, FMUS2 must

**Algorithm 2.** Merge($i, M_{i-1}$)

**Input:** the set of all MUSes after merging $i - 1$ clauses of $F$
**Output:** the set of all MUSes after merging $i$ clauses of $F$
1:  $M_i' := \emptyset$
2:  $N_i := \{\phi \mid \phi \in M_{i-1} \text{ and } \phi \cap \{L_1^i, \dots, L_{m_i}^i\} = \emptyset\}$
3:  $S_i := \{(\Phi_1^i, \dots, \Phi_{m_i}^i) \mid \Phi_j^i \in M_{i-1}, L_j^i \in \Phi_j^i, j \in [1, m_i]\}$
4:  **for all** $(\Phi_1^i, \dots, \Phi_{m_i}^i) \in S_i$ **do**
5:     **if** $(\Phi_1^i, \dots, \Phi_{m_i}^i)$ can merge **then**
6:        $M_i' := M_i' \cup \left\{ \{C^i\} \cup \bigcup_{j=1}^{m_i} (\Phi_j^i - \{L_j^i\}) \right\}$
7:     **end if**
8:  **end for**
9:  $M_i := \texttt{MS}(N_i \cup M_i')$
10: **return** $M_i$

terminate in finite steps. The output of FMUS2 will be the set which consists of all MUSes of the input. Besides, FMUS2 can be altered to a partial MUS enumerating algorithm by simply outputting all MUSes newly found after merging every clause. This is based on the fact that if there is a MUS $\{1, 3\}$ after merging clauses 1 to 3, $\{1, 3\}$ is also a MUS of the whole set of clauses 1 to $n$, where $n \geq 3$.

# 4  Optimization Strategies

In this section, we will discuss some optimization strategies used to improve the performance of FMUS2. The strategies can be divided into two categories. One is concerned with the order used in the merging procedure, and the other is concerned with pruning, i.e. reducing the number of intermediate results.

## 4.1  Merging Strategies

For FMUS2, the merging procedure is the most important and time-consuming part. Though different orders of merging do not affect the correctness of the algorithm, they do affect the number of intermediate results significantly. Thus the efficiency of the algorithm will be affected. A good order may solve an input rapidly while a bad order may timeout for the same input. We propose a simple heuristic merging strategy to determine the merging order.

The heuristic merging strategy is based on the theoretical maximum number $M(C)$ of intermediate results for each clause $C$ when it is the first to merge. In detail, $M(C^i) = \prod_{j=1}^{m_i} n_j^i$. The $m_i$ denotes the number of literals of $C^i$, and the numbers of contradictory literals of $C_1^i, \cdots, C_{m_i}^i$ are $n_1^i, \cdots, n_{m_i}^i$. In order to rein in the potentially exponential growth of intermediate results as much as possible, before merging, $M(C^i)$ will be calculated for every clause and then arranged from least to most which is the merging order. For the consideration of comparison, a completely opposite order and a random order are implemented as contrast strategies.

Except for deciding the order of merging, the heuristic strategy will also renumber the clauses opposite to the merging order. The reasons are as follows.

While merging, we should decide whether two intermediate results can be merged. The principle is that, when merging clause $i$, if two intermediate results contains two different literals that come from the same clause $j(j \neq i)$ separately, they can not be merged. If they are merged, the unsatisfiability of the newly generated intermediate result can not be maintained.

*Example 2.* Considering

$$F = \{\overset{1.1}{x_1} \vee \overset{1.2}{x_2}, \overset{2.1}{\neg x_1} \vee \overset{2.2}{\neg x_2}\}.$$

The $x.y$ labels on the top of literals are identifiers. The $x$ denotes the clause number which this literal belongs to, and the $y$ denotes the literal number in clause. In particular, the $x.0$ label denotes the whole $x$-th clause.

It is obvious that $F$ is satisfiable. Before merging, there are two intermediate results, $I_1 = \{\overset{1.1}{x_1}, \overset{2.1}{\neg x_1}\}$ and $I_2 = \{\overset{1.2}{x_2}, \overset{2.2}{\neg x_2}\}$. According to the principle above, $I_1$ and $I_2$ can not be merged. If they are merged, the result is $I_3 = \{x_1 \overset{1.0}{\vee} x_2, \neg x_1 \overset{2.0}{\vee} \neg x_2\}$, which is incorrect, in other words, satisfiable.

Because we should check whether two intermediate results can be merged, we should traverse and check all possible clauses, for which one or more literals are contained in these two intermediate results. The larger the $M(C)$ for a clause is, the more likely different literals of this clause will be contained in two different intermediate results. Thus when merging two intermediate results, the clause for which one or more literals are contained in these two intermediate results and $M(C)$ is larger, will be checked first.

We give an example to show how the merging strategy works and why we renumber the clauses opposite to the merging order.

*Example 3.* Considering

$$F = \{\overset{1.1}{x_1} \vee \overset{1.2}{\neg x_4}, \overset{2.1}{x_4} \vee \overset{2.2}{\neg x_3} \vee \overset{2.3}{x_2}, \overset{3.1}{\neg x_1} \vee \overset{3.2}{x_2} \vee \overset{3.3}{x_3}, \overset{4.1}{\neg x_2} \vee \overset{4.2}{x_4}, \overset{5.0}{\neg x_1}\}.$$

$F$ is satisfiable. Before merging, there are 7 intermediate results $I_1$ to $I_7$.

$$I_1 = \{\overset{1.1}{x_1}, \overset{3.1}{\neg x_1}\}, I_2 = \{\overset{1.1}{x_1}, \overset{5.0}{\neg x_1}\},$$
$$I_3 = \{\overset{1.2}{\neg x_4}, \overset{2.1}{x_4}\}, I_4 = \{\overset{1.2}{\neg x_4}, \overset{4.2}{x_4}\},$$
$$I_5 = \{\overset{2.2}{\neg x_3}, \overset{3.3}{x_3}\}, I_6 = \{\overset{2.3}{x_2}, \overset{4.1}{\neg x_2}\}, I_7 = \{\overset{3.2}{x_2}, \overset{4.1}{\neg x_2}\}.$$

Because $C^5$ is a clause with only one literal, we just need to compute $M(C^1)$ to $M(C^4)$.

$$M(C^1) = n_1^1 \times n_2^1 = 2 \times 2 = 4,$$
$$M(C^2) = n_1^2 \times n_2^2 \times n_3^2 = 1 \times 1 \times 1 = 1,$$
$$M(C^3) = n_1^3 \times n_2^3 \times n_3^3 = 1 \times 1 \times 1 = 1,$$
$$M(C^4) = n_1^4 \times n_2^4 = 2 \times 1 = 2.$$

Thus the merging order is $2, 3, 4, 1$.

After merging $C^2$, we will get $I_8 = \{\overset{1.2}{\neg x_4}, x_4 \vee \overset{2.0}{\neg x_3} \vee x_2, \overset{3.3}{x_3}, \overset{4.1}{\neg x_2}\}$. And the set of all intermediate results is $\Gamma = \{I_1, I_2, I_4, I_7, I_8\}$.

Then we will merge $C^3$. $I_1$, $I_7$ and $I_8$ involve $L_1^3$, $L_2^3$ and $L_3^3$ separately. First we get $I_9 = \{\overset{1.1}{x_1}, \overset{3.1}{\neg x_1}, \overset{3.2}{x_2}, \overset{4.1}{\neg x_2}\}$ by merging $I_1$ and $I_7$. Then we try to merge $I_8$ and $I_9$. Because there are literals of $C^1$ and $C^4$ in $I_8$ and $I_9$, we should check whether $I_8$ and $I_9$ can be merged. Because $M(C^1)$ is larger than $M(C^4)$, we first check literals of $C^1$. $I_8$ contains $L_2^1$ and $I_9$ contains $L_1^1$, thus they can not be merged. $I_8$ and $I_9$ will be discarded. The remaining $\Gamma = \{I_2, I_4\}$.

If we do not check opposite to the merging order, we could first check literals of $C^4$, and we will see that $I_8$ and $I_9$ both contain $L_1^4$. Then we should also check literals of $C^1$. As a result, a useless check is processed.

The remaining $I_2$ and $I_4$ can be merged, but there is no another intermediate result contains $L_1^4$. Thus no MUS is found, and the original $F$ is satisfiable.

## 4.2    Pruning Strategies

Two strategies are applied to prune the search space of MUSes, i.e., eliminate useless intermediate results. The core of these two strategies is keeping every intermediate result $I$ *minimal*, in other words, $I$ is not a superset of any other intermediate result $I_2$ or any already obtained MUS. If $I$ is a superset of a MUS, it is obvious that $I$ can never be merged (expanded) to a MUS. If $I$ is a superset of another intermediate result $I_2$, for any larger intermediate result $I'$ that contains $I$ by merging it with some other intermediate result $I_3, I_4, \ldots$, there will be another $I_2'$ that contains $I_2$ by merging it with the same $I_3, I_4, \ldots$. So $I'$ is not minimal, and so it cannot be a MUS.

Strategy 1 focuses on eliminating useless intermediate results after merging every literal. The merging operation processes literal by literal and clause by clause. After merging every literal, each newly generated intermediate result will be checked whether it is a superset of any other intermediate result that is not used while merging this literal. And after merging every clause, each remaining newly generated intermediate result will be checked whether it is a superset of any already obtained MUS.

The ideal situation is that no useless intermediate result will be generated. But Strategy 1 cannot prevent the appearance of useless intermediate results. It can only discard them after their appearance. Though it can benefit the following merging steps, time and space are spent to generate the useless intermediate results and check whether they are useless. So we propose the next strategy to partly prevent the appearance of useless intermediate results.

Strategy 2 is recording an affirmative propositions set and a negative propositions set for every intermediate result, and then these sets will be used to decide whether two intermediate results can merge. For every intermediate result, its affirmative propositions set is a set that contains every *single* affirmative proposition which belongs to this intermediate result, and its negative propositions set is a set that contains every *single* negative proposition which belongs to this intermediate result. The *single* proposition means proposition contained in this intermediate result, and not the whole clause which this proposition belongs to is contained in this intermediate result.

While merging, the intersection of two intermediate results' affirmative propositions sets $P'$ and the intersection of two intermediate results' negative propositions sets $N'$ will be computed first. Then, we will find literals contained in both intermediate results, and remove their corresponding propositions in $P'$ or $N'$. Finally, if $P'$ and $N'$ are both $\emptyset$, these two intermediate results can be merged. If not, these two intermediate results can not be merged.

$P'$ contains propositions in these two intermediate results' affirmative propositions sets both. If it is not empty, it means there are propositions with the same name but coming from different clauses, i.e., duplicate propositions. If we merge these two intermediate results and get a new intermediate result $I$, $I$ cannot be *minimal*. Because there are duplicate propositions in $I$, we at least can remove one duplicate proposition without changing the unsatisfiability of $I$.

Example 4 shows how Strategy 2 works.

*Example 4.*

$$F = (\overset{1.1}{x_1} \vee \overset{1.2}{x_2}) \wedge (\overset{2.1}{\neg x_1} \vee \overset{2.2}{\neg x_2}) \wedge (\overset{3.1}{\neg x_3} \vee \overset{3.2}{\neg x_2}) \wedge (\overset{4.1}{\neg x_3} \vee \overset{4.2}{x_2}) \wedge (\overset{5.0}{x_3}).$$

After merging the first clause, we get an intermediate result $I_1$ with its affirmative propositions set $\emptyset$ and its negative propositions set $\{\neg x_1, \neg x_2\}$.

$$I_1 = \{\overset{1.0}{x_1 \vee x_2}, \overset{2.1}{\neg x_1}, \overset{3.2}{\neg x_2}\}.$$

And then, we merge the second clause, i.e., merge $I_1$ and $I_2 = \{\overset{2.2}{\neg x_2}, \overset{4.2}{x_2}\}$. The affirmative propositions set of $I_2$ is $\{x_2\}$, and the negative propositions set is $\{\neg x_2\}$. The proposition $\neg x_2$ appears in both $I_1$ and $I_2$, but it comes from different literals of different clauses (clause 2 and clause 3). So we choose not to merge $I_1$ and $I_2$.

If we merge $I_1$ and $I_2$, we will get an intermediate result $I_3$.

$$I_3 = \{\overset{1.0}{x_1 \vee x_2}, \overset{2.0}{\neg x_1 \vee \neg x_2}, \overset{3.2}{\neg x_2}, \overset{4.2}{x_2}\}.$$

$I_3$ is a superset of another intermediate result $I_4 = \{\overset{3.2}{\neg x_2}, \overset{4.2}{x_2}\}$. According to the reason described above, $I_3$ will be discarded.

As a result, this strategy prevents $I_1$ and $I_2$ merging, instead of merging and discarding the newly generated intermediate result. Because it will not carry out the merging and discarding process, which needs to traverse the intermediate result set to decide whether one should be discarded or not, runtime will be saved. When the original formula becomes complex, the situation similar to Example 4 will occur many times. So a lot of runtime will be saved.

# 5    Experiments

In this section, a series of experiments are performed to evaluate the general performance of FMUS2 by comparing it with the state-of-the-art algorithms and verify the effectiveness of the heuristic merging and pruning strategies we adopted. All experiments were performed on a Ubuntu 16.04 LTS Linux server with an Intel Xeon E5-4607 v2 2.6 GHz CPU and 15 GB main memory. Timeout is set to 300 s for all test cases. For timeout instances, we use the Penalized Average Runtime (PAR-10) [9], where a timeout counts 10 times the time limit. That is, the runtime for every timeout instance is set to 3000 s.

## 5.1    Performance

As mentioned earlier, the FEF fragment is a special case of EPR. Since many implementations of MUS enumeration algorithms only deal with propositional logic, we shall first evaluate the performance of FMUS2 on FEF by comparing its performance on industrial benchmarks with one of the state-of-the-art partial MUSes enumerators—MARCO [11], which support enumerating MUSes in the EPR fragment by using Z3 [5] as a SAT-solver. In this experiment, MARCO and Z3 are both open source and the version of MARCO is 2.0.1. The evaluation is performed on 100 instances from the EPR division of the TPTP Problem Library [18]. The majority of instances considered are originally from realistic problems, including geometry, puzzles, and software verification.

Figure 1 shows the the runtime of FMUS2 and MARCO for each instance. The $x$-coordinate represents the number of solved instances, and the $y$-coordinate

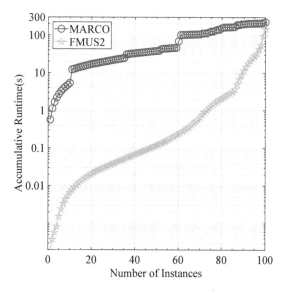

**Fig. 1.** Comparing FMUS2 against MARCO on industrial benchmarks.

represents the accumulative runtime spent by MARCO or FMUS2 when solving these instances. The line for FMUS2 is always below the line for MARCO, which implies that FMUS2 is faster than MARCO in general for these instances. Although FMUS2 does not spend less time than MARCO for every instance, the accumulative runtime, in other words, the average runtime for FMUS2 is less than MARCO (the average runtime for FMUS2 and MARCO are 2.230 s and 1.460 s respectively).

The experimental results also reveal that FMUS2 is still not optimized enough to compete with methods which utilize highly optimized SAT-solvers when dealing with large-scale formula sets which have complex inconsistency relations between clauses of formulas, i.e. hard instances for FMUS2.

Since FMUS2 is a complete MUSes enumeration algorithm, we shall do a further comparison with one of the state-of-the-art complete MUSes enumeration algorithm CAMUS [12]. Because CAMUS only supports propositional logic and the above industrial instances for comparing with MARCO are relatively scattered and smaller in their scales, randomly generated benchmarks in propositional logic are adopted to further comparison on large-scale instances. Note that in this experiment, MARCO uses its built-in SAT solver—MiniSAT [6].

The randomly generated benchmarks are divided into classes such that all instances in each class have the same number of formulas, which can be found in https://github.com/luojie-sklsde/MUS_Random_Benchmarks. Each class contains 200 unsatisfiable formulas, denoted as the form "mus$x$-$y$", where the first number $x$ stands for the number of clauses of instances in this class and the second number $y$ stands for the average number of literals in each instance. For example, class "mus400-798" is composed of instances (formulas) containing 400 clauses, where the average number of literals in these instances is 798. Although the number of clauses (i.e. $x$) is fixed in each class, the number of literals within clauses can vary (so $y$ is an average number), which allows us to simulate as many cases as possible.

Table 1 shows experimental results of CAMUS, MARCO and FMUS2 on the randomly generated benchmarks.

**Table 1.** Comparing among CAMUS, MARCO and FMUS2

| Benchmarks | CAMUS | | | MARCO | | | FMUS2 | | |
|---|---|---|---|---|---|---|---|---|---|
| | $N_{TO}$ | $N_{best}$ | $T_{Ave}$ | $N_{TO}$ | $N_{best}$ | $T_{Ave}$ | $N_{TO}$ | $N_{best}$ | $T_{Ave}$ |
| mus100-200 | 25 | 10 | 379.344 | 12 | 0 | 183.483 | 3 | 187 | 45.798 |
| mus200-401 | 121 | 0 | 1822.921 | 76 | 0 | 1150.496 | 16 | 184 | 242.506 |
| mus400-798 | 194 | 0 | 2911.875 | 182 | 0 | 2736.391 | 34 | 166 | 511.340 |
| mus600-1200 | 200 | 0 | 3000 | 200 | 0 | 3000 | 56 | 144 | 841.770 |
| mus800-1601 | 200 | 0 | 3000 | 200 | 0 | 3000 | 64 | 136 | 961.467 |
| mus1000-2002 | 200 | 0 | 3000 | 200 | 0 | 3000 | 69 | 131 | 1035.995 |

The first column of Table 1 are different classes of the benchmarks, followed by statistical runtime data for CAMUS, MARCO and FMUS2. $N_{TO}$ is the number of instances which are timeout after 300s, $N_{best}$ is the number of instances which get the best runtime among the 3 approaches, and $T_{Ave}$ is the average runtime (in seconds) of all instances. The bold number in each row represents the best results among different approaches. It is clear that FMUS2 outperforms CAMUS and MARCO in all three numbers, i.e. $N_{TO}$, $N_{best}$, and $T_{Ave}$. FMUS2 has the smallest number of timeout instances in all six classes and gets the best runtime for most of instances in each class of benchmarks, which means the performance of FMUS2 is stable among different instances. Based on a detailed analysis of the experimental data, we find that FMUS2 is especially efficient when dealing with instances that contain multiple MUSes, which are exactly the ideal targeting input of the MUSes enumeration problem.

The performance experiment shows the competitive power of FMUS2, that is, FMUS2 can perform better than the state-of-the-art algorithms MARCO and CAMUS in some industrial and randomly generated cases.

## 5.2   Effectiveness of the Optimization Strategies

To evaluate whether the strategies are effective or not, we carried out a series of experiments. Table 2 shows experimental results of FMUS2 and FMUS on the same benchmarks with Table 1.

**Table 2.** Comparing FMUS2 with FMUS on randomly generated benchmarks

| Benchmarks | FMUS | | | FMUS2 | | |
|---|---|---|---|---|---|---|
| | $N_{TO}$ | $N_{best}$ | $T_{Ave}$ | $N_{TO}$ | $N_{best}$ | $T_{Ave}$ |
| mus100-200 | **3** | 2 | 46.713 | **3** | **195** | **45.798** |
| mus200-401 | 19 | 3 | 287.875 | **16** | **182** | **242.506** |
| mus400-798 | 36 | 0 | 542.194 | **34** | **166** | **511.340** |
| mus600-1200 | 57 | 1 | 860.813 | **56** | **143** | **841.770** |
| mus800-1601 | 68 | 1 | 1026.261 | **64** | **135** | **961.467** |
| mus1000-2002 | 71 | 0 | 1070.605 | **69** | **131** | **1035.995** |

The result shows that these optimizations adopted are effective in this randomly generated benchmark.

To evaluate the impact of different merging strategies on the performance of the proposed FMUS2 algorithm, a series of experiments are performed on the industrial benchmarks from TPTP Problem Library.

Table 3 shows statistical data of experimental results. Note that the contrast merging strategy adopts an opposite strategy to the heuristic merging strategy. In Table 3, $N_{TO}$, $T_{Ave}$ are the same as Table 1, while $T'_{Ave}$ is the average runtime of all instances which are solved in time. From the average runtime data, we can see that different merging strategies greatly affect the performance

**Table 3.** Statistical data for different merging strategies on industrial benchmarks

| Benchmarks | Random | | | Heuristic | | | Contrast | | |
|---|---|---|---|---|---|---|---|---|---|
| | $N_{TO}$ | $T_{Ave}$ | $T'_{Ave}$ | $N_{TO}$ | $T_{Ave}$ | $T'_{Ave}$ | $N_{TO}$ | $T_{Ave}$ | $T'_{Ave}$ |
| TPTP instances (100) | 6 | 185.685 | 6.047 | **0** | **1.460** | **1.460** | 9 | 271.479 | 1.626 |

of our algorithm. It is obvious that the heuristic strategy yields the best performance overall, especially obvious when timeout instances are also counted. Hence, adopting the proposed heuristic merging strategy greatly improves the performance of FMUS2 on practical problems in general, which we view as a reasonable metric of its effectiveness.

However, there are still some cases where the heuristic merging strategy is beaten by the random strategy, and the runtime for some instances can be shortened, which means there is still a lot of potential for further optimizing of the heuristic merging strategy. Figure 2 demonstrates the change of the numbers of intermediate results for different merge orders while running the "HWV003-3" instance from TPTP. The $x$-axis represents the number of merged clauses, and the $y$-axis represents the number of intermediate results during each after each merging. More specifically, there are 61 clauses that need to be merged in this test case, thus the $y$-value becomes zero when the $x$-value increases to 61, indicating the end of the merging. The line labeled with order3 represents the status of our current heuristic merging strategy. On the one hand, a slight change to order3 can result in order4, which maintains a large amount of intermediate results from merging 26 clauses to merging 54 clauses such that the runtime increases dramatically. Further change to order4 can lead to the merge order order5, which triggers a visible explosion of intermediate results and run out

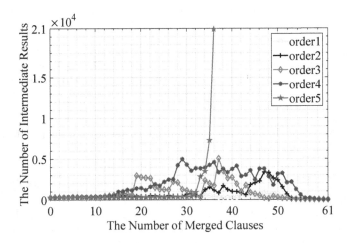

**Fig. 2.** Variation trends of intermediate results for different merge orders.

of memory at the end. This is one of main reasons for the timeout of some instances in these benchmarks. On the other hand, changes to order3 may also lead to merge orders such as order2 and order1 which is the best merge order we obtained for the HWV003-3 instance. Hence there is still much room left for the optimization of the merging strategy, especially when dealing with hard instances for FMUS2.

## 6  Conclusions

In this paper, we proposed a "decompose-merge" algorithm to enumerate all minimal unsatisfiable subsets for a CNF formula in the field of propositional logic and FEF fragment of first-order logic. A heuristic merging strategy and two pruning strategies are adopted to improve the performance of the algorithm. Experimental results show that our algorithm FMUS2 is competitive, and can perform better on some industrial and randomly generated cases when compared with two other state-of-the-art MUS enumerating algorithms. And the optimization strategies adopted has proved to be effective.

For future work, further improvements to FMUS2 will be one of our focuses. As mentioned above, there are still some weaknesses in FMUS2 when dealing with hard instances, i.e. large-scale formulas which have a complex inconsistency relations between clauses. The experimental results also show that the current heuristic merging strategy can be optimized. There is still a lot of room to improve. For instance, it would be interesting to explore better merging strategies and techniques to intelligently select a strategy according to the characteristics of the input set. Besides, we would like to investigate whether our algorithm can be applied to larger fragments of first-order logic in future work.

**Acknowledgments.** This work was supported by National Natural Science Foundation of China (Grand No. 61502022) and State Key Laboratory of Software Development Environment (Grand No. SKLSDE-2017ZX-17).

## References

1. Bacchus, F., Katsirelos, G.: Finding a collection of MUSes incrementally. In: Quimper, C.-G. (ed.) CPAIOR 2016. LNCS, vol. 9676, pp. 35–44. Springer, Cham (2016). https://doi.org/10.1007/978-3-319-33954-2_3
2. Bailey, J., Stuckey, P.J.: Discovery of minimal unsatisfiable subsets of constraints using hitting set dualization. In: Hermenegildo, M.V., Cabeza, D. (eds.) PADL 2005. LNCS, vol. 3350, pp. 174–186. Springer, Heidelberg (2005). https://doi.org/10.1007/978-3-540-30557-6_14
3. de la Banda, M.G., Stuckey, P.J., Wazny, J.: Finding all minimal unsatisfiable subsets. In: Proceedings of the 5th ACM SIGPLAN International Conference on Principles and Practice of Declaritive Programming, pp. 32–43. ACM (2003)
4. Belov, A., Lynce, I., Marques-Silva, J.: Towards efficient MUS extraction. AI Commun. **25**(2), 97–116 (2012)

5. de Moura, L., Bjørner, N.: Z3: an efficient SMT solver. In: Ramakrishnan, C.R., Rehof, J. (eds.) TACAS 2008. LNCS, vol. 4963, pp. 337–340. Springer, Heidelberg (2008). https://doi.org/10.1007/978-3-540-78800-3_24

6. Eén, N., Sörensson, N.: An extensible SAT-solver. In: Giunchiglia, E., Tacchella, A. (eds.) SAT 2003. LNCS, vol. 2919, pp. 502–518. Springer, Heidelberg (2004). https://doi.org/10.1007/978-3-540-24605-3_37

7. Gomes, C.P., Kautz, H., Sabharwal, A., Selman, B.: Satisfiability solvers. Found. Artifi. Intell. **3**, 89–134 (2008)

8. Hou, A.: A theory of measurement in diagnosis from first principles. Artif. Intell. **65**(2), 281–328 (1994)

9. Hutter, F., Hoos, H.H., Leyton-Brown, K.: Tradeoffs in the empirical evaluation of competing algorithm designs. Ann. Math. Artif. Intell. **60**(1–2), 65–89 (2010)

10. Li, W., Shen, N., Wang, J.: R-calculus: a logical approach for knowledge base maintenance. Int. J. Artif. Intell. Tools **4**(01n02), 177–200 (1995)

11. Liffiton, M.H., Previti, A., Malik, A., Marques-Silva, J.: Fast, flexible MUS enumeration. Constraints **21**(2), 223–250 (2016)

12. Liffiton, M.H., Sakallah, K.A.: Algorithms for computing minimal unsatisfiable subsets of constraints. J. Autom. Reason. **40**(1), 1–33 (2008)

13. Luo, J., Li, W.: R-calculus without the cut rule. Sci. China Inf. Sci. **54**(12), 2530–2543 (2011)

14. Previti, A., Marques-Silva, J.: Partial MUS enumeration. In: AAAI (2013)

15. Ramsey, F.P.: On a problem of formal logic. In: Gessel, I., Rota, G.C. (eds.) Classic Papers in Combinatorics. MBC, pp. 1–24. Springer, Boston (2009). https://doi.org/10.1007/978-0-8176-4842-8_1

16. Ryvchin, V., Strichman, O.: Faster extraction of high-level minimal unsatisfiable cores. In: Sakallah, K.A., Simon, L. (eds.) SAT 2011. LNCS, vol. 6695, pp. 174–187. Springer, Heidelberg (2011). https://doi.org/10.1007/978-3-642-21581-0_15

17. Stern, R.T., Kalech, M., Feldman, A., Provan, G.M.: Exploring the duality in conflict-directed model-based diagnosis. In: AAAI, vol. 12, pp. 828–834 (2012)

18. Sutcliffe, G.: The TPTP problem library and associated infrastructure. J. Autom. Reason. **59**(4), 483–502 (2017)

19. Xiao, G., Ma, Y.: Inconsistency measurement based on variables in minimal unsatisfiable subsets. In: European Conference on Artificial Intelligence 2012 (ECAI 2012) (2012)

20. Xie, H., Luo, J.: An algorithm to compute minimal unsatisfiable subsets for a decidable fragment of first-order formulas. In: 2016 IEEE 28th International Conference on Tools with Artificial Intelligence (ICTAI), pp. 444–451. IEEE (2016)

# Deciding Extended Modal Logics by Combining State Space Generation and SAT Solving

Martin Strecker[✉][iD]

IRIT, University of Toulouse, Toulouse, France
martin.strecker@irit.fr

**Abstract.** This paper presents a method of deciding extended modal formulas that arise, in particular, when reasoning about transformations of relational structures and graphs. The method proceeds by first unwinding a structure that is an over-approximation of potential models, and then selecting effective models with the aid of a SAT solver.

**Keywords:** Automated theorem proving · Modal logic
Graph transformations · Program verification

## 1 Introduction

Modal logics have a relatively long history in computer science, and nevertheless, they are still an active research area. This is due to the wide spectrum of variants and possible application areas of modal logics. Basic modal logics have mainly been conceived for reasoning about possibility and necessity or related modalities, such as obligation and knowledge. Temporal logics such as LTL and CTL are variants that can capture properties of a system at different time instants and, thus, characterize how a system may evolve as time passes. An important practically relevant application is the verification of concurrent and reactive systems. Description Logics, used as the foundation of semantic databases and knowledge representation formalisms, have been recognized to be variants of modal logics.

The work presented here has arisen out of an effort to verify graph transformations, which are themselves important for reasoning about processes that modify graph-like structures, for example pointer-manipulating imperative programs. We will discuss this application area and resulting proof problems in Sect. 2. The essential feature of the formulas to be verified is that they give rise to models that are genuine graph and not tree structures, which is a considerable complication. The logic used in this paper will be defined in Sect. 3.

We here restrict our attention to propositional multi-modal logics. Syntactically, they are made up of propositional formulas to which modal operators indexed by binary relations, traditionally $\Box_r$ and $\Diamond_r$, can be applied. Semantically, these formulas are interpreted in Kripke structures, *i.e.* sets of possible

© Springer Nature Switzerland AG 2018
J. Fleuriot et al. (Eds.): AISC 2018, LNAI 11110, pp. 119–135, 2018.
https://doi.org/10.1007/978-3-319-99957-9_8

worlds linked by binary accessibility relations. In each of these worlds, different combinations of elementary predicates may hold.

When it comes to proofs methods, several approaches are possible:

- Many modal logics can easily be embedded into first order logic, so that one can use first-order provers for attempting a proof. This approach is rather straightforward and easily adaptable to different variants of modal logic, but it has a severe drawback: many modal logics have pleasant meta-theoretic properties, such as the finite model property, and many of them are decidable. But there are no guarantees that a standard first-order prover will come to a halt when started on the translation of a modal logic formula. Indeed, it can be expected that the difficulties to be discussed in this article, like containing the number of worlds created during model exploration, have an analogy in first-order proof search, like preventing the generation of useless instances of universally quantified formulas.
- Tableaux are a standard method for trying to construct a model of a formula. They proceed by decomposing connectors until only elementary propositions are left and it is evident whether a model exists or not. Modal operators complicate the picture, because they lead to the creation of new worlds, or to copies of formulas from one world into another, and it is not evident that this procedure stops, in particular in the presence of graph structures. In Sect. 4, we will sketch tableau methods with the purpose of highlighting their difficulties, and for preparing the ground for the state space generation techniques to be introduced subsequently.
- SAT solvers have become impressively efficient for finding models of propositional proof problems. The difficulty consists in deriving a propositional formula corresponding to a modal formula, and this is the main topic of this paper. Our approach works in two phases: we traverse the modal formula a first time. By exploring the structure of modal operators in the formula, we derive an over-approximation of the graph that will yield the Kripke structure. Using this graph structure, we can traverse the modal formula a second time, in order to generate a propositional formula that can then be submitted to the SAT solver and determine the propositions true in each world.

*Related work:* There have been previous efforts to use SAT solvers for deciding modal [6] and description logics [10], and SMT solvers to come to terms with number restrictions in description logics [7]. As compared to the work considered here, the models constructed only have tree form, which is makes state space generation (*cf.* Sect. 5.2) substantially simpler. In [1], state space generation and model checking are interleaved, whereas we generate the potential state space in one run and then search for a possible model. There is a growing interest in quantifier instantiation [9] in conjunction with SAT/SMT solvers; the modal operators considered here pose the problem of quantifier instantiation (and the number of instances to be considered) in a specific context. There exist enumerative techniques, using bounded model checking, that are for example used in the Alloy analyzer [8] but that, contrary to the work presented here, are not complete.

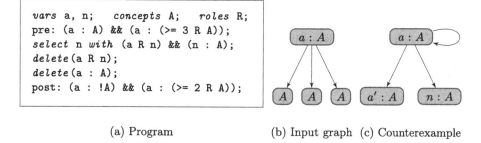

```
vars a, n;   concepts A;   roles R;
pre: (a : A) && (a : (>= 3 R A));
select n with (a R n) && (n : A);
delete(a R n);
delete(a : A);
post: (a : !A) && (a : (>= 2 R A));
```

(a) Program                    (b) Input graph  (c) Counterexample

**Fig. 1.** An example transformation (Color figure online)

## 2  Intended Application

Our intention is to verify programs for graph transformations, automatically and in a sound and complete fashion. The idea is best illustrated by an example program, as in Fig. 1a. The program consists of executable code (in black) and pre- and post-conditions (in blue, marked with **pre:** and **post:**). The pre-conditions describe the shape an input graph is assumed to have; the post-conditions describe the shape after the transformation.

The pre-condition states that the graph has a node a belonging to concept A (for this terminology, see Sect. 3; roughly, concepts are like types or classes) and that a has at least 3 successors with relation R which also belong to concept A. This latter requirement is expressed by a: (>= 3 R A). A typical input graph is displayed in Fig. 1b, where the successors of a are not named, but only their concept membership is shown. The program selects non-deterministically a node n linked to a with relation R and that is also of concept A. It then deletes the arc between a and n and removes a from concept A. The post-condition to be satisfied in the end states that a does not belong to concept A and a has at least 2 R-successors of concept A, which is correct for the input graph Fig. 1b, However, when running our verifier, it comes up with the counterexample of Fig. 1c, which is a graph also satisfying the pre-condition but violating the post-condition after completing the transformation.

We do not spell out the details of the program verification methodology here, see for example [5]. For our purposes, it suffices to say that correctness statements of the program (and in part also selection conditions as in the **select** statement) are formulas of a Description Logic, an extension of $\mathcal{ALCQ}$ whose main ingredients are individual variables (such as a), relations (R) and concepts (simple ones such as A and complex ones such as (>= 3 R A), called number restrictions). The logic to be presented in Sect. 3 does not include number restrictions, see there. The example also highlights the fact that for our verification purposes, dealing with genuine graph structures is essential.

# 3  Logic

The logic considered in this paper can be understood as an extension of description logic $\mathcal{ALC}$ [2] or, alternatively, as an extension of a multi-modal logic. It consists of a hierarchy of three syntactic categories: Concepts $C$ (see Fig. 2a) are Boolean combinations of set expressions built up from elementary concepts. Facts $fact$ (see Fig. 2b) correspond to set membership statements of the form $i : C$ or being an instance of a role $i \, r \, i$. Differently said, we can reason with unary and binary relations (concepts resp. roles). On the last level of the hierarchy, there are formulas $fm$, which are Boolean combinations of facts.

The syntax is not minimal. Role complement $i \, (\neg r) \, i$ has been introduced for stating clash conditions in the tableau calculus, see Sect. 4. We often require formulas to be in negation normal form, obtained by recursively pushing negations inside, thereby swapping Boolean connectives, such as $\neg(C \sqcap D) = \neg C \sqcup \neg D$ and modal operators, such as $\neg(\Diamond_r C) = (\Box_r \neg C)$, or by feeding them into the next level of the syntactic hierarchy, such as $\neg(x : C) = (x : \neg C)$.

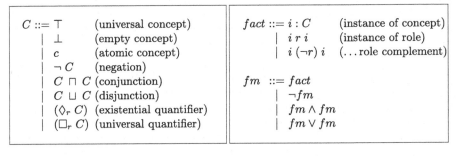

(a) Concepts                    (b) Facts and formulas

**Fig. 2.** Syntax of the logic

For defining the semantics, we assume an interpretation to be a quadruple consisting of (1) a domain of elements $\Delta_{\mathcal{I}}$; (2) an interpretation function mapping elementary concepts $c$ to sets $c^{\mathcal{I}}$ of elements; (3) an interpretation function mapping roles $r$ to sets of pairs $r^{\mathcal{I}}$ of elements; and (4) an interpretation function mapping individual variables $x$ to elements $x^{\mathcal{I}}$. We use this notion of interpretation both for quantifier-free first-order formulas with unary and binary predicate symbols, as introduced in Sect. 5.3, and for the modal logic of this section. Interpretations are extended to concepts as defined in Fig. 3a: The concept constructors $\neg, \sqcap, \sqcup$ are translated by their set-theoretic counterparts; $\Diamond_r C$ is interpreted as the set of elements having an $r$-successor with property $C$, and $\Box_r C$ as the set of elements all of whose $r$-successors have property $C$. Facts are interpreted to produce a truth value (see Fig. 3b); the extension to formulas (not shown) is then straightforward.

$$
\begin{aligned}
\top^{\mathcal{I}} &= \Delta_{\mathcal{I}} \\
\bot^{\mathcal{I}} &= \emptyset \\
(\neg C)^{\mathcal{I}} &= \Delta_{\mathcal{I}} - C^{\mathcal{I}} \\
(C \sqcap D)^{\mathcal{I}} &= C^{\mathcal{I}} \cap D^{\mathcal{I}} \\
(C \sqcup D)^{\mathcal{I}} &= C^{\mathcal{I}} \cup D^{\mathcal{I}} \\
(\Diamond_r C)^{\mathcal{I}} &= \{ x \in \Delta_{\mathcal{I}} \mid \exists y.(x,y) \in r^{\mathcal{I}} \wedge y \in C^{\mathcal{I}} \} \\
(\Box_r C)^{\mathcal{I}} &= \{ x \in \Delta_{\mathcal{I}} \mid \forall y.(x,y) \in r^{\mathcal{I}} \rightarrow y \in C^{\mathcal{I}} \}
\end{aligned}
$$

$$
\begin{aligned}
(x : C)^{\mathcal{I}} &= x^{\mathcal{I}} \in C^{\mathcal{I}} \\
(x \ r \ y)^{\mathcal{I}} &= (x^{\mathcal{I}}, y^{\mathcal{I}}) \in r^{\mathcal{I}} \\
(x \ (\neg r) \ y)^{\mathcal{I}} &= (x^{\mathcal{I}}, y^{\mathcal{I}}) \notin r^{\mathcal{I}}
\end{aligned}
$$

(a) Semantics of concepts                    (b) Semantics of facts

**Fig. 3.** Semantics

As usual, a *model* is an interpretation satisfying a formula. An alternative view on models is as graphs such as the one in Fig. 1c, where nodes are tagged uniquely by variable names and decorated by the elementary concepts which are true for the corresponding variable, and directed arcs decorated by relation names.

## 4    Tableau Methods

We give an overview of the tableau method, with a particular emphasis on the way the modal operators are handled. We will see shortly that $\Diamond$ permits to "generate new worlds", and that there is a subtle interplay between $\Diamond$ and $\Box$ that the SAT method will have to simulate. We will first describe the calculus in Sect. 4.1 and then state essential properties in Sect. 4.2.

### 4.1    Calculus

The procedure manipulates a tableau, which is a set $\mathcal{A}$ of formulas. The formulas in $\mathcal{A}$ are decomposed depending on their shape, according to the rules in Fig. 4, giving rise to a new tableau $\mathcal{A}'$. This process is formally modelled by a transition relation $\mathcal{A} \xrightarrow{f} \mathcal{A}'$. This relation is non-deterministic, as witnessed by the rules DISJC and DISJF. All formulas in $\mathcal{A}$ are supposed to be in negation normal form, an invariant maintained by the procedure; for this reason, there are no explicit rules for negated formulas.

We briefly comment on the rules: The rules for decomposing binary connectives (CONJC/DISJC for concepts and CONJF/DISJF for formulas) are standard for tableau procedures and directly reflect the semantics. For example, CONJC states that if $x$ is member of the intersection of concepts $C_1 \sqcap C_2$, then it is member of each of $C_1$ and $C_2$. The side conditions of the rule ensure the termination of the calculus.

The modal rules are best explained with the graph view of models: the statement $x : (\Diamond_r C)$ expresses that node $x$ in the graph has an $r$-successor marked $C$.

$$\text{CONJC} \dfrac{(x:(C_1 \sqcap C_2)) \in \mathcal{A} \quad \neg((x:C_1) \in \mathcal{A} \wedge (x:C_2) \in \mathcal{A})}{\mathcal{A} \xrightarrow{(x:(C_1 \sqcap C_2))} \mathcal{A} \cup \{x:C_1, x:C_2\}}$$

$$\text{DISJC} \dfrac{\substack{(x:(C_1 \sqcup C_2)) \in \mathcal{A} \quad (x:C_1) \notin \mathcal{A} \quad (x:C_2) \notin \mathcal{A} \\ \mathcal{A}' = \mathcal{A} \cup \{x:C_1\} \vee \mathcal{A}' = \mathcal{A} \cup \{x:C_2\}}}{\mathcal{A} \xrightarrow{(x:(C_1 \sqcup C_2))} \mathcal{A}'}$$

$$\Diamond\text{C} \dfrac{(x:(\Diamond_r C)) \in \mathcal{A} \quad \forall y.\neg((x\ r\ y) \in \mathcal{A} \wedge (y:C) \in \mathcal{A}) \quad z \notin fv(\mathcal{A})}{\mathcal{A} \xrightarrow{(z,(x:(\Diamond_r C)))} \mathcal{A} \cup \{z:C, x\ r\ z\}}$$

$$\Box\text{C} \dfrac{(x:(\Box_r C)) \in \mathcal{A} \quad \exists y.(x\ r\ y) \in \mathcal{A} \wedge (y:C) \notin \mathcal{A}}{\mathcal{A} \xrightarrow{(x:(\Box_r C))} \mathcal{A} \cup \{g.(\exists y.(g = (y:C)) \wedge (x\ r\ y) \in \mathcal{A})\}}$$

$$\text{CONJF} \dfrac{f_1 \wedge f_2 \in \mathcal{A} \quad \neg(f_1 \in \mathcal{A} \wedge f_2 \in \mathcal{A})}{\mathcal{A} \xrightarrow{f_1 \wedge f_2} (\mathcal{A} - \{f_1 \wedge f_2\}) \cup \{f_1, f_2\}}$$

$$\text{DISJF} \dfrac{\substack{f_1 \vee f_2 \in \mathcal{A} \quad f_1 \notin \mathcal{A} \quad f_2 \notin \mathcal{A} \\ \mathcal{A}' = (\mathcal{A} - \{f_1 \vee f_2\}) \cup \{f_1\} \vee \mathcal{A}' = (\mathcal{A} - \{f_1 \vee f_2\}) \cup \{f_2\}}}{\mathcal{A} \xrightarrow{f_1 \vee f_2} \mathcal{A}'}$$

**Fig. 4.** Tableau rules

Provided such a node does not yet exist, we create a new node $z$, an arc $(x\ r\ z)$ and mark node $z$ with $C$. In this case, we also record the variable that has been created in the transition relation: $\mathcal{A} \xrightarrow{(z,f)} \mathcal{A} \cup \{z : C, x\ r\ z\}$. The statement $x : (\Box_r C)$ postulates that all $r$-successors of $x$ are marked with $C$, *i.e.* for all existing arcs $(x\ r\ y)$, the nodes $y$ are marked with $C$. The applicability condition ensures that not all $r$-successors of $x$ are already marked, because then we would not make progress.

The aim of the procedure is to derive a contradiction, called a *clash*: A tableau $\mathcal{A}$ contains a clash if, for $C$ a concept, $r$ a role and $x, y$ individual variables, $(x : \bot) \in \mathcal{A}$ or $\{x : C, x : \neg C\} \subseteq \mathcal{A}$ or $\{(x\ r\ y), (x\ (\neg r)\ y)\} \subseteq \mathcal{A}$. For determining whether an initial tableau $\mathcal{A}$ is satisfiable, the tableau procedure explores all complete tableaux reachable via the rule relation (a tableau is *complete* if no further rule is applicable). If all complete reachable tableaux contain a clash, then the initial tableau is unsatisfiable; otherwise, there is a complete clash-free tableau from which a model of the original tableau can be derived.

We note that most of the rules can only be applied at most once to a particular formula, which blocks the rule for further application. The $\Box$C rule is the only exception – and the essential difficulty of the calculus: application of other

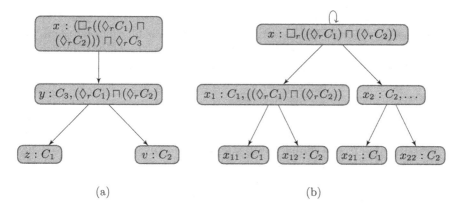

**Fig. 5.** (a) See Example 1. (b) See Example 2

rules in the tableau can lead to the generation of new arcs $(x\ r\ z)$ that may trigger rule $\Box C$ again. We consider two examples to illustrate the situation.

*Example 1.* The first example is of the kind typically considered in the literature for Description Logics. The aim is to ascertain whether a concept $C$ is consistent, which can be done by verifying that the tableau $\{x : C\}$ is satisfiable. An instance of this situation is the following: $\mathcal{A}_0 = \{x : (\Box_r((\Diamond_r C_1) \sqcap (\Diamond_r C_2))) \sqcap \Diamond_r C_3\}$.

Decomposing the conjunction yields $\mathcal{A}_1 = \mathcal{A}_0 \cup \{x : \Box_r((\Diamond_r C_1) \sqcap (\Diamond_r C_2)), x : \Diamond_r C_3\}$. Applying the $\Diamond C$ rule introduces a new variable $y$, such that $\mathcal{A}_2 = \mathcal{A}_1 \cup \{(x\ r\ y), y : C_3\}$. With the $\Box C$ rule, we get $\mathcal{A}_3 = \mathcal{A}_2 \cup \{y : (\Diamond_r C_1) \sqcap (\Diamond_r C_2)\}$. Another round of decompositions and $\Diamond C$ rule applications gives the complete, satisfiable tableau $\mathcal{A}_4 = \mathcal{A}_3 \cup \{(y\ r\ z), z : C_1, (y\ r\ v), v : C_2\}$.

It should be noted that the resulting graph is a tree (see Fig. 5a, here annotated with atomic and composite concepts), and an easy induction shows that this is indeed so for all models derived from an initial tableau of the form $\{x : C\}$. Furthermore, from the syntax tree of the original concept, one can read off the structure of the resulting model: Each direct $\Diamond_r$ subtree in the syntax (direct = not separated by a modal operator) gives potentially rise to a new child node in the model, and these are the only children. It is therefore possible to index the generated nodes by positions in the original formula. This is the approach taken in [6] to identify all nodes in the model to be generated.

*Example 2.* The initial tableau $\mathcal{A} = \{(x\ r\ x), x : \Box_r((\Diamond_r C_1) \sqcap (\Diamond_r C_2))\}$ already starts out with a partial model (the node $x$ with a self-loop) that is not tree-shaped. Without the arc $(x\ r\ x)$, the tableau would be complete; with it, we apply the $\Box C$ rule and add $x : (\Diamond_r C_1) \sqcap (\Diamond_r C_2)$ to the tableau. Decomposition and rule $\Diamond C$ yield two nodes $x_1 : C_1$ and $x_2 : C_2$ and arcs $(x\ r\ x_1)$ and $(x\ r\ x_2)$. This triggers the $\Box C$ rule twice, and we obtain $x_1 : (\Diamond_r C_1) \sqcap (\Diamond_r C_2)$ and similarly for $x_2$. After a renewed decomposition, we obtain still another set of nodes $x_{11}, x_{21} : C_1$ and $x_{12}, x_{22} : C_2$.

There are two main differences compared to Example 1: Indeed, the resulting model is not a tree (see Fig. 5b) but, in a sense, tree-like: all the newly generated nodes only have one parent; and the new nodes cannot be indexed by the syntactic structure of the formula in a straightforward way. In the node generating algorithm of Sect. 5.2, we have therefore avoided to do so.

## 4.2  Properties

We sketch soundness and completeness arguments for the tableau calculus (which are relatively standard) and present more in detail a novel termination result that will also be instrumental for the discussion in Sect. 5.2.

Let us designate by $\mathcal{A} \overset{*}{\hookrightarrow} \mathcal{A}'$ that tableau $\mathcal{A}'$ is reachable by a sequence of applications of tableau rules from tableau $\mathcal{A}$. Let $\overset{+}{\hookrightarrow}$ be the strict part of $\overset{*}{\hookrightarrow}$.

**Theorem 1 (Soundness).** *The tableau calculus is sound: if $\mathcal{A} \overset{*}{\hookrightarrow} \mathcal{A}'$ and $\mathcal{A}'$ is satisfiable, then so is $\mathcal{A}$.*

*Proof.* The proof is by induction on the derivation, showing that each rule application does not create new models.    □

**Theorem 2 (Completeness).** *The tableau calculus is complete: if $\mathcal{A}'$ is unsatisfiable for all $\mathcal{A}'$ with $\mathcal{A} \overset{*}{\hookrightarrow} \mathcal{A}'$, then so is $\mathcal{A}$.*

*Proof.* The proof is by induction on the derivation, showing that the model property is preserved by at least one of the alternatives of each derivation.    □

As far as termination of the calculus is concerned, the situation is considerably more complex for graph-like structures than for tree structures, as motivated in the above examples. When starting with a proof problem $x : C$, a tableau derivation generates a tree with new nodes of the form $x' : C'$, where $C'$ is a strict subconcept of $C$. The well-foundedness of the subconcept order is the essential ingredient for the termination argument in the case of tree structures.

However, in a graph structure with nodes $x_1 : C_1$ and $x_2 : C_2$ and relation $(x_1 \ r \ x_2)$, the node $x_2$ may be decorated with more complex concepts $C$ than $C_2$ in the course of the proof, for example if $C_1$ is $\square_r C$, with $C$ more complex than $C_2$.

The termination argument developed here uses the notion of *bound*, which is a set of formulas with which a variable can potentially be decorated. In contrast, the *annotation* of a variable $x$ in a tableau $\mathcal{A}$ is the set $\{C . (x : C) \in \mathcal{A}\}$ with which the node is actually annotated. To take a simple example, the node decorated with $x : (A \sqcup B) \sqcap C$ will be bounded by the set $\{(A \sqcup B) \sqcap C, A \sqcup B, A, B, C\}$. After a decomposition with the CONJC rule, additional annotations will be $x : A \sqcup B$ and $x : C$, all of which remain within the bounds.

We will define an order on pairs $(bound, annotation)$ as follows: $(b_1, a_1) < (b_2, a_2) := (finite(b_2) \wedge b_1 \subseteq b_2 \wedge b_2 \supseteq a_1 \wedge a_1 \supset a_2)$. In a tableau proof, when deriving a new node from an existing node, the bound may become smaller

$(b_1 \subseteq b_2)$ whereas the actual annotation should increase $(a_1 \supset a_2)$ and come closer to the former bound without exceeding it $(b_2 \supseteq a_1)$. It can be shown that this order is well-founded. With these ingredients, we can now state and prove that the tableau rules do not permit infinite derivations:

**Theorem 3 (Termination).** *Relation $\overset{+}{\hookrightarrow}$ is well-founded.*

*Proof.* We begin by defining a variant of the calculus that, apart from formulas, also manages bounds. In the initial tableau, all the variables have the same bound, *viz.* the set of all subconcepts occurring in the tableau. In general, nodes keep their bounds as the result of a rule application; the only exception is the $\Diamond C$ rule, which is the only rule that creates new nodes. If $x$ is the node that the $\Diamond C$ rule is applied to and it has a bound containing modal operators $\Box_{r_1} B_1, \ldots, \Box_{r_m} B_m, \Diamond_{s_1} D_1, \ldots, \Diamond_{s_n} D_n$, then the new node $z$ that is generated (*cf.* Fig. 4) will have a bound that is the subconcept-closure of $\{B_1, \ldots, B_m, D_1, \ldots, D_n\}$.

We now define the *potential of a node $x$* as the pair

- *(bound, annotation)* of node $x$ with the order defined above;
- number of facts $x : \Diamond_r C$ in the tableau to which the rule $\Diamond C$ is still applicable;

endowed with a lexicographic order. The *potential of a formula* is the tuple

- size of the formula, defined as the number of formula constructors of the syntax trees, where facts have size 0;
- node potential of the node $x$ for facts of the form $x : C$;

endowed with a lexicographic order. The potential of a tableau is then the multiset of the potentials of the formulas it contains.

Obviously, the rules CONJF and DISJF decrease the potential of a tableau, by decreasing the potential of a formula. The rules CONJC and DISJC decrease the potential of a node, by increasing the *annotation* component while keeping the bound constant. The $\Diamond C$ rule replaces the potential of node $x$ by two smaller potentials (the node $x$ with the number of $\Diamond C$ decreased, and the node $z$ with lower bounds). The $\Box C$ rule, when applied to a node $x$ linked to a node $y$ (*cf.* Fig. 4), will increase the annotation of node $y$. It may at the same increase the number of possible $\Diamond C$ applications of $y$, but the net effect is to decrease the potential of $y$.                                                                    □

## 5    Translation to Propositional Logic

### 5.1    Principles

As has been seen in Sect. 4, a tableau interleaves decomposition of formulas and generation of new variables corresponding to nodes of the model. The disjunctive rules are non-deterministic. In practice, different strategies exist for exploring these alternatives, among them depth-first search with backtracking or breadth-first search. In both cases, there is a risk of duplicate work, *i.e.* of testing over and over again which combination among a set of mutually incompatible choices (such as $x : C$ or $x : \neg C$) leads to a satisfiable formula.

The principal hypothesis of our approach is that it is more efficient to lay out the whole space of possible worlds in a first step. A central ingredient for this is the notion of *node*, which is the representative of a variable $x$ in a tableau and which collects all the concept membership information $x : C_1, \ldots x : C_n$ available in a tableau. For this, we generate in parallel all the nodes a tableau procedure might create, and then carry out the combinatorial exploration by a procedure optimized for that purpose, namely a SAT solver.

The node generation phase (Sect. 5.2) mimics the tableau algorithm, as far as the modal operators are concerned: for $\Diamond$, this means to create new successor nodes, and for $\Box$, enrich nodes with new concepts, and this is repeated until reaching a fixpoint. In this process, we have to memorize which instances have already been created, and various other information. In all of the following discussion, we assume formulas, facts and concepts in negation normal form.

*Terminology:* Before describing the structure of nodes, we fix some terminology. The set of *subconcepts* of a concept $C$ is defined as consisting of $C$ and recursively of all immediately constituent subconcepts of $C$. The set of *Boolean subconcepts* is the set of all subconcepts except for those below modal operators. Thus, the set of Boolean concepts of $\Box_r C_1 \sqcap (C_2 \sqcup \Diamond_r(C_3 \sqcap C_4))$ is $\{\Box_r C_1, C_2, \Diamond_r(C_3 \sqcap C_4)\}$.

We denote $y$ as a *successor instance* of a fact $x : \Box_r C'$ if $y$ is an $r$-successor of $x$ and $y : C'$. We now describe the record structure *node* that contains this information. It consists of the fields

- *name*: a unique identifier for the node, whose precise structure is immaterial.
- *new*: a list of newly added subconcepts that still have to be processed.
- *old*: keeps track of subconcepts that already have been processed; necessary for ensuring termination.
- *somec*: in principle, a list of Boolean $\Diamond$-subconcepts of the initial concept $C$ of this node, *i.e.* the Boolean subconcepts of $C$ that have the form $\Diamond_r C'$. Instead of storing a list of concepts $\Diamond_r C'$, we record a list of tuples $(r, C')$.
- *allc*: in principle, a list of Boolean $\Box$-subconcepts of the initial concept $C$ of this node, *i.e.* the Boolean subconcepts of $C$ that have the form $\Box_r C'$. As for $\Diamond$, instead of a list of concepts $\Box_r C'$, we record a list of tuples $(r, C')$.

A note on syntax: The description of the algorithms presented in the following has been derived from an implementation in Ocaml, from which we have borrowed the syntax for lists: [] for the empty list and :: for consing. Duplicate-free concatenation is written $\cup$. We enumerate record components within banana brackets, and write record updates as $r(\!|c := v|\!)$ (update of component $c$ of record $r$ with value $v$). Selection of component $c$ of record $r$ is written as $r.c$. Adding an element $e$ to a component $c$ in a record $r$ is written as $r(\!|c+ = e|\!)$, shorthand for $r(\!|c := \{e\} \cup r.c|\!)$.

$$\text{END} \quad \frac{(nds', wns') = partition\_allc\_instances(nds, rels) \qquad wns' = []}{\mathcal{N}_c(nds, rels, []) = (nds', rels)}$$

$$\text{REL} \quad \frac{(nds', wns') = partition\_allc\_instances(nds, rels) \qquad wns' \neq []}{\mathcal{N}_c(nds, rels, []) = \mathcal{N}_c(nds', rels, wns')}$$

$$\text{POPN} \quad \frac{wn.new = []}{\mathcal{N}_c(nds, rels, wn :: wns) = \mathcal{N}_c(wn :: nds, rels, wns)}$$

$$\text{SK}\Diamond \quad \frac{wn.new = \Diamond_r C :: cs \quad (r, C) \in wn.somec}{\mathcal{N}_c(nds, rels, wn :: wns) = \mathcal{N}_c(nds, rels, (wn(\!new := cs; old+ = \Diamond_r C)\! ) :: wns))}$$

$$\text{DEC}\Diamond \quad \frac{\begin{array}{c} wn.new = \Diamond_r C :: cs \quad (r, C) \notin wn.somec \\ (nnd, nrel) = create\_somec\_successor(wn.name, r, C) \\ wn' = wn(\!new = cs; old+ = \Diamond_r C; somec+ = (r, C)\!) \end{array}}{\mathcal{N}_c(nds, rels, wn :: wns) = \mathcal{N}_c(nds, nrel :: rels, nnd :: wn' :: wns)}$$

$$\text{DEC}\square \quad \frac{wn.new = \square_r C :: cs \qquad wn' = wn(\!new = cs; old+ = \square_r C; allc+ = (r, C)\!)}{\mathcal{N}_c(nds, rels, wn :: wns) = \mathcal{N}_c(nds, rels, wn' :: wns)}$$

$$\text{SKB} \quad \frac{wn.new = c :: cs \quad c = (C_1 \sqcap C_2) \vee c = (C_1 \sqcup C_2) \quad (C_1 \,\epsilon\, wn \wedge C_2 \,\epsilon\, wn)}{\mathcal{N}_c(nds, rels, wn :: wns) = \mathcal{N}_c(nds, rels, (wn(\!new := cs; old+ = c)\!) :: wns)}$$

$$\text{DECB} \quad \frac{\begin{array}{c} wn.new = c :: cs \quad c = (C_1 \sqcap C_2) \vee c = (C_1 \sqcup C_2) \\ \neg(C_1 \,\epsilon\, wn \wedge C_2 \,\epsilon\, wn) \quad wn' = (wn(\!new := \{C_1, C_2\} \cup cs; old+ = c)\!) \end{array}}{\mathcal{N}_c(nds, rels, wn :: wns) = \mathcal{N}_c(nds, rels, wn' :: wns)}$$

**Fig. 6.** Computing the node set

## 5.2  Generating the Set of Nodes

*Nodes of a Concept.* We now present function $\mathcal{N}_c()$ which computes the set of nodes for a concept. The function manipulates a work list of nodes (record structure *node*) that still have to be processed; once a node is finished, it is added to the list of nodes whose processing is complete. Manipulating the work list amounts to manipulating in turn the nodes that will make up the Kripke structure. Apart from recursing over the list of nodes, we also consider the *new* subconcepts of the current node, which corresponds to decomposing the concepts of the current node. We also keep track of the relations between these nodes, with a relation represented as a list of triples $(r, x, y)$: relation name $r$, source $x$ and target $y$ node identifier. Thus, altogether, function $\mathcal{N}_c()$ takes as arguments a list of completed nodes $nds$, a list of relations $rels$ between nodes, and the worklist $wns$. The function returns the list of completed nodes and the relations.

The function can be written straightforwardly as a tail-recursive function, but in order to better distinguish the different patterns, we present the function in a rule format in Fig. 6. Our actual implementation differs in one detail: the implemented function takes an additional counter for keeping track of node names. To avoid clutter, we assume here that generation of fresh names happens behind the scenes.

*Rules:* Let us now comment on the rules.

- END and REL are the cases when the worklist is empty. For deciding what to do, we split the existing node set $nds$ into a set of definitely finished nodes $nds'$ and nodes $wns'$ that have to be reprocessed. If $wns'$ is empty, we are done (END) and return the nodes and relations accumulated so far. Otherwise, we relaunch (REL) the function with the new worklist. The auxiliary function *partition_allc_instances* will be described further below.
- POPN transfers a worklist node with an empty *new* component to the list of completed nodes, because it contains no more concepts to be processed.
- The remaining rules all assume that the *new* list of the current node $wn$ is not empty, and manipulate its first element. If this first element is $\Diamond_r c$, there are two cases:
  - SK$\Diamond$: there already exists an $r$-successor node containing $C$ attached to the current node $((r, C) \in wn.somec)$. This corresponds to the situation when the tableau rule $\Diamond C$ is not applicable; in this case, we skip $\Diamond_r C$ and continue with the rest of the *new* list.
  - DEC$\Diamond$: no such successor exists. This is the case when the rule $\Diamond C$ is applicable, and we decompose the operator. So assume that the current node has name $x$, then function *create_somec_successor* generates a node $nnd$ with a fresh name, say $z$, and a relation $nrel = (r, x, z)$. We memorize that we now have an appropriate $r$-successor of concept $C$ for node $x$, so that we do not create one again (the case handled by SK$\Diamond$). The new node and the modified current node are added to the worklist.
- DEC$\Box$: decomposition of the box is applied if the first element is $\Box_r C$. Nothing interesting happens at this point: we record in the *allc* component that the current node has a $\Box$ modal operator. It is during partitioning with function *partition_allc_instances* that this information will be propagated.
- SKB and DECB for handling Boolean connectors. Let us look at conjunction $C_1 \sqcap C_2$ first. The skip rule SKB corresponds to the case when the CONJC tableau rule is not applicable (we write $C \epsilon wn$ for $C \in wn.new \cup wn.old$), and the concept $c = C_1 \sqcap C_2$ is simply marked as old. Otherwise, the decompose rule DECB adds $C_1$ and $C_2$ to the new nodes, in analogy to the CONJC rule. Perhaps surprisingly, disjunction is handled the same way, corresponding to

following simultaneously two different evolutions of the tableau as of rule DisJC. The fact the set of concepts accumulated in a node now possibly becomes inconsistent is not relevant at this stage and will be taken care of by formula translation in Sect. 5.3.

*Partition Instances.* Without giving a full definition, we now describe the idea of function *partition_allc_instances*. To motivate the special treatment reserved to the $\Box$ operator, let us remark that all the rules of the tableau calculus remain inhibited after one application to a particular instance; only the $\Box C$ rule is an exception, where an application to instance $x : \Box_r C$ can be reactivated whenever a new link to a node $y$ is created. So *partition_allc_instances(nds, rels)* does the following: For each node $x$ of $nds$ with $x : \Box_r C$ as recorded in the *allc* field, and for each relation $(r\ x\ y)$ in $rels$, it checks whether $y : C$ (*i.e.*concept $C$ is in the *old* field of node $y$). If this is so, node $y$ remains inactive. Otherwise, $C$ is added to the *new* component of $y$, and node $y$ is put back into the worklist.

*Nodes of a Formula.* Function $\mathcal{N}_c()$ is by far the most complex function. The analogous function on formulas, $\mathcal{N}_f(nds, rels, fm)$, traverses formula $fm$ recursively, gathering all the nodes and relations it finds. For a fact $x : C$, it either creates a node $x$ and adds concept $C$ to its *new* list, or simply adds $C$ to *new* if node $x$ already exists. We will write $\mathcal{N}_f(fm)$ instead of $\mathcal{N}_f([], [], fm)$.

## 5.3   Translating Concepts and Formulas

Given a formula, once we have computed an over-approximation of the nodes and relations of the Kripke structure of this formula, we can translate it to propositional logic. Checking the satisfiability of this translated formula either demonstrates its unsatisfiability (in which case the original formula is unsatisfiable as well) or yields a model.

We now define this translation, first with function $\mathcal{T}_c()$ for concepts which takes as additional argument the set of nodes $nds$ which are a (not necessarily strict) superset of the nodes of the model to be constructed. The translation rules are displayed in Fig. 7.

The rules for $\top, \bot$ and elementary (positive or negative) concepts $c$ are straightforward. The rules for conjunction and disjunction reflect the semantics of the respective connective. As to the translation of the modal operators $\Diamond$ and $\Box$, remember that the set of nodes $nds$ represents an over-approximation of the domain of the model to be constructed. For both operators, we project out the names of the nodes, to obtain the set $Y$ of all node names. We then quantify existentially resp. universally over this set. This reflects the semantics of the operators (*cf.* Fig. 3a) with $(x \in (\Diamond_r C)^{\mathcal{I}}) = \exists y \in \varDelta_{\mathcal{I}}. (x, y) \in r^{\mathcal{I}} \wedge y \in C^{\mathcal{I}}$ and $(x \in (\Box_r C)^{\mathcal{I}}) = \forall y \in \varDelta_{\mathcal{I}}. (x, y) \in r^{\mathcal{I}} \longrightarrow y \in C^{\mathcal{I}}$. The translation of formulas, $\mathcal{T}_f(nds, fm)$, is then a recursive function traversing formula $fm$, such that $\mathcal{T}_f(nds, (x : C)) = \mathcal{T}_c(nds, x, C)$. Both $\mathcal{T}_c()$ and $\mathcal{T}_f()$ are defined by simple structural recursion, so their termination is evident.

$$\mathcal{T}_c(nds, x, \top) = true$$
$$\mathcal{T}_c(nds, x, \bot) = false$$
$$\mathcal{T}_c(nds, x, c) = c(x)$$
$$\mathcal{T}_c(nds, x, \neg c) = \neg c(x)$$
$$\mathcal{T}_c(nds, x, C_1 \sqcap C_2) = \mathcal{T}_c(nds, x, C_1) \wedge \mathcal{T}_c(nds, x, C_2)$$
$$\mathcal{T}_c(nds, x, C_1 \sqcup C_2) = \mathcal{T}_c(nds, x, C_1) \vee \mathcal{T}_c(nds, x, C_2)$$
$$\mathcal{T}_c(nds, x, \Diamond_r C') = \bigvee_{y \in Y} r(x, y) \wedge \mathcal{T}_c(nds, y, C')$$
$$\mathcal{T}_c(nds, x, \Box_r C') = \bigwedge_{y \in Y} (r(x, y) \longrightarrow \mathcal{T}_c(nds, y, C'))$$
$$\text{where } Y = \{y. \ y = nd.name \wedge nd \in nds\}$$

**Fig. 7.** Translating concepts to propositional logic

## 6   Soundness and Completeness

We summarize the essential steps of our verification framework and then demonstrate that it effectively provides a decision procedure for the formulas defined in Sect. 3. Given a formula $fm$:

- Generate the set of nodes and use these to translate the formula to propositional logic: $\mathcal{T}_f(\mathcal{N}_f(fm), fm)$.
- Submit the resulting formula to a SAT solver which will either state its unsatisfiability or provide a model, yielding a model of the original formula.

**Lemma 1 (Termination of node generation).** $\mathcal{N}_f()$ and $\mathcal{N}_c()$ terminate.

*Proof.* $\mathcal{N}_f()$ is a simple structurally recursive function whose termination is immediate. For $\mathcal{N}_c()$, we can essentially use the concept of *potential* introduced for the proof of Theorem 3, and show that recursive calls of the function lead to a decrease in the multiset of potentials of $nds$. □

We first show that the translation function $\mathcal{T}_c()$ preserves models, under certain circumstances. To get an intuition, take the formula $x : (\Diamond_r c) \wedge x : (\Diamond_r (\neg c))$. It stipulates that $x$ has a successor $y_1$ where $c$ holds, and a successor $y_2$ where $\neg c$ holds. The formula, translated to predicate logic, is $(\bigvee_{y \in Y} .r(x, y) \wedge c(y)) \wedge (\bigvee_{y \in Y} .r(x, y) \wedge \neg c(y))$. This formula is only satisfiable for a set $Y$ consisting of at least two nodes, thus if $y_1$ and $y_2$ are not forced to be the same. We conclude that $\mathcal{T}_c()$ preserves models if the node set $nds$ that is a parameter of $\mathcal{T}_c()$ is "sufficiently large". In a sense, the main difficulty of the soundness and completeness result resides in showing that $\mathcal{N}_c()$ expands to a universe with enough elements.

**Lemma 2 (Model preservation of translation)**

1. *If $\mathcal{I}$ is a model of $\mathcal{T}_c(nds, x, C)$, then also of $x : C$.*
2. *Let $\mathcal{I}$ be a model of $x : C$, and let there be an injective mapping from $\Delta_\mathcal{I}$ into nds. Then there exists a model $\mathcal{I}'$ of $\mathcal{T}_c(nds, x, C)$.*

*Analogous results hold for the translation function $\mathcal{T}_f()$.*

*Proof.*

1. A model of $\mathcal{T}_c(nds, x, C)$ immediately gives a model of $x : C$ (*cf.* semantics in Fig. 3a).
2. We construct inductively a model $\mathcal{I}'$ whose domain $\Delta_{\mathcal{I}'}$ is a superset of $\Delta_\mathcal{I}$. The cases where the concept $C$ is a constant or an elementary concept are immediate. The case of disjunction is easy. For the case of a conjunction $C_1 \sqcap C_2$, the construction inductively yields two models $\mathcal{I}'_1$ and $\mathcal{I}'_2$ of $C_1$ and $C_2$ respectively which might be mutually incompatible (as seen in the above example). By a remapping of variables, we can construct two models that coincide on the interpretation of $x$ and otherwise map variables to disjoint elements of $\Delta_\mathcal{I}$. By joining these two models into a single interpretation $\mathcal{I}'$, we obtain a model of the translation of $C_1 \sqcap C_2$.

   As to the modal operators: Let $\mathcal{I}$ be a model of $x : \Diamond_r C'$, so $x^\mathcal{I} \in \Delta_\mathcal{I}$ and there exists $y^\mathcal{I} \in \Delta_\mathcal{I}$ such that $(x^\mathcal{I}, y^\mathcal{I}) \in r^\mathcal{I} \wedge y^\mathcal{I} \in C^\mathcal{I}$. Let $m$ be the injective mapping from $\Delta_\mathcal{I}$ into $nds$. The interpretation of the translated formula $\bigvee_{y \in Y} r(x, y) \wedge \mathcal{T}_c(nds, y, C')$ is possibly over a larger domain (the set $Y$ may contain more elements than $\Delta_\mathcal{I}$), so we decide to interpret any $\mathcal{I}'(y)$ for $y$ in the image of $m$ as $\mathcal{I}(m^{-1}(y))$, and any other $y$ arbitrarily. Similarly, relation $r(x, y)$ for $y$ in the image of $m$ is interpreted as for $\mathcal{I}$ and as false otherwise. This interpretation satisfies the translated formula. The argument for the $\square$ operator is analogous.

   $\square$

**Theorem 4 (Soundness).** *The proof method, when applied to a formula $fm$, is sound: A model found by the solver for $\mathcal{T}_f(fm)$ is also a model of $fm$.*

*Proof.* We assume that the SAT solver is sound, so a model that the solver claims to be one for $\mathcal{T}_f(fm)$ is indeed one. With Lemma 2(1), we thus obtain a model of $fm$. Remark that this is independent of the correctness of $\mathcal{N}_c()$ and $\mathcal{N}_f()$. $\square$

**Theorem 5 (Completeness).** *The proof method, when applied to a formula $fm$, is complete: Whenever there exists a model for $fm$, then the solver will find one for $\mathcal{T}_f(\mathcal{N}_f(fm), fm)$.*

*Proof.* Suppose $fm$ has a model. According to Theorems 2 and 3, starting from tableau $\mathcal{A} = \{fm\}$, there exists a terminating run $\mathcal{A} \overset{*}{\hookrightarrow} \mathcal{A}'$ such that $\mathcal{A}'$ permits to construct a model $\mathcal{I}$ of $fm$ with domain $\Delta_\mathcal{I} = vars(\mathcal{A}')$. It is easy to show that $\mathcal{N}_c()$ returns a set of nodes $nds$ such that $vars(\mathcal{A}') \subseteq nds$. According to Lemma 2(2), there exists a model $\mathcal{I}'$ of $\mathcal{T}_f(nds, fm)$, and since the solver is assumed to be complete, it will find a model of $\mathcal{T}_f(nds, fm)$. $\square$

## 7    Conclusions

We have presented a decision procedure for an extension of a modal logic that is particularly appropriate for reasoning about graphs and their transformations, and thus constitutes an essential increase of expressivity *w.r.t.* logics restricted to tree models.

The approach described here has been implemented in a prototype in Ocaml, using alternatively CVC4 [3] or veriT [4] as SAT solvers. The prototype shows a good response time for formulas from the application scenario it is intended for, but it has not been exercised on performance benchmarks. The models obtained by this method are sometimes surprizing, yielding (often very compact) genuine graphs where a tree model would also exist. Sometimes, however, the model contains a great number of nodes that are spurious in the sense that already a subgraph would be a model. This indicates a potential for optimizations. Further work to be considered are more expressive logics, for example including number restrictions (as in Sect. 2) or particular properties of relations, like transitivity.

**Acknowledgements.** Most of the work described here has been carried out while the author was at Inria Nancy during a research stay financially supported by Inria and hosted by the VeriDis team. I am grateful to Stephan Merz for inviting me, and to him and to Pascal Fontaine for discussions about and suggestions for improvement of this paper.

## References

1. Areces, C., Fontaine, P., Merz, S.: Modal satisfiability via SMT solving. In: De Nicola, R., Hennicker, R. (eds.) Software, Services, and Systems. LNCS, vol. 8950, pp. 30–45. Springer, Cham (2015). https://doi.org/10.1007/978-3-319-15545-6_5
2. Baader, F., Lutz, C.: Description logic. In: Blackburn, P., van Benthem, J., Wolter, F. (eds.) The Handbook of Modal Logic, pp. 757–820. Elsevier (2006)
3. Barrett, C., et al.: CVC4. In: Gopalakrishnan, G., Qadeer, S. (eds.) CAV 2011. LNCS, vol. 6806, pp. 171–177. Springer, Heidelberg (2011). https://doi.org/10.1007/978-3-642-22110-1_14
4. Bouton, T., Caminha B. de Oliveira, D., Déharbe, D., Fontaine, P.: veriT: an open, trustable and efficient SMT-solver. In: Schmidt, R.A. (ed.) CADE 2009. LNCS (LNAI), vol. 5663, pp. 151–156. Springer, Heidelberg (2009). https://doi.org/10.1007/978-3-642-02959-2_12
5. Brenas, J.H., Echahed, R., Strecker, M.: A hoare-like calculus using the SROIQ$^\sigma$ logic on transformations of graphs. In: Diaz, J., Lanese, I., Sangiorgi, D. (eds.) TCS 2014. LNCS, vol. 8705, pp. 164–178. Springer, Heidelberg (2014). https://doi.org/10.1007/978-3-662-44602-7_14
6. Giunchiglia, E., Tacchella, A., Giunchiglia, F.: SAT-based decision procedures for classical modal logics. J. Autom. Reason. **28**(2), 143–171 (2002). https://doi.org/10.1023/A:1015071400913
7. Haarslev, V., Sebastiani, R., Vescovi, M.: Automated reasoning in $\mathcal{ALCQ}$ via SMT. In: Bjørner, N., Sofronie-Stokkermans, V. (eds.) CADE 2011. LNCS (LNAI), vol. 6803, pp. 283–298. Springer, Heidelberg (2011). https://doi.org/10.1007/978-3-642-22438-6_22

8. Jackson, D.: Software Abstractions. MIT Press, Cambridge (2011)
9. Reynolds, A., Barbosa, H., Fontaine, P.: Revisiting enumerative instantiation. In: Beyer, D., Huisman, M. (eds.) TACAS 2018. LNCS, vol. 10806, pp. 112–131. Springer, Cham (2018). https://doi.org/10.1007/978-3-319-89963-3_7
10. Sebastiani, R., Vescovi, M.: Automated reasoning in modal and description logics via SAT encoding: the case study of K (m)/ALC-satisfiability. J. Artif. Intell. Res. **35**(1), 343 (2009)

# Symbolic and Numerical Computation

# What Does Qualitative Spatial Knowledge Tell About Origami Geometric Folds?

Fadoua Ghourabi[1]([✉]) and Kazuko Takahashi[2]

[1] Ochanomizu University, Tokyo, Japan
ghourabi.fadoua@ocha.ac.jp
[2] Kwansei Gakuin University, Sanda, Japan
ktaka@kwansei.ac.jp

**Abstract.** Origami geometry is based on a set of 7 fundamental folding operations. By applying a well-chosen sequence of the operations, we are able to solve a variety of geometric problems including those impossible by using Euclidean tools. In this paper, we examine these operations from spatial qualitative point of view, i.e. a common-sense knowledge of the space and the relations between its objects. The qualitative spatial representation of the origami folds is suitable for human cognition when practicing origami by hand. We analyze the spatial relations between the parameters of the folding operations using some existing spatial calculus. We attempt to divide the set of possible values of the parameters into disjoint spatial configurations that correspond to a specific number of fold lines. Our analyses and proofs use the power of a computer algebra system and in particular the Gröbner basis algorithm.

**Keywords:** Origami geometry · Huzita-Justin folding operations
Region Connection Calculus · Relative distance

## 1 Introduction

Origami is the art of paper folding and can serve as framework for solving geometric problems. Seven fundamental operations have been defined by Huzita [8] and Justin [9] to show how to fold the origami and make variety of geometric objects and in particular objects that require solving cubic equations. Origami is simple as only hands are involved in the folding process, affordable as paper is abundant and powerful as it solves problems unsolvable by using straightedge and compass. These advantages give grounds for incorporating origami in a lesson of geometry. Are the fundamental operations of origami geometry suitable for human (or a pupil) cognition?

We thank Prof. Isao Nakai from Ochanomizu University for our discussions on the 6th fold operation. We also thank Prof. Tetsuo Ida from Tsukuba University for giving the permission to use the computational origami system EOS to produce the origami figures. This work is supported by JSPS KAKENHI Grant No. 18K11453.

© Springer Nature Switzerland AG 2018
J. Fleuriot et al. (Eds.): AISC 2018, LNAI 11110, pp. 139–154, 2018.
https://doi.org/10.1007/978-3-319-99957-9_9

Research on the fundamental origami operations focused on their possibilities or increasing their power [1,10,11]. However, anyone who has ever struggled with the challenge of making a geometrical origami object by hand, is familiar with the difficulty of applying the 6th operation. The operation goes as follow (as originally stated by Huzita).

(6) Given two distinct points and two distinct lines, you can fold super-posing the first point onto the first line and the second point onto the second line at the same time.

The 6th operation requires superposing two points on two lines simultaneously. We invite the reader to try it with a piece of paper. Martin advised to use a transparent paper or to fold in front of a lightbulb [12]. Others hinted that the fold includes sliding a point on a line to bring the other point on the other line [14].

A diagram in Euclidean geometry or a shape in origami geometry is, in the first place, a collection of spatial objects such as lines, points, segment lines, circles, etc. The spatial knowledge is given by relations that describe a common sense understanding of the space and its objects. Examples of such relations are on, inside, outside, to the left, to the right, etc. These relations are rudimentary in the sense they can be described by the naked eye without further calculations or reasoning and thus suitable for human cognition. We attempt at providing a qualitative representation of the fundamental fold operations.

In this paper, we build on the first author's previous work [7]. The paper [7] analyzes the fundamental fold operations by identifying the degenerate cases, enumerating the cases where some operations can be derived from others, among other things. The degenerate cases are configurations of points and lines on the origami where the fold operation is not well-defined because of infinite possibilities. By excluding these cases, the fold operation has a finite number of solutions and thus well defined. In this paper, we further develop this analysis. We divide the origami space into disjoint configurations that give an exact number $n$ of fold lines, where $0 \leq n \leq 3$. To that end, we present a mapping of the spatial relations and the fold operations into algebraic terms. We also present a systematic proof strategy to show the statements on the number of fold lines.

The rest of the paper is organized as follows. We first introduce origami geometry based on the fundamental fold operations in Sect. 2. Then, in Sect. 3, we explain the various qualitative calculi that we use. The spatial configuration of well definedness are listed in Sect. 4 and the configurations on the number of solutions are explained in Sects. 5 and 6. Finally, in Sect. 7, we conclude.

# 2    Origami Spatial Objects and Their Construction

## 2.1    Origami Shape

We work with a square origami paper. By hand, we can fold the origami paper and make a crease. A crease leaves a trace on the origami paper, a line segment

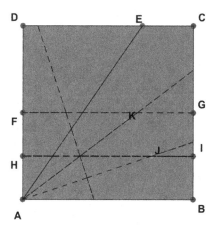

**Fig. 1.** An origami geometric shape: lines $AK$ and $AJ$ are trisectors of $\angle EBA$

whose endpoints are on the edges of the origami square paper. The extension of the line segment of a crease is a line that we call *fold line*. The intersection of two non-parallel fold lines is an origami point that can be outside the origami square.

An origami construction is a sequence of folds (and unfolds). When the collection of points and lines, constructed by folds, have a geometric meaning, we say that we constructed an origami geometric shape. For instance, the origami geometric shape in Fig. 1 depicts two line trisectors $AK$ and $AJ$ of angle $\angle EAB$. The remaining points and line segments on the origami, e.g. $F$, $G$, $H$, $I$, are constructed during the intermediate steps and used to make the trisectors.

## 2.2   Origami Fold

How to obtain a meaningful origami shape such as the one in Fig. 1? We need, foremost, a rigorous definition of the origami folds in the way Euclid's Elements define constructions with a compass and a straightedge. Let $\mathcal{O}$ be an origami square $\square ABCD$. An origami shape is obtained by applying the following fundamental fold operations [8,9].

**(O1)** Given two distinct points $P$ and $Q$, fold $\mathcal{O}$ along the unique line that passes through $P$ and $Q$.

**(O2)** Given two distinct points $P$ and $Q$, fold $\mathcal{O}$ along the unique line to superpose $P$ and $Q$.

**(O3)** Given two distinct lines $m$ and $n$, fold $\mathcal{O}$ along a line to superpose $m$ and $n$.

**(O4)** Given a line $m$ and a point $P$, fold $\mathcal{O}$ along the unique line passing through $P$ to superpose $m$ onto itself.

**(O5)** Given a line $m$, a point $P$ not on $m$ and a point $Q$, fold $\mathcal{O}$ along a line passing through $Q$ to superpose $P$ and $m$.

**(O6)** Given two lines $m$ and $n$, a point $P$ not on $m$ and a point $Q$ not on $n$, where $m$ and $n$ are distinct or $P$ and $Q$ are distinct, fold $\mathcal{O}$ along a line to superpose $P$ and $m$, and $Q$ and $n$.

**(O7)** Given two lines $m$ and $n$ and a point $P$ not on $m$, fold $\mathcal{O}$ along the unique line to superpose $P$ and $m$, and $n$ onto itself.

The fold line described by operation (O1) is the line passing through two distinct points. The fold line described by operation (O2) is the perpendicular bisector of the line segment $PQ$ as shown in Fig. 2. Operation (O3) gives rise to at most two fold lines, which are the interior and exterior bisectors of the angle formed by the two lines $m$ and $n$. To perform operation (O4), we drop a line perpendicular to $m$ and passing through $P$. The fold line of operation (O5) is the line tangent to the parabola of focus $P$ and directrix $m$, denoted by $\mathcal{P}(P, m)$, and passing through point $Q$. This operation is shown in Fig. 3. The operation (O6) in Fig. 4 is about finding a common tangent to the parabolas $\mathcal{P}(P, m)$ and $\mathcal{P}(Q, n)$. Finally, to perform operation (O7), we fold along the tangent to the parabola $\mathcal{P}(P, m)$ and perpendicular to the line $n$.

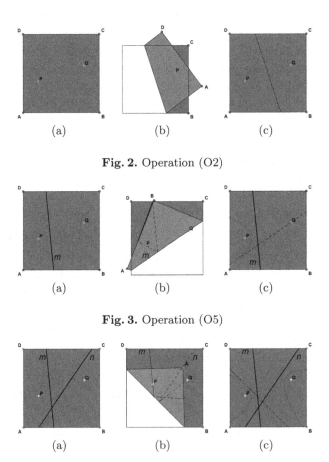

|     |     |     |
| :-: | :-: | :-: |
| (a) | (b) | (c) |

**Fig. 2.** Operation (O2)

|     |     |     |
| :-: | :-: | :-: |
| (a) | (b) | (c) |

**Fig. 3.** Operation (O5)

|     |     |     |
| :-: | :-: | :-: |
| (a) | (b) | (c) |

**Fig. 4.** Operation (O6)

# 3    Qualitative Spatial Relations in Origami

The following common sense concepts of connection, orientation and distance will be used to describe origami folds qualitatively.

## 3.1    Object Connection

There is a finite number of situations on the way objects are put together. For instance, whether they are connected and, if true, in which way they are connected. Such situations are described using Region Connection Calculus known in literature by RCC [3]. RCC defines a set of spatial relations. The commonly used ones are either 5 or 8 relations depending on the topology of the spatial objects and the purpose of the representation. Nevertheless, the set of relations must satisfy an important property: pairwise disjoint and jointly exhaustive, which means exactly one relation holds between two arbitrary objects.

RCC5 works for an object equal to its topological closure, in other words its boundary and interior coincide. This is the case of origami points and lines. Table 1 describes all possible connections between points and lines without ambiguities. Note that the 5 spatial relations are equivalent to basic geometric properties in the 2D plane. For instance, the relation **proper-part** stands for the geometric property that a point is on a line, two lines are **disjoint** when they are parallel, etc.

**Table 1.** RCC5 relations between origami points and lines

| | equal | proper-part | intersect | proper-part$^{-1}$ | disjoint |
|---|---|---|---|---|---|
| Point×Point | $\overset{\bullet}{P_1, P_2}$ | – | – | – | $\overset{\bullet}{P_1}$  $\overset{\bullet}{P_2}$ |
| Line×Line | $m, n$ | – | $\dfrac{m}{n}$ | – | $\dfrac{m}{n}$ |
| Point×Line | – | $\overset{\bullet}{P}\ m$ | – | – | $\overset{m}{\underset{\bullet\,P}{}}$ |
| Line×Point | – | – | – | $\overset{m}{\underset{\bullet\,P}{}}$ | $\overset{m}{\underset{\bullet\,P}{}}$ |

Circles are more complex objects. The RCC5 is limited since we cannot distinguish between a line tangent to a circle and a line not intersecting a circle, or when two circles are tangent or disjoint. We use two of RCC8 relations, namely relations **disconnected** and **externally-connected**, to improve the expressiveness as shown in Fig. 5 [13].

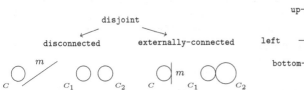

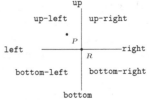

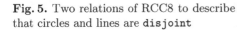

**Fig. 5.** Two relations of RCC8 to describe that circles and lines are `disjoint`

**Fig. 6.** Relations of relative orientation with $(P, R) \in$ `up-left`

### 3.2 Relative Orientation

A well-known qualitative description of the positions of objects relative to each other is Freska's calculus for points on the 2D plane [6]. To describe the position of an origami point $P$ with respect to a reference point $R$, we divide the origami plane into 8 regions intersecting in $R$ as shown in Fig. 6.

### 3.3 Relative Distance

Several approaches have been defined to compare lengths of intervals which can be regarded as distance between ending points [4]. However, these approaches do not make a good use of the possibilities that the space may offer. The origami plane, for instance, is a dynamic medium. By means of folds, points can be moved by reflection while preserving the length.

**Example.** We want to compare the distances $d(P, Q)$ and $d(R, S)$ in Fig. 7(a), where $d$ is the conventional Euclidean distance. First, we perform an (O3) fold to bring points $R$ and $S$ on the line $PQ$ as shown in Fig. 7(b). In Fig. 7(c), $R1$ and $S1$ are the reflections of $R$ and $S$ by the fold.[1] Next, we perform an (O2) fold to bring $R1$ onto $P$. Figures 7(d) and (e) show this operation, where $R2 = P$ is the reflection of $R1$ and $S2$ the reflection of $S1$. Finally, we perform along the line that passes through $R2$ and perpendicular to line $PQ$, i.e. (O4) fold. The operation is depicted in Fig. 7(f) and (g) shows a new point $S3$ obtained by the reflection of $S2$ by the fold. Since folding preserves the distance, we have

$$d(R, S) = d(R1, S1) = d(R2, S2) = d(P, S2) = d(P, S3).$$

The points $P$, $S3$ and $Q$ are aligned consecutive points in a homogeneous distance system where any given interval is bigger or equal than the previous one [4]. Since $S3$ and $Q$ are disjoint, $d(P, Q) > d(P, S3) = d(R, S)$.

Distance between a point and a line, e.g. $d(P, m)$, is in essence a distance between points. For instance, in the case of $d(P, m)$, we perform (O4) along the point $P$ and perpendicular to $m$. Let $Q$ be the intersection of the fold line and

---

[1] The reflection point can be easily obtained by (O1)–(O7) folds. We omit the steps in Fig. 7.

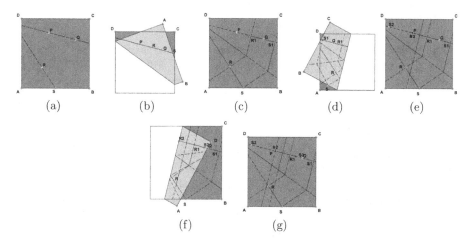

**Fig. 7.** Origami folds to deduce that $d(P, Q) \geq d(R, S)$

$m$. Then, $d(P, m) = d(P, Q)$. Similarly, the distance between two parallel lines, e.g. $d(m, n)$, is defined as the distance between two points $M$ and $N$ on the lines, where $MN \perp m$.

## 4   Well-Definedness of Fold Operations

The statements of (O1)–(O7) in Sect. 2.2 include conditions like "two distinct points $P$ and $Q$", "two distinct lines $m$ and $n$", "a point $P$ not on $m$". These are the conditions to eliminate degenerate configurations or incidence configurations. The degenerate situations are configurations of points and lines where there are infinite possibilities for the fold line. The incidence configurations occurs when we superpose a point $P$ and a line $m$ and $(P, m) \in$ proper-part. The operation becomes solvable with simpler operations, i.e. operations that solve lower degree equations. See [7] for a discussion on the configurations of degeneracy and incidence.

These conditions are intuitive and can be expressed qualitatively. First, we use the following lemma. Distinct points (respectively distinct lines) means not equal (respectively lines not equal).

**Lemma 1.**  – $(P, Q) \notin$ equal *if and only if* $(P, Q) \in$ disjoint.
– $(m, n) \notin$ equal *if and only if* $(m, n) \in$ intersects $\cup$ disjoint.
– $(P, m) \notin$ proper-part *if and only if* $(P, m) \in$ disjoint.

The proof follows from the fact that the RCC5 relations between points and lines are jointly exhaustive and pairwise distinct.

– The Operation (O1) is well defined when $(P, Q) \in$ disjoint.
– The Operation (O2) is well defined when $(P, Q) \in$ disjoint.

- The Operation (O3) is well defined when $(m, n) \in$ intersects $\cup$ disjoint.
- The Operation (O4) is always well defined.
- The Operation (O5) is well defined when $(P, m) \in$ disjoint.
- The Operation (O6) is well defined when $(P, m) \in$ disjoint and $(Q, m) \in$ disjoint and $((m, n) \in$ intersects $\cup$ disjoint or $(P, Q) \in$ disjoint.
- The Operation (O7) is well defined when $(P, m) \in$ disjoint.

### 4.1  Spatial Conditions of the Solutions of (O1)–(O4)

Non-degenerate configurations are further processed by identifying the number of solutions. The solutions of operations (O1), (O2), (O3) and (O4) are straightforward. We can easily show that (O1) and (O2) have a unique solution if and only if the points are not equal, whereas (O4) always has a unique solution independently from the spatial configuration of the parameters. Operation (O3) has two solutions if the lines parameters are in relation intersects and one solution if the lines are in disjoint.

## 5  Spatial Conditions of the Solutions of (O5)–(O7)

### 5.1  A Systematic Approach

**Objects.** To analyze the solutions of (O5)–(O7), we use an algebraic approach. We consider a Cartesian Coordinate system. Points are defined by pairs of their coordinates. We denote the coordinates of a point $P$ by $(x_p, y_p)$. A well defined line has an equation of the form $ax + by + c = 0$, where $a \neq 0 \vee b \neq 0$. We denote by $a_m, b_m$ and $c_m$ the coefficients of a line $m$. A circle $\mathcal{C}(P, r)$, whose center and radius are $P$ and $r > 0$, has the equation $\sqrt{(x - x_p)^2 + (y - y_p)^2} = r$.

The determination of the (exact) domain of the coordinates and coefficients is tricky. $\mathbb{Q}$ is too small since it doesn't include $\sqrt{x}$ numbers and $\mathbb{R}$ is too much. An algebraic extension of $\mathbb{Q}$ would be a good candidate since the origami fundamental fold operations allow the construction of rational numbers plus numbers of the form $\sqrt{x}$ and $\sqrt[3]{x}$ [5].

**Algebraic Relations and Functions.** Table 2 shows the algebraic relations of the qualitative spatial relations explained in Sect. 3. The algebraic forms are self-explanatory.

Furthermore, in our analysis of the fold operations, specifically operations (O5)–(O7), we work with parabolas $\mathcal{P}(P, m)$ represented by the following equation $f(x, y)$.

$$f(x, y) := (x - x_p)^2 + (y - y_p)^2 - \frac{(a_m x + b_m y + c_m)^2}{a_m{}^2 + b_m{}^2} = 0 \qquad (1)$$

Let $t$ be a tangent to the parabola $\mathcal{P}(P, m)$ at a point $(x_1, y_1)$. Also, let $\lambda$ be the slope of $t$. Then the following equation $g(x_1, y_1)$ defines the tangent $t$.

$$g(x_1, y_1) := \frac{\partial f}{\partial x}(x_1, y_1) + (\frac{\partial f}{\partial y}(x_1, y_1))\lambda = 0 \qquad (2)$$

**Table 2.** Algebraic forms of the qualitative spatial relations and functions

| | Spatial relation/function | Algebraic relation/function |
|---|---|---|
| Object connection | $(P, Q) \in$ `equal` | $x_p = x_q \wedge y_p = y_q$ |
| | $(P, Q) \in$ `disjoint` | $x_p \neq x_q \wedge y_p \neq y_q$ |
| | $(m, n) \in$ `equal` | $\exists k.\ a_n = ka_m \wedge b_n = kb_m \wedge c_n = kc_m$ |
| | $(m, n) \in$ `disjoint` | $(a_n b_m - a_m b_n = 0) \wedge \neg((m, n) \in$ `equal`$)$ |
| | $(m, n) \in$ `intersects` | $a_n b_m - a_m b_n \neq 0$ |
| | $(\mathcal{C}(P, r_1), \mathcal{C}(Q, r_2)) \in$ `equal` | $(P, Q) \in$ `equal` $\wedge\, r_1 = r_2$ |
| | $(\mathcal{C}(P, r_1), \mathcal{C}(Q, r_2)) \in$ `intersects` | $((P, Q) \in$ `disjoint`$) \wedge (\sqrt{(x_p - x_q)^2 + (y_p - y_q)^2} < r_1 + r_2)$ |
| | $(\mathcal{C}(P, r_1), \mathcal{C}(Q, r_2)) \in$ `externally-connected` | $((P, Q) \in$ `disjoint`$) \wedge (\sqrt{(x_p - x_q)^2 + (y_p - y_q)^2} = r_1 + r_2)$ |
| | $(\mathcal{C}(P, r_1), \mathcal{C}(Q, r_2)) \in$ `disconnected` | $((P, Q) \in$ `disjoint`$) \wedge (\sqrt{(x_p - x_q)^2 + (y_p - y_q)^2} > r_1 + r_2)$ |
| | $(Q, \mathcal{C}(P, r)) \in$ `proper-part` | $\sqrt{(x_q - x_p)^2 + (y_q - y_p)^2} = r$ |
| | $(Q, \mathcal{C}(P, r)) \in$ `disjoint` | $(\sqrt{(x_q - x_p)^2 + (y_q - y_p)^2} < r) \vee (\sqrt{(x_q - x_p)^2 + (y_q - y_p)^2} > r)$ |
| | $(m, \mathcal{C}(P, r)) \in$ `intersects` | $\frac{\lvert a_m x_p + b_m y_p + c_m \rvert}{\sqrt{a_m^2 + b_m^2}} < r$ |
| | $(m, \mathcal{C}(P, r)) \in$ `externally-connected` | $\frac{\lvert a_m x_p + b_m y_p + c_m \rvert}{\sqrt{a_m^2 + b_m^2}} = r$ |
| | $(m, \mathcal{C}(P, r)) \in$ `disconnected` | $\frac{\lvert a_m x_p + b_m y_p + c_m \rvert}{\sqrt{a_m^2 + b_m^2}} > r$ |
| Orientation | $(P, Q) \in$ `left` | $x_p < x_q \wedge y_p = y_q$ |
| | $(P, Q) \in$ `right` | $x_p > x_q \wedge y_p = y_q$ |
| | $(P, Q) \in$ `up` | $x_p = x_q \wedge y_p > y_q$ |
| | $(P, Q) \in$ `bottom` | $x_p = x_q \wedge y_p < y_q$ |
| | $(P, Q) \in$ `up-left` | $x_p < x_q \wedge y_p > y_q$ |
| | $(P, Q) \in$ `up-right` | $x_p > x_q \wedge y_p > y_q$ |
| | $(P, Q) \in$ `bottom-left` | $x_p < x_q \wedge y_p < y_q$ |
| | $(P, Q) \in$ `bottom-right` | $x_p > x_q \wedge y_p < y_q$ |
| Distance | $d(P, Q)$ | $\sqrt{(x_p - x_q)^2 + (y_p - y_q)^2}$ |
| | $d(P, m)$ | $\frac{\lvert a_m x_p + b_m y_p + c_m \rvert}{\sqrt{a_m^2 + b_m^2}}$ |
| | $d(m, n)$, where $m \parallel n$ | $\frac{\lvert c_m - c_n \rvert}{\sqrt{a_m^2 + b_m^2}}$ |

**Proof Strategy.** To prove the number of the solutions of (O5)–(O7), we perform the following steps.

1. Define a system $S$ of the algebraic relations that describe the fold line. These relations are (1) and (2) as well as well-established algebraic form of geometric properties that we will explain when used.
2. Compute Gröbner basis of $S$. This step attempts to eliminate some of the variables and obtain one equation in the slope of the fold line.
   - If one polynomial is obtained, then compute its discriminant.
   - If more than one polynomial are obtained, then solve for some of the dependent variables.
3. Analyze the obtained polynomial expressions to identify the cases with real solutions. Specifically, we watch out for the appearance of relations of Table 2.

We use the power of the computer algebra system *Mathematica* to perform the computations in the above steps. We use Buchberger's algorithm to generate Gröbner bases. The computations are performed symbolically, thus we prove our results in the general case.

## 5.2   Spatial Conditions of the Solutions of (O5)

**Theorem 2.** *Let $P$, $Q$ and $m$ be on the origami where $(P, m) \in$ disjoint. We perform the (O5) operation along the fold line passing through $Q$ to superpose $P$ and $m$.*

- *If $d(Q, P) = d(Q, m)$, then there is a unique fold line.*
- *If $d(Q, P) > d(Q, m)$, then there are two distinct fold lines.*
- *If $d(Q, P) < d(Q, m)$, then there is no fold line.*

*Proof.* Since $(P, m) \in$ disjoint, we know that the solutions of (O5) are the lines passing through $Q$ and tangent to the parabola $\mathcal{P}(P, m)$. Let $\lambda$ be the slope of the tangent to the parabola $\mathcal{P}(P, m)$ at point $(x_1, y_1)$. Hence, we have the system of equations $S = \{f(x_1, y_1), g(x_1, y_1), (y_1 - y_q) - \lambda(x_1 - x_q) = 0\}$. $f$ and $g$ are given in (1) and (2), respectively. The equation $(y_1 - y_q) - \lambda(x_1 - x_q) = 0$ means that the tangent passes through the points $(x_1, y_1)$ and $Q$.

We compute the Gröbner basis of $S$. We obtain a 2nd degree polynomial in $\lambda$ whose discriminant is the following.

$$4(a_m^2 + b_m^2)^2(c_m + a_m x_p + b_m y_p)^2 \times \tag{3}$$

$$(a_m^2 + b_m^2)((x_p - x_q)^2 + (y_p - y_q)^2) - (a_m x_q + b_m y_q + c_m)^2 \tag{4}$$

Line $m$ is well defined, then $a_m$ and $b_m$ cannot vanish at the same time and $a_m^2 + b_m^2 > 0$. Also, since $(P, m) \in$ disjoint, $(c_m + a_m x_p + b_m y_p)^2 > 0$. Thus, the factors in (3) are always strictly positive and the sign of the polynomial in (4) determines the number of solutions of $\lambda$. (4) is the expression of $d(Q, P)^2 - d(Q, m)^2$ in algebraic terms. If strictly positive, we have two solutions for $\lambda$, i.e. two fold lines. If strictly negative, then there is no solution. If equal to 0 then there is a unique fold line.                                                                      □

From a geometric point of view, a tangent $t$ to a parabola is the perpendicular bisector of the line segment joining $P$ and a point on $m$ that we name $P'$. Since $t$ passes through $Q$, the circle whose center is $Q$ and radius $\overline{QP}$ intersects $m$ in $P'$. In Fig. 8, we show the situation where the circle intersects $m$ in two points and we have two tangents or two fold lines $t_1$ and $t_2$. If the circle does not intersect $m$, then no tangent exists. If the circle is tangent to $m$ ($Q$ is a point on the parabola) then there is one tangent.

## 5.3   Spatial Conditions of the Solutions of (O7)

**Theorem 3.** *Let $P$, $m$ and $n$ be on the origami where $(P, m) \in$ disjoint. We perform operation (O7) along the fold line $t$ perpendicular to $n$ to superpose $P$ and $m$.*

- *If $(m, n) \in$ equal $\cup$ disjoint, then there is no fold line.*
- *If $(m, n) \in$ intersects, then there is one fold line.*

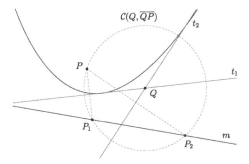

**Fig. 8.** $d(Q, P) > d(Q, m)$: two fold lines $t_1$ and $t_2$ for (O5)

*Proof.* We solve for $x_1$ and $y_1$ in $\{f(x_1, y_1), g(x_1, y_1), a_n\lambda - b_n = 0\}$, where $\lambda$ is the slope of the fold line, and $f$ and $g$ are defined in (1) and (2). The former equation $a_n\lambda - b_n = 0$ states that $n$ and the tangent are perpendicular. We obtain one solution of the following form:

$$x_1 \rightarrow \frac{\text{some polynomial expression of } a_m, b_m, c_m, a_n, b_n, c_n, x_p \text{ and } y_p}{2(a_n b_m - a_m b_n)^2}$$

$$y_1 \rightarrow \frac{\text{some polynomial expression of } a_m, b_m, c_m, a_n, b_n, c_n, x_p \text{ and } y_p}{2(a_n b_m - a_m b_n)^2}$$

However, the solutions are undefined when the denominator is null, i.e. $a_n b_m - a_m b_n = 0$. This is the algebraic relations of two lines that are `disjoint` or `equal` according to Table 2. □

### 5.4    Spatial Conditions of the Solutions of (O6) with Disjoint Lines

**Theorem 4.** *Let points $P$ and $Q$ and `disjoint` lines $m$ and $n$ be on origami, where $(P, m) \in$ `disjoint` and $(Q, m) \in$ `disjoint`. We perform operation (O6) to superpose $P$ and $m$ and $Q$ and $n$.*

- *If $d(m, n) > d(P, Q)$, then there is no fold line.*
- *If $d(m, n) = d(P, Q)$, then there is a unique fold line.*
- *If $d(m, n) < d(P, Q)$, then there are two fold lines.*

*Proof.* We proceed similarly to the proof of Theorem 2. We know that the fold line is a common tangent to parabolas $\mathcal{P}(P, m)$ and $\mathcal{P}(Q, n)$. We compute the Gröbner basis of $\{f_1(x_1, y_1), g_1(x_1, y_1), f_2(x_2, y_2), g_2(x_2, y_2), (y_1 - y_2) - (x_1 - x_2)\lambda = 0, a_n b_m - a_m b_n = 0\}$, where $\lambda$ is the slope of the common tangent of $\mathcal{P}(P, m)$ and $\mathcal{P}(Q, n)$ at $(x_1, y_1)$ and $(x_2, y_2)$, respectively. Note that $f_1, g_1, f_2$ and $g_2$ are Eqs. (1) and (2) defined for the first parabola $\mathcal{P}(P, m)$ and the second parabola $\mathcal{P}(Q, n)$. The discriminant of the result of Gröbner basis computation gives

$$(a_m^2 + b_m^2)((x_p - x_q)^2 + (y_p - y_q)^2 - (c_m - c_n)^2),$$

which stands for the algebraic form of $d(P, Q)^2 - d(m, n)^2$. □

# 6    Analysis of (O6) with Intersecting Lines

## 6.1    Simplification Without Loss of Generality

Operation (O6) contributes to geometry by solving problems that are impossible by classical Euclidean tools. The operation has the merits of solving any cubic equation of the general form $ax^3 + bx^2 + cx + d = 0$. Coefficients $a$, $b$, $c$ and $d$ are in the field of origami constructible numbers, i.e. an algebraic extension of $\mathbb{Q}$ with square root and cubic square root [5].

To simplify our analysis of operation (O6), we use lines parallel to $xy$-axes. This reduces the number of parameters that come from the lines $m$ and $n$ and simplifies the Gröbner basis computation, which is, in the worst case, double exponential in the number of variables [2].

**Lemma 5.** *Any cubic equation $ax^3 + bx^2 + cx + d = 0$, where $a \neq 0$, can be solved with lines $m$ and $n$ perpendicular and parallel to $xy$-axes, respectively.*

*Proof.* We apply (O6) to superpose $P$ and $m$, and $Q$ and $n$, simultaneously. The fold line is a common tangent to the parabolas $\mathcal{P}(P, m)$ and $\mathcal{P}(Q, n)$. Let $\lambda$ be the slope of the common tangent. We take $m$ and $n$ to be of equations $x + c_m = 0$ and $y + c_n = 0$. We compute the Gröbner basis of

$$\{f_1(x_1, y_1), f_2(x_2, y_2), g_1(x_1, y_1), g_2(x_2, y_2), (y_2 - y_1) - \lambda(x_2 - x_1) = 0\},$$

where the former equation $(y_2 - y_1) - \lambda(x_2 - x_1) = 0$ states that the tangent passes through the points $(x_1, y_1)$ and $(x_2, y_2)$ on the parabolas. The result is a cubic polynomial in $\lambda$.

$$(c_n + y_q)\lambda^3 + (c_m - x_p + 2x_q)\lambda^2 + (c_n + 2y_p - x_q)\lambda + c_m + x_p \qquad (5)$$

We match the coefficient of the above polynomial with $a$, $b$, $c$ and $d$. We solve for the coordinates of $P$ and $Q$ and obtain:

$$\{x_p \rightarrow -c_m + d, y_p \rightarrow (a + c - 2c_n)/2, x_q \rightarrow (b - 2c_m + d)/2, y_q \rightarrow a - c_n\} \quad (6)$$

□

We can further simplify by taking $c_m = c_n = 0$. In that case, solutions in (6) gives rise to $P(d, (a + c)/2)$ and $Q((b + d)/2, a)$. For instance, to solve the cubic $x^3 - 3x^2 + \frac{27}{8} = 0$ with (O6), we can take the lines $m : x = 0$ and $n : y = 0$ and the points $P(\frac{27}{8}, \frac{1}{2})$ and $Q(\frac{3}{16}, 1)$.

## 6.2    The Discriminant Function

**Lemma 6.** *Let $\Delta$ be the discriminant of the polynomial (5). We have the following result about the number of real roots.*

*(i) If $\Delta < 0$, then polynomial (5) has a single real root.*
*(ii) If $\Delta > 0$, then polynomial (5) has three distinct real roots.*

*(iii) If $\Delta = 0$, then polynomial (5) has either a triple real root, or one double real root and one single real root.*

*Proof.* The proof is a result of solving cubic equations with radicals.    □

The description of $\Delta$ in term of spatial relations between (O6) parameters $m$, $n$, $P$ and $Q$ is not straightforward. So, we observe how point $Q$ would relate to $P$. We fix $P$ to be the point $(3, 4)$, for instance, and take $m$ and $n$ to be the $y$-axis and the $x$-axis, respectively. We plot $\Delta(x_q, y_q) = 0$. Figure 9(a) depicts the 3 regions defined by the curve $\Delta(x_q, y_q) = 0$. If $Q$ is on the blue region then we are in case (ii), i.e. there are 3 distinct fold lines, if on the white region then case (i), i.e. one fold line, if on the curve then case (iii), i.e. either one or two fold lines.

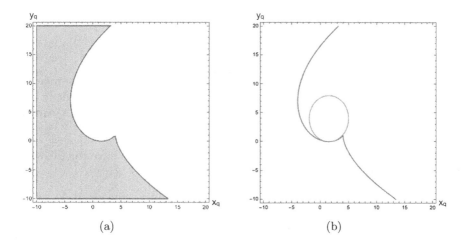

(a)                              (b)

**Fig. 9.** The curve $\Delta(x_q, y_q) = 0$ (Color figure online)

Hereafter, we give a geometric explanation of the curve $\Delta(x_q, y_q)$. In Fig. 10, we take two parabolas tangent at point $S$. Obviously, a possible fold line is the tangent passing through point $S$. We move the point $S$ on the parabola $\mathcal{P}(P, m)$ and trace the point $Q$. The locus of point $Q$ is $\Delta(x_q, y_q) = 0$. Since the parabolas are tangent, then 2 or 3 fold lines coincide and thus correspond to double or triple real roots.

## 6.3   Spatial Case 1: $Q$ is Equal to the Cusp Point

**Lemma 7.** *If $Q$ is the cusp point of $\Delta(x_q, y_q) = 0$, then (O6) has a unique fold line.*

*Proof.* Another useful constant of a cubic $ax^3 + bx^2 + cx + d = 0$ is $\Delta_0 = b^2 - 3ac$. When $\Delta_0 = \Delta = 0$, there exists a triple real root. In our example $\Delta_0(x_q, y_q) = 0$

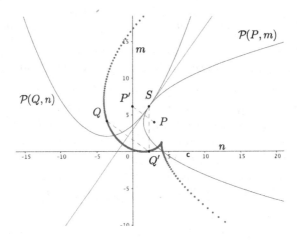

**Fig. 10.** The locus of $Q$ (blue curve) when $P$, $m$ and $n$ are fixed and $\mathcal{P}(P,m)$, and $\mathcal{P}(Q,m)$ are tangent at a point $S$ (Color figure online)

is shown in Fig. 9(b). The cusp point is an intersection point of the two curves and correspond to a situation where we have one fold line of multiplicity 3. A second intersection point is on $n$ excluded by the condition $(Q,n) \in \mathtt{disjoint}$ of (O6).                                                                                                          $\square$

Figure 11 shows the circles when $Q$ is the cusp point. We have the following result based on spatial observation.

**Lemma 8.** *Let point $O$ be the intersection of $m$ and $n$, and point $M$ be the middle point of the line segment $PQ$. Furthermore, let $\mathcal{C}_1$ and $\mathcal{C}_2$ be the circles $\mathcal{C}(M, \overline{MP})$ and $\mathcal{C}(O, \overline{PQ})$. If $(\mathcal{C}_1, \mathcal{C}_2) \in \mathtt{externally\text{-}connected}$, then $Q$ is the cusp point.*

*Proof.* We provide the sketch of the proof. Using *Mathematica*:

1. Solve for the coordinates $x_q$ and $y_q$ of the cusp point using

$$\frac{\partial \Delta(x_q, y_q)}{\partial y_q} = \frac{\partial \Delta(x_q, y_q)}{\partial x_q} = 0.$$

2. Show that the circles $\mathcal{C}(M, \overline{MP})$ and $\mathcal{C}(O, \overline{PQ})$ are **externally-connected**using the appropriate relation from Table 2.

                                                                                                          $\square$

### 6.4   Spatial Case 2: $P$ and $Q$ are on Opposite Half-Planes

The origami plane is divided by lines $m$ and $n$ into the 8 regions of relative orientation (see Sect. 3.2). The curve $\Delta(x_q, y_q) = 0$ intersects only 3 half-planes.

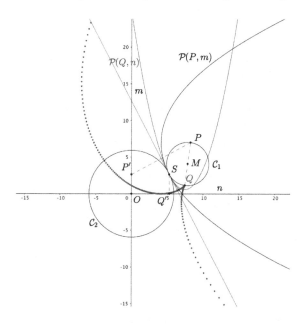

**Fig. 11.** $(\mathcal{C}_1, \mathcal{C}_2) \in$ **externally-connected** when $Q$ is the cusp of $\Delta(x_q, y_q) = 0$

Referring to Fig. 10, for instance, the curve $\Delta(x_q, y_q) = 0$ intersects the **up-left**, **up-right** and **bottom-right** half-planes. The remaining **bottom-left** half-plane is a subset of the region $\Delta(x_q, y_q) > 0$. Therefore, if we take $Q$ to be any point on the **bottom-left** half-plane, then we have 3 distinct fold lines for (O6).

## 7   Conclusion

We analyzed the fundamental folding operations. Based on Gröbner basis and other computer algebra methods, we proved the conditions on the number of fold lines. The conditions are described using qualitative relations between points and lines parameters of the fold operations. This approach worked for operations (O1)–(O5), (O7) and (O6) with disjoint lines. In the case of (O6) with intersecting lines, the spatial configurations cannot be described in a simple qualitative language. To tackle this operation, we identified some spatial cases that are easy to recognize when performing origami by hands.

# References

1. Alperin, R.: A mathematical theory of origami constructions and numbers. N. Y. J. Math. **6**, 119–133 (2000)
2. Becker, T., Kredel, H., Weispfenning, V.: Gröbner Bases: A Computational Approach to Commutative Algebra, pp. 511–514. Springer, New York (1993). https://doi.org/10.1007/978-1-4612-0913-3
3. Chen, J., Cohn, A., Liu, D., Wang, S., Ouyang, J., Yu, Q.: A survey of qualitative spatial representations. Knowl. Eng. Rev. **30**(1), 106–136 (2013)
4. Clementini, E., Di Felice, P., Hernández, D.: Qualitative representation of positional information. Artif. Intell. **95**(2), 317–356 (1997)
5. Cox, D.: Galois Theory, pp. 274–279. Wiley-Interscience, Hoboken (2004)
6. Freksa, C.: Using orientation information for qualitative spatial reasoning. In: Frank, A.U., Campari, I., Formentini, U. (eds.) GIS 1992. LNCS, vol. 639, pp. 162–178. Springer, Heidelberg (1992). https://doi.org/10.1007/3-540-55966-3_10
7. Ghourabi, F., Kasem, A., Kaliszyk, C.: Algebraic analysis of Huzita's Origami operations and their extensions. In: Ida, T., Fleuriot, J. (eds.) ADG 2012. LNCS (LNAI), vol. 7993, pp. 143–160. Springer, Heidelberg (2013). https://doi.org/10.1007/978-3-642-40672-0_10
8. Huzita, H.: Axiomatic development of origami geometry. In: Proceedings of the First International Meeting of Origami Science and Technology, pp. 143–158 (1989)
9. Justin, J.: Résolution par le pliage de l'équation du troisième degré et applications géométriques. In: Proceedings of the First International Meeting of Origami Science and Technology, pp. 251–261 (1989)
10. Haga, K.: Origamics Part I: Fold a Square Piece of Paper and Make Geometrical Figures. Nihon Hyoron Sha (1999). (in Japanese)
11. Kasem, A., Ghourabi, F., Ida, T.: Origami axioms and circle extension. In: Proceedings of the 26th Symposium on Applied Computing (SAC 2011), pp. 1106–1111. ACM Press (2011)
12. Martin, G.: Geometric Constructions, pp. 145–159. Springer, New York (1998). https://doi.org/10.1007/978-1-4612-0629-3
13. Randell, D.A., Cui, Z., Cohn, A.G.: A spatial logic based on regions and connection. In: Proceedings of the 3rd International Conference on Knowledge Representation and Reasoning, pp. 165–176 (1992)
14. Robu, J., Ida, T., Ţepeneu, D., Takahashi, H., Buchberger, B.: Computational origami construction of a regular heptagon with automated proof of its correctness. In: Hong, H., Wang, D. (eds.) ADG 2004. LNCS (LNAI), vol. 3763, pp. 19–33. Springer, Heidelberg (2006). https://doi.org/10.1007/11615798_2

# Discovering Geometry Theorems
# in Regular Polygons

Zoltán Kovács[(✉)] ⓘ

The Private University College of Education of the Diocese of Linz,
Salesianumweg 3, 4020 Linz, Austria
zoltan@geogebra.org

**Abstract.** By using a systematic, automated way, we discover a large
amount of geometry statements on regular polygons. Given a regular
$n$-gon, its diagonals are taken, two pairs of them may determine a pair
of intersection points that define a segment. By considering all possible segments defined in this way, we can compute the lengths of them
symbolically, and, depending on the simplicity of the symbolic result we
classify the segment either as "interesting" or "not interesting".

Among others, we prove that in a regular 11-gon with unit sides the
only rational lengths appearing the way described above, are 1 and 2, and
the only quadratic surd is $\sqrt{3}$. The applied way of proving is exhaustion,
by using the freely available software tool RegularNGons, programmed
by the author. The combinatorial explosion, however, calls for future
improvements involving methods in artificial intelligence.

The symbolic method being used is Wu's algebraic geometry approach
[1], combined with the discovery algorithm communicated by Recio and
Vélez [2]. The heavy computations are performed by a recent version of
the Giac computer algebra software, running in a web browser with the
support of the recent technology WebAssembly. Visual communication
of the obtained results is operated by the dynamic geometry software
GeoGebra.

**Keywords:** Automated theorem proving · Computer algebra
Regular polygons · WebAssembly · GeoGebra

## 1 Introduction

Obtaining interesting mathematical theorems automatically is a usual dream of
many mathematicians. By defining a formal language (with its logical axioms)
on a research field, and a set of (non-logical) axioms, one can deduce various
statements only by repeating the axioms. In principle, proofs for all propositions
in a research field can be traced back to consecutive uses of the axioms.

Several axiomatizations are available for many research fields in mathematics, however, interesting theorems (with proofs) are more difficult to find.
One problem is that combining the axioms consecutively usually produces an

© Springer Nature Switzerland AG 2018
J. Fleuriot et al. (Eds.): AISC 2018, LNAI 11110, pp. 155–169, 2018.
https://doi.org/10.1007/978-3-319-99957-9_10

unmanagable big database of propositions, including trivial or uninteresting ones. The other problem is to identify which propositions are interesting enough to call them theorems [3].

In this paper we limit our considerations to planar Euclidean geometry, namely to find interesting properties in a regular polygon. The literature on listing such properties is, however, huge, including constructible polygons (by compass and straightedge or origami, for example). From the very start of the availability of computer algebra systems (CAS) and dynamic geometry software (DGS), namely, the 1990s, however, non-constructable polygons can also be better observed, either numerically or symbolically.

In this study we limit the available axioms to very simple operations on a regular $n$-gon. Its diagonals (including the sides) can be taken, two pairs of them may determine a pair of intersection points which define a segment. By considering all possible segments given in this way, we can compute the lengths of them symbolically, and, depending on the simplicity of the symbolic result we classify the segment either as "interesting" or "not interesting". This is, of course, somewhat subjective, but this approach can be slightly modified by allowing other results interesting enough, or to define some other points as well for the domain of interest.

The paper consists of the following parts: In Sect. 2 the mathematical background is explained on computing an appearing segment symbolically. Section 3 presents a manually obtained new result. Section 4 demonstrates how the mathematical computations can be automated by using the tool `RegularNGons`. Finally, Sect. 5 depicts some future ideas.

## 2     Mathematical Background

In this section first we refer to two classic theorems on constructibility. An algebraic formula will be then shown by using former work.

### 2.1     Constructibility

Algebraization of the setup of a planar geometry statement is a well known process since the revolutional book [4] of Chou's. It demonstrates on 512 mathematical statements how an equation system describe a geometric construction, and by performing some manipulations on the equation system, a mechanical proof can be obtained. Chou's work focuses mainly on constructible setups, that is, mostly on such constructions that can be created only by using the classic approach, namely by compass and straightedge.

It is well known (since 1801, according to Gauß, and since 1837, according to Wantzel, see [5,6]) that a regular $n$-gon is constructible by using compass and straightedge if and only if $n$ is the product of a power of 2 and any number of distinct Fermat primes (including none). We recall that a Fermat prime is a prime number of the form $2^{2^m} + 1$. A generalization of this result (Pierpont see [7]) by allowing an angle trisector as well (for example, origami folding steps), is that a regular $n$-gon is constructible if and only if

$$n = 2^r \cdot 3^s \cdot p_1 \cdot p_2 \cdots p_k,$$

where $r, s, k \geq 0$ and the $p_i$ are distinct primes of form $2^t \cdot 3^u + 1$ [8]. The first constructible regular $n$-gons of this kind are

$$n = 3, 4, 5, 6, 7, 8, 9, 10, 12, 13, 14, 15, 16, 17, 18, 19, 20, \ldots$$

From the list the case $n = 11$ is missing, and, as a natural consequence, there are much less scientific results known on regular 11-gons than for other polygons. In this paper, therefore, we will focus on obtaining results on a regular 11-gon.

## 2.2    An Algebraic Formula for the Vertices

In this part of the paper we derive a formula for the coordinates of the vertices of a regular $n$-gon.

From now on we assume that $n \geq 1$. The cases $n = 1, 2$ have no real geometrical meaning, but they will be useful from the algebraic point of view.

In Chou's book—which is based on Wu's algebraic geometry method [1], and in the following it will be mentioned therefore as *Wu's approach*—the usual way to describe the points of a construction is to assign coordinates $(x_i, y_i)$ for a given point $P_i$ ($i = 0, 1, 2, \ldots$). When speaking about a polygon, in many cases the first vertices are put into coordinates $P_0 = (0, 0)$ and $P_1 = (1, 0)$, and the other coordinates are described either by using exact rationals, or the coordinates are expressed as possible solutions of algebraic equations.

For example, when defining a square, for instance, $P_2 = (1, 1)$ and $P_3 = (0, 1)$ seem to make sense, but for a regular triangle two equations for $P_2 = (x_2, y_2)$ are required, namely $x_2^2 + y_2^2 = 1$ and $(x_2 - 1)^2 + y_2^2 = 1$. It is easy to see that this equation system has two solutions, namely $x_2 = \frac{1}{2}, y_2 = \frac{\sqrt{3}}{2}$ and $x_2 = \frac{1}{2}, y_2 = -\frac{\sqrt{3}}{2}$. It is well known that there is no way in Wu's approach to avoid such duplicates, unless the coordinates are rational. In other words, if both minimal polynomials of the coordinates are linear (or constant), then the duplicates can be avoided, otherwise not. Here, for $x_2$ we have $2x_2 - 1(= 0)$, but for $y_2$ the minimal polynomial is $4y_2^2 - 3(= 0)$. We remark that the minimal polynomials are irreducible over $\mathbb{Z}$.

Clearly, minimal polynomials of a regular $n$-gon with vertices $P_0 = (0, 0)$ and $P_1 = (1, 0)$ can play an important role here. The paper [9] (based on Lehmer's work [10]) suggests an algorithm to obtain the minimal polynomial $p_c(x)$ of $\cos(2\pi/n)$, based on the Chebyshev polynomials $T_j(x)$ of the first kind (see Algorithm 1).

Now, by adding the equation $p_c(x)^2 + p_s(y)^2 = 1$ to the equation system, we have managed to describe a polynomial $p_s(y)$ such that $p_s(\sin(2\pi/n)) = 0$. Table 1 shows the minimal polynomials for $n \leq 17$.

It is clear, that—not considering the cases $n = 1, 2, 3, 4, 6$—the number of roots of $p_c$ is more than one, therefore the solution of the equation system $\{p_c(x) = 0, p_s(x) = 0\}$ is not unique. The number of solutions for $p_c(x) = 0$ depends on the degree of $p_c$, and—not considering the cases $n = 1, 2$—the number of solutions for $p_s(x) = 0$ is two for each root of $p_c(x)$, therefore the number of solutions for $\{p_c(x) = 0, p_s(y) = 0\}$ is usually $2 \cdot \deg(p_c)$. As a result, the point

**Algorithm 1.** Computing the minimal polynomial of $\cos(2\pi/n)$

1: **procedure** COS2PIOVERNMINPOLY($n$)
2:     $p_c \leftarrow T_n - 1$
3:     **for all** $j \mid n \ \wedge \ j < n$ **do**
4:         $q \leftarrow T_j - 1$
5:         $r \leftarrow \gcd(p_c, q)$
6:         $p_c \leftarrow p_c/r$
7:     **return** SquarefreeFactorization($p_c$)

**Table 1.** List of minimal polynomials of $\cos(2\pi/n)$, $n \leq 7$

| $n$ | Minimal polynomial of $\cos(2\pi/n)$ |
|---|---|
| 1 | $x - 1$ |
| 2 | $x + 1$ |
| 3 | $2x + 1$ |
| 4 | $x$ |
| 5 | $4x^2 + 2x - 1$ |
| 6 | $2x - 1$ |
| 7 | $8x^3 + 4x^2 - 4x - 1$ |

$$P = (\cos(2\pi/n), \sin(2\pi/n))$$

can be exactly determined by an algebraic equation in Wu's approach only in case $n = 4$, as shown in Table 2.

**Table 2.** Degree of ambiguity for $(\cos(2\pi/n), \sin(2\pi/n))$, $3 \leq n \leq 13$

| $n$ | 3 | 4 | 5 | 6 | 7 | 8 | 9 | 10 | 11 | 12 | 13 |
|---|---|---|---|---|---|---|---|---|---|---|---|
| Degree | 2 | 2 | 4 | 2 | 6 | 4 | 6 | 4 | 10 | 4 | 12 |

It seems to make sense that the degree of ambiguity (not considering the case $n = 4$) can be computed with Euler's totient function, that is, the degree equals to $\varphi(n)$. Later we will give a short proof on this.

Now we are ready to set up additional formulas to describe the coordinates of the vertices of a regular $n$-gon, having its first vertices $P_0 = (0,0)$ and $P_1 = (1,0)$, and the remaining vertices $P_2 = (x_2, y_2), \ldots, P_{n-1} = (x_{n-1}, y_{n-1})$ are to be found. By using consecutive rotations and assuming $x = \cos(2\pi/n), y = \sin(2\pi/n)$, we can claim that

$$\begin{pmatrix} x_i \\ y_i \end{pmatrix} - \begin{pmatrix} x_{i-1} \\ y_{i-1} \end{pmatrix} = \begin{pmatrix} x & -y \\ y & x \end{pmatrix} \cdot \left( \begin{pmatrix} x_{i-1} \\ y_{i-1} \end{pmatrix} - \begin{pmatrix} x_{i-2} \\ y_{i-2} \end{pmatrix} \right)$$

and therefore

$$x_i = -xy_{i-1} + x_{i-1} + xx_{i-1} + yy_{i-2} - xx_{i-2}, \tag{1}$$
$$y_i = y_{i-1} + xy_{i-1} + yx_{i-1} - xy_{i-2} - yx_{i-2} \tag{2}$$

for all $i = 2, 3, \ldots, n - 1$.

## 3    A Manual Result on a Regular Pentagon

In this section we present a well-known statement on a regular 5-gon that can be obtained by using the formulas from the previous section.

**Theorem 1.** *Consider a regular pentagon (Fig. 1) with vertices $P_0, P_1, \ldots, P_{n-1}$. Let $A = P_0, B = P_2, C = P_1, D = P_3, E = P_0, F = P_2, G = P_1, H = P_4$. Let us define diagonals $d = AB, e = CD, f = EF, g = GH$ and intersection points $R = d \cap e, S = f \cap g$. Now, when the length of $P_0P_1$ is 1, then the length of $RS$ is $\frac{3-\sqrt{5}}{2}$.*

This result is well-known from elementary geometry, but here we provide a proof that uses the developed formulas from Sect. 2. We will use the variables $x_0, x_1, x_2, x_3, x_4$ for the $x$-coordinates of the vertices, $y_0, y_1, y_2, y_3, y_4$ for the $y$-coordinates, and $x$ and $y$ for the cosine and sine of $2\pi/5$, respectively. Points $P_0$ and $P_1$ will be put into $(0,0)$ and $(1,0)$.

By using Table 1 and Eqs. (1) and (2), we have the following hypotheses:

$$h_1 = 4x^2 + 2x - 1 = 0,$$
$$h_2 = x^2 + y^2 - 1 = 0,$$
$$h_3 = x_0 = 0,$$
$$h_4 = y_0 = 0,$$
$$h_5 = x_1 - 1 = 0,$$
$$h_6 = y_1 = 0,$$
$$h_7 = -x_2 + -xy_1 + x_1 + xx_1 + yy_0 - xx_0 = 0,$$
$$h_8 = -y_2 + y_1 + xy_1 + yx_1 - xy_0 - yx_0 = 0,$$
$$h_9 = -x_3 + -xy_2 + x_2 + xx_2 + yy_1 - xx_1 = 0,$$
$$h_{10} = -y_3 + y_2 + xy_2 + yx_2 - xy_1 - yx_1 = 0,$$
$$h_{11} = -x_4 + -xy_3 + x_3 + xx_3 + yy_2 - xx_2 = 0,$$
$$h_{12} = -y_4 + y_3 + xy_3 + yx_3 - xy_2 - yx_2 = 0.$$

Since $R \in d$ and $R \in e$, we can claim that

$$h_{13} = \begin{vmatrix} x_0 & y_0 & 1 \\ x_2 & y_2 & 1 \\ x_r & y_r & 1 \end{vmatrix} = 0, h_{14} = \begin{vmatrix} x_1 & y_1 & 1 \\ x_3 & y_3 & 1 \\ x_r & y_r & 1 \end{vmatrix} = 0,$$

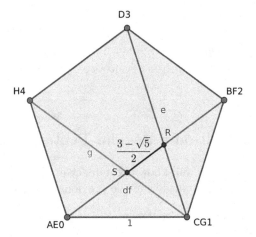

**Fig. 1.** A well-known theorem on a regular pentagon (for convenience we use only the indices of the points in the figure, that is, $0, 1, \ldots, n - 1$ stand for $P_0, P_1, \ldots, P_{n-1}$, respectively)

where $R = (x_r, y_r)$. Similarly,

$$h_{15} = \begin{vmatrix} x_0 & y_0 & 1 \\ x_2 & y_2 & 1 \\ x_s & y_s & 1 \end{vmatrix} = 0, h_{16} = \begin{vmatrix} x_1 & y_1 & 1 \\ x_4 & y_4 & 1 \\ x_s & y_s & 1 \end{vmatrix} = 0,$$

where $S = (x_s, y_s)$. Finally we can define the length $|RS|$ by stating

$$h_{17} = |RS|^2 - \left( (x_r - x_s)^2 + (y_r - y_s)^2 \right) = 0.$$

From here we can go ahead with two methods:

1. We directly prove that $|RS| = \frac{3-\sqrt{5}}{2}$. As we will see, this actually does not follow from the hypotheses, because they describe a different case as well, shown in Fig. 2. That is, we need to prove a weaker thesis, namely that $|RS| = \frac{3-\sqrt{5}}{2}$ or $|RS| = \frac{3+\sqrt{5}}{2}$, which is equivalent to

$$\left( |RS| - \frac{3 - \sqrt{5}}{2} \right) \cdot \left( |RS| - \frac{3 + \sqrt{5}}{2} \right) = 0.$$

Unfortunately, this form is still not complete, because $|RS|$ is defined implicitly by using $|RS|^2$, that is, if $|RS|$ is a root, also $-|RS|$ will appear. The correct form for $t$ is therefore

$$t = \left( |RS| - \frac{3 - \sqrt{5}}{2} \right) \cdot \left( |RS| - \frac{3 + \sqrt{5}}{2} \right) \cdot$$

$$\left( -|RS| - \frac{3 - \sqrt{5}}{2} \right) \cdot \left( -|RS| - \frac{3 + \sqrt{5}}{2} \right) = 0,$$

that is, after expansion,

$$t = (|RS|^2 - 3|RS| + 1) \cdot (|RS|^2 + 3|RS| + 1) = |RS|^4 - 7|RS|^2 + 1 = 0.$$

Proving the thesis can be done by denying $t$, with inserting $t \cdot z - 1 = 0$ into the equation system $\{h_1, h_2, \ldots, h_{17}\}$ and obtaining a contradiction. This approach is based on the Rabinowitsch trick, introduced by Kapur in 1986 (see [11]).

2. We can also discover the exact value of $|RS|$ by eliminating all variables from the ideal $\langle h_1, h_2, \ldots, h_{17} \rangle$, except $|RS|$. We will follow this second method, suggested by Recio and Vélez in 1999 (see [2]).

Let us emphasize that the first method can be used only *after* one has a conjecture already. In contrast, the second method can be used *before* having a conjecture, namely, to find a conjecture *and* its proof at the same time.

For the first method we must admit that in Wu's approach there is no way to express that the length of a segment is $\frac{3-\sqrt{5}}{2}$. Instead, we need to use its minimal polynomial, having integer (or rational) coefficients. Actually, $|RS|^2 - 3|RS| + 1$ is a minimal polynomial of both $\frac{3-\sqrt{5}}{2}$ and $\frac{3+\sqrt{5}}{2}$, and $|RS|^2 + 3|RS| + 1$ is of $-\frac{3-\sqrt{5}}{2}$ and $-\frac{3+\sqrt{5}}{2}$. In fact, given a length $|RS|$ in general, we need to prove that the equation $t = t_1 \cdot t_2 = 0$ is implied where $t_1$ and $t_2$ are the minimal polynomials of the expected $|RS|$ and $-|RS|$, respectively. Even if geometrically $t_1$ is implied, from the algebraic point of view $t_1 \cdot t_2$ is to be proven.

Also, we remark that $|RS|$ always appears to an even power in $t$.

Finally, when using the second method, by elimination (here we utilize computer algebra), we will indeed obtain that

$$\langle h_1, h_2, \ldots, h_{17} \rangle \cap \mathbb{Q}[|RS|] = \langle |RS|^4 - 7|RS|^2 + 1 \rangle.$$

## 3.1   Star-Regular Polygons

Before going further, we need to explain the situation with the star-regular pentagon in Fig. 2. Here we need to mention that the equation $h_1 = 4x^2 + 2x - 1 = 0$ describes not only $\cos(2\pi/5)$ but also $\cos(2 \cdot 2\pi/5)$, $\cos(3 \cdot 2\pi/5)$ and $\cos(4 \cdot 2\pi/5)$, however, because of symmetry, the first and last, and the second and third values are the same. (We can think of these values as the projections of $z_1, z_2, z_3, z_4$ on the real axis, where

$$z_j = (\cos(2\pi/5) + i\sin(2\pi/5))^j = \cos(j \cdot 2\pi/5) + i\sin(j \cdot 2\pi/5),$$

$j = 1, 2, 3, 4$.)

That is, in this special case (for $n = 5$) $h_1$ is a minimal polynomial of $\operatorname{Re} z_1 (= \operatorname{Re} z_4)$ and $\operatorname{Re} z_2 (= \operatorname{Re} z_3)$. By considering the formulas (1) and (2) we can learn that the rotation is controlled by the vector $(x, y)$, where $2\pi/n$ is the external angle of the regular $n$-gon. When changing the angle to a double, triple, ..., value,

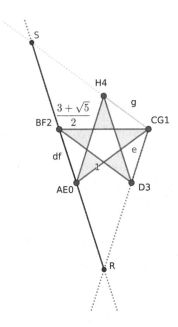

**Fig. 2.** A variant of the theorem in a star-regular pentagon

we obtain star-regular $n$-gons, unless the external angle describes a regular (or star-regular) $m$-gon ($m < n$).

This fact is well-known in the theory of regular polytopes [12], but let us illustrate this property by another example. When choosing $n = 6$, we have $h_1' = 2x - 1 = 0$ that describes $\cos(2\pi/6) = \cos(5 \cdot 2\pi/6)$. Now by considering $z_1', z_2', z_3', z_4', z_5'$ where

$$z_j' = \cos(j \cdot 2\pi/6) + i\sin(j \cdot 2\pi/6),$$

$j = 1, 2, 3, 4, 5$, we can see that $z_2'$ can also be considered as a generator for $\cos(1 \cdot 2\pi/3)$ (when projecting it on the $x$-axis) since $2 \cdot 2\pi/6 = 1 \cdot 2\pi/3$. That is, $z_2'(= \overline{z_4'})$ is not used when generating the minimal polynomial of $\cos(2\pi/6)$ (it occurs at the creation of the minimal polynomial of $\cos(2\pi/3)$), and this is the case also for $z_3'$ (because it is used for the minimal polynomial of $\cos(2\pi/2)$).

An immediate consequence is that $z_j'$ is used as a generator in the minimal polynomial of $\cos(2\pi/6)$ if and only if $j$ and 6 are coprimes, but since $\cos(2\pi/6) = \cos(5 \cdot 2\pi/6)$, only the first half of the indices $j$ play a technical role. In general, when $n$ is arbitrary, the number of technically used generators are $\varphi(n)/2$ (the other $\varphi(n)/2$ ones produce the same projections).

Finally, when considering the equation $x^2 + y^2 = 1$ as well, if $n \geq 3$, there are two solutions in $y$, hence the hypotheses describe *all* cases when $j$ and $n$ are coprimes (not just for the half of the cases, that is, for $1 \leq j \leq n/2$). Practically, the hypotheses depict not just the regular $n$-gon case, but also *all* star-regular $n$-gons. It is clear, after this chain of thoughts, that the number of cases is $\varphi(n)$ (which is the number of positive coprimes to $n$, less than $n$). From this immediately follows that the degree of ambiguity for $(\cos(2\pi/n), \sin(2\pi/n))$ is *exactly* $\varphi(n)$.

Also, it is clear that there exists essentially only one regular 5-gon and one star 5-gon (namely, $\{5/2\}$, when using the Schäfli symbol, see [12]). But these are just two different cases. The other two ones, according to $\varphi(5)$, are symmetrically equivalent cases. The axis of symmetry is the $x$-axis in our case.

On the other hand, by using our method, it is not always possible to distinguish between these $\varphi(n)$ cases:

1. $t = |RS|^2 - c$ where $c$ is a rational. In this case clearly $|RS| = \sqrt{c}$ follows.
2. Otherwise, the resulting polynomial $t$ is a product of two polynomials $t_1, t_2 \in \mathbb{Q}[|RS|]$, and the half of the union of their roots are positive, while the others are negative. On the other hand, the positive roots can be placed in several combinations in $t_1$ and $t_2$ in general:
   (a) In our concrete example there are two positive roots in $t_1$ and two negative ones in $t_2$. When considering similar cases, the positive roots can always occur in, say $t_1$, and the negative roots then in $t_2$. Albeit the elimination delivers the product $t = t_1 \cdot t_2$, clearly $t_2$ cannot play a geometrical role, therefore $t_1$ can be concluded.

   However, if $t_1$ contains more than one (positive) root, those roots cannot be distinguished. This is the case in our concrete example as well.
   (b) In general, $t_1$ may contain a few positive solutions, but $t_2$ may also contain some other ones. In such cases the positive solutions in $t_1$ and $t_2$ cannot be distinguished from each other.

   Such an example is the polynomial $t = t_1 \cdot t_2$ where $t_1 = |RS|^2 - |RS| - 1$ and $t_2 = |RS|^2 + |RS| - 1$. It describes the length of the diagonal of a regular (star-) pentagon, namely both lengths $\frac{\sqrt{5}\pm 1}{2}$. Here $t_1$ contains one of the positive roots, namely $\frac{\sqrt{5}+1}{2}$, while $t_2$ the other one, $\frac{\sqrt{5}-1}{2}$. At the end of the day, only $t$ can be concluded, none of its factors can be dropped because both contain geometrically useful data.

## 4    Automated Discovery of Theorems

Obtaining new results randomly is one of the possible aims when observing regular polygons. But, luckily, this kind of discovery can be systematic when the different setups $\mathcal{S}$ are numbered consecutively. If there is a bijective map

$$S : \{0, 1, 2, \ldots, s - 1\} \to \mathcal{S},$$

there are some programmatical benefits for the processing of the cases:

1. A database $D : \{0, 1, 2, \ldots, s - 1\} \to \{\text{true, false}\}$ can be maintained. Here for each $k \in \mathbb{N}_0, k < s$ there is an explicitly defined setup $S(k) \in \mathcal{S}$, and it can be saved as a database entry $D(k)$ if the check was already performed or not. If the computation loop needs to be suspended or stopped due to the high amount of computations for a given $k$, it can be restarted at the same value $k$ in a next loop, independently from the first run.

2. This also supports parallel or distributed computing. The number of cases $k$ can be then split and the setups can be divided among several processors or computers.
3. The distributed computation can also be controlled via a centralized Internet application that communicates with the clients, assigns the task to them, collects the results, and updates the central database. Of course, not only the success of the performed computations should be stored, but also their results, by using a map $D' : \{0, 1, 2, \ldots, s - 1\} \to \ldots$ that has a sophisticated output data structure.

### 4.1    A Bijective Mapping

In our approach we assume that a regular $n$-gon is to be studied. It has $\binom{n}{2}$ diagonals (including the sides). From these we select two different ones, $d$ and $e$ (the order of selection does not matter) to designate their intersection point $R$. That is, the number of possible selections are $\binom{\binom{n}{2}}{2}$. On the other hand, to designate another intersection point $S$ from another combination of the diagonals, we finally have

$$\binom{\binom{\binom{n}{2}}{2}}{2} \tag{3}$$

different selections for the segment $RS$. When expanding the formula (3) we learn that the number of cases is

$$\frac{n^8 - 4n^7 + 2n^6 + 8n^5 - 15n^4 + 12n^3 + 12n^2 - 16n}{128} \sim \frac{n^8}{128},$$

that is, $s$ is equal to $n^8/128$ asymptotically.

It would be useful to find a formula for $S(k)$ to compute $RS$ quickly. For the first step we will construct another map

$$c : \{0, 1, 2, \ldots, \binom{m}{2} - 1\} \to \binom{\{0, 1, 2, \ldots, m - 1\}}{2}$$

where $\binom{\{0,1,2,\ldots,m-1\}}{2}$ stands for the set of 2-combinations of the set $\{0, 1, 2, \ldots, m - 1\}$. Here we will assume that

$$c(0) = \{0, 1\},\ c(1) = \{0, 2\},\qquad c(2) = \{0, 3\}, \ldots,\ c(m - 2) = \{0, m - 1\},$$
$$c(m - 1) = \{1, 2\},\ c(m) = \{1, 3\},\ c(m + 1) = \{1, 4\}, \ldots, c(2m - 4) = \{1, m - 1\},$$
$$c(2m - 3) = \{2, 3\}, \ldots,$$

..., and finally $c\left(\binom{m}{2} - 1\right) = \{m - 2, m - 1\}$. To compute $c$ quickly, we consider the inverse map $c^{-1}$. It is clear that $c^{-1}(k, k+1) = (m-1)+(m-2)+\ldots+(m-k)$, that is, $\frac{(m-1)+(m-k)}{2} \cdot k = -\frac{1}{2}k^2 + k \cdot \frac{2m-1}{2} = p.$

Let us now assume that $p$ is given, and $k$ is to be computed. Clearly $-\frac{1}{2}k^2 + k \cdot \frac{2m-1}{2} - p = 0$, and using the quadratic equation solver formula,

$$k = \frac{\frac{1-2m}{2} \pm \sqrt{\left(\frac{2m-1}{2}\right)^2 - 2p}}{-1} = m - \frac{1}{2} \mp \sqrt{\left(m - \frac{1}{2}\right)^2 - 2p}.$$

Here obviously the subtraction should be chosen. By some further simple calculations finally we obtain the formula $c(p) = \{k, l\}$ where

$$k = \left\lfloor m - \frac{1}{2} - \sqrt{\left(m - \frac{1}{2}\right)^2 - 2p} \right\rfloor, \tag{4}$$

$$l = \frac{2p + k^2 - (2m - 3) \cdot k}{2} + 1. \tag{5}$$

This formula can be used then multiple times for $m = \left(\binom{n}{2}\right)$, $m = \binom{n}{2}$ and $m = n$.

**Example.** Let $n = 5$, then $s = \left(\binom{\binom{5}{2}}{2}\right) = 990$. We are interested in, say, the 678th case when observing all possible segments $RS$.

1. First we compute $\binom{5}{2} = 45 = m_1$. That is, we search for $c(678)$. By using formulas (4) and (5), we get $k = 19$ and $l = 33$.
2. Now we search for the 19th and 33th combinations of a set with $\binom{5}{2} = 10 = m_2$ elements. Using the same formulas, we get $k = 2, l = 5$ and $k = 4, l = 8$ values for $p = 19$ and $p = 33$, respectively.
3. Finally we search for the 2nd, 5th, 4th and 8th combinations of a set with $5 = m_3$ elements. Using the same formulas again, we get $k = 0, l = 3$, $k = 1, l = 3$, $k = 1, l = 2$ and $k = 2, l = 4$ values for $p = 2, 5, 4$ and $8$, respectively.

Lastly we conclude that the 678th case describes when $A = P_0$, $B = P_3$, $C = P_1$, $D = P_3$, $E = P_1$, $F = P_2$, $G = P_2$, $H = P_4$.

## 4.2   An Implementation

This automated algorithm has been recently implemented in the software tool RegularNGons [13].

The following input parameters can be used to fine tune its output:

- $n = \ldots$ defines the number of vertices in the regular polygon.
- $s$ and $e$ define the starting and ending cases (both are non-negative integers, less than the formula (3)).
- By adding $m = \ldots$ or $M = \ldots$ the minimal and maximal degrees of outputs can be controlled, respectively. By default $m = 1$ and $M = 2$, that is, either linear results or quadratic surds are collected.
- The parameter $u$ will force searching for results given as parameters. For example, $u = 2$ considers only the outputs that are of $|RS| = 2$.
- The option $S = 0$ tries to avoid checking cases that were already checked in a symmetrically equivalent position. When this is set, only the $A = 0$, $B \leq n/2$ cases will be checked. (Here, and from now on, we will use the indices of the points, that is, 0 stands for $P_0$, 1 for $P_1$, and so on.)

– When using parameter $f = 1$, once a length is found, no more results will be printed that have the same length.

The software tool runs in a modern web browser, for example, *Google Chrome* 64. It uses the *Giac* computer algebra system to compute eliminations (its *Web-Assembly* [14] version is used in an embedded way), and *GeoGebra* to visualize the obtained results on-the-fly—finally (or during the run) the results can be saved as a GeoGebra file.

The timing for a complete run for a given $n$-gon depends on the magnitude of $n$. For smaller $n$ values the complete run can be performed in seconds or minutes. For bigger $n$ values, a complete run may take several hours, or days, or even more. Some, yet unresolved memory issues in Giac may require multiple runs for bigger $n$ values.

Parts of a typical output of `RegularNGons` look like the following, when using inputs $n = 7$, $S = 0$ and parameter $f = 1$:

```
Welcome to RegularNGons (https://github.com/kovzol/RegularNGons)...
s can be incremented until 21945
n=7, s=4: A=0, B=1, C=0, D=2, E=0, F=1, G=1, H=2: {RS^2-1}, {{RS=1}}
n=7, s=124: A=0, B=1, C=0, D=2, E=1, F=3, G=2, H=6: {RS^2-2}, {{RS=(√2)}}
n=7, s=2113: A=0, B=1, C=2, D=3, E=0, F=5, G=1, H=6: {RS^2-4}, {{RS=2}}
Elapsed time: 0h 28m 40s
Finished after finding 3 solutions
11627 cases were not checked to ignore symmetry
```

This result will be recalled later in Theorem 3.

### 4.3    Some Results

We will find the following definition useful when presenting the statements that can be collected by using `RegularNGons`.

**Definition 1.**    – **Points of the first kind** of a regular $n$-gon are its vertices. We denote this set by $\mathcal{P}_1$.
–  **Segments of the first kind** of a regular $n$-gon are its sides and diagonals. We denote this set by $\mathcal{S}_1$.
–  **Points of the $k$-th kind** of a regular $n$-gon are the intersection points of its segments of the $(k-1)$-th kind. We denote this set by $\mathcal{P}_k$.
–  **Segments of the $k$-th kind** of a regular $n$-gon are the segments defined by its points of the $(k)$-th kind. We denote this set by $\mathcal{S}_k$.

By using this notion, in this paper we consider *segments of the second kind* of a regular $n$-gon. We remark that it makes sense to study segments of higher kinds in a regular $n$-gon. It is easy to see that a recursive formula can be given to determine the number of possible cases for the various kinds of points and segments of a regular $n$-gon:

**Proposition 1.**    – $|\mathcal{P}_1| = n$.
–  $|\mathcal{S}_1| = \binom{|\mathcal{P}_1|}{2}$.
–  $|\mathcal{P}_k| = \binom{|\mathcal{S}_{k-1}|}{2}$.
–  $|\mathcal{S}_k| = \binom{|\mathcal{P}_k|}{2}$.

*Proof.* By construction, these formulas are obvious.                          □

Now we present some results that were obtained by `RegularNGons`.

**Theorem 2.** *Given a regular 7-gon, there are 42 segments of its second kind that are of length 2, shown in Fig. 3.*

**Fig. 3.** Some properties of a regular heptagon (Color figure online)

*Proof.* By exhausting all $|S_2| = 21945$ cases, there exist exactly the cases as presented. (The running time on a modern PC was about 1 h and 15 min.)    □

The 42 different cases can be classified into 3 substantially different groups, shown in green, red and magenta in Fig. 3. Because of symmetry, each substantially different segment have 6 rotated copies and a mirrored copy with 6 other rotated copies. In total there are $7 + 7 = 14$ elements of the groups. In the figure only 2 representants are colored in each group (they are mirror images), the others are all blue.

**Theorem 3.** *Let us consider all segments of the second kind in regular 7-, 9- and 11-gons, having side lengths 1. Then the following hold:*

- *In a regular heptagon the only rational lengths in $S_2$ are 1 and 2, and the only quadratic surd is $\sqrt{2}$.*
- *In a regular nonagon the only rational lengths in $S_2$ are 1, 2 and 3, and the only quadratic surds are $\sqrt{3}$ and $\sqrt{7}$.*

– *In a regular 11-gon the only rational lengths in $S_2$ are 1, 2, and the only quadratic surd is $\sqrt{3}$.*

*Proof.* By exhaustion, using a computer.

Some other results can be found at https://www.geogebra.org/m/AXd5ByHX. The software tool `RegularNGons` can be launched on-line at http://prover-test.geogebra.org/~kovzol/RegularNGons/. An example run can be started to request solving the case $n = 5$ by invoking the URL http://prover-test.geogebra.org/~kovzol/RegularNGons/?n=5.

## 5   Conclusion and Future Work

We presented an automated way on obtaining various new theorems on regular polygons, based on the work of [1,2,9]. Some results may be not elegant from some perspectives, but others can be, especially those where rational numbers or quadratic surds appear.

On the other hand, classifying results to be "elegant" or not, can be very subjective. For this reason our implementation `RegularNGons` is able to filter the results by various criteria. Extending the existing filters by other ones, for example, by searching for good approximations of a given real number, could be another step forward in this direction.

We emphasize that enumerating the possible cases is a crucial detail in our work. We found a simple way to map the first non-negative numbers to the possible cases bijectively, however, some cases in our definitions still yield the same segment $RS$. This case occurs when $R$ or $S$, or both, are among the vertices of the $n$-gon. We will try to address this issue in the future.

Automatizing theorems on regular $n$-gons can be further developed by considering segments of higher kinds, not just of the second. The number of cases to check—according to Proposition 1—grows rapidly. For the third kind, the number of cases is asymptotic to $n^{16}/2^{15}$, for instance, for $n = 5$ the number of segments of the third kind is $119,831,804,235$. That is, there can be lots of new theorems to explore! Of course, many of them may be uninteresting, but some of them may be of interest. (Here we did not mention that considering the diagonals of $n$- and $m$-gons at once, may lead to a very high amount of new statements as well.)

The high number of cases calls for distributed computing. Our further plan is to extend our software tool to be a centralized system that assigns interesting tasks to the contributors' computers. By this way the idle computer time could be used to "mine" new, interesting geometry theorems.

Also, using efficient methods on filtering the high amount of cases may avoid combinatorial explosion. Algorithms in artificial intelligence, in particular, in machine learning, may help future versions of the tool to obtain useful results also for higher $n$ in a reasonable time.

**Acknowledgments.** The author was partially supported by a grant MTM2017-88796-P from the Spanish MINECO (Ministerio de Economía y Competitividad) and the ERDF (European Regional Development Fund). Many thanks to Tomás Recio and Francisco Botana for their valuable comments and suggestions. The JavaScript implementation in `RegularNGons` was supported by Gábor Ancsin and the GeoGebra Team.

# References

1. Wu, W.T.: On the decision problem and the mechanization of theorem-proving in elementary geometry (1984)
2. Recio, T., Vélez, M.P.: Automatic discovery of theorems in elementary geometry. J. Autom. Reason. **23**, 63–82 (1999)
3. Colton, S., Bundy, A., Walsh, T.: On the notion of interestingness in automated mathematical discovery. Int. J. Hum.-Comput. Stud. **53**, 351–375 (2000)
4. Chou, S.C.: Mechanical Geometry Theorem Proving. Springer, Dordrecht (1987)
5. Wantzel, P.: Recherches sur les moyens de reconnaître si un problème de géométrie peut se résoudre avec la règle et le compas. J. Math. Pures Appl. **1**, 366–372 (1837)
6. Sethuraman, B.: Rings, Fields, and Vector Spaces: An Introduction to Abstract Algebra via Geometric Constructibility. Springer, New York (1997). https://doi.org/10.1007/978-1-4757-2700-5
7. Pierpont, J.: On an undemonstrated theorem of the disquisitiones arithmeticæ. Bull. Am. Math. Soc. **2**, 77–83 (1895)
8. Gleason, A.M.: Angle trisection, the heptagon, and the triskaidecagon. Am. Math. Mon. **95**, 185–194 (1988)
9. Watkins, W., Zeitlin, J.: The minimal polynomial of $\cos(2\pi/n)$. Am. Math. Mon. **100**, 471–474 (1993)
10. Lehmer, D.H.: A note on trigonometric algebraic numbers. Am. Math. Mon. **40**, 165–166 (1933)
11. Kapur, D.: Using Gröbner bases to reason about geometry problems. J. Symb. Comput. **2**, 399–408 (1986)
12. Coxeter, H.S.M.: Regular Polytopes, 3rd edn. Dover Publications, New York (1973)
13. Kovács, Z.: RegularNGons. A GitHub project (2018). https://github.com/kovzol/RegularNGons
14. Haas, A., et al.: Bringing the web up to speed with WebAssembly. In: Proceedings of the 38th ACM SIGPLAN Conference on Programming Language Design and Implementation. Association for Computing Machinery, pp. 185–200 (2017)

# Revealing Bistability in Neurological Disorder Models By Solving Parametric Polynomial Systems Geometrically

Changbo Chen[1,2] and Wenyuan Wu[1,2]

[1] Chongqing Key Laboratory of Automated Reasoning and Cognition,
Chongqing Institute of Green and Intelligent Technology,
Chinese Academy of Sciences, Chongqing, China
{chenchangbo,wuwenyuan}@cigit.ac.cn
[2] University of Chinese Academy of Sciences, Beijing, China

**Abstract.** Understanding the mechanisms of the brain is a common theme for both computational neuroscience and artificial intelligence. Machine learning technique, like artificial neural network, has been benefiting from a better understanding of the neuronal network in human brains. In the study of neurons, mathematical modeling plays a vital role. In this paper, we analyze the important phenomenon of bistability in neurological disorders modeled by ordinary differential equations in virtue of our recently developed method for solving bi-parametric polynomial systems. Unlike the algebraic symbolic approach, our numeric method solves parametric systems geometrically. With respect to the classical bifurcation analysis approach, our method naturally has good initial points thanks to the critical point technique in real algebraic geometry.

Special heuristic strategies are proposed for addressing the multi-scale problem of parameters and variables occurring in biological models, as well as taking into account the fact that the variables representing concentrations are non-negative. Comparing with its symbolic algebraic counterparts, one merit of this geometrical method is that it may compute smaller boundaries.

## 1 Introduction

Due to the tremendous increase in computing power, a machine learning approach named artificial neural network is enjoying a renaissance. Design of artificial neural network has been being inspired by advances in neuroscience and it is believed that a better understanding of the mechanisms of human brain and nervous system will play a vital role for the advent of more powerful artificial intelligence technology [13]. The study of human brain and nervous system is also the main subject of computational neuroscience [1,2,11,21]. Different from the "black-box" approach in machine learning, explicit mathematical models were built and analyzed in this discipline.

© Springer Nature Switzerland AG 2018
J. Fleuriot et al. (Eds.): AISC 2018, LNAI 11110, pp. 170–180, 2018.
https://doi.org/10.1007/978-3-319-99957-9_11

Historically, different models were proposed for studying neurons [11]. In this paper, we are particularly interested in neurological disorder models [21]. A study of the underlying mechanism is important for understanding various modalities of learning, such as long-term memory [20,23].

In general, dynamical system defined by ordinary different equations (ODEs) is a powerful tool for modeling biochemical networks [10]. The dominant approach for solving these systems is numerical simulation [24], which can handle large systems of equations. For small size problems, bifurcation analysis [12,16] is a very useful tool for understanding the role that parameters play.

The study of equilibria and their stability can be cast to an algebraic problem [19,26], which creates opportunities for symbolic tools for solving parametric polynomial systems, such as cylindrical algebraic decomposition [6,8], the border polynomial approach [29], the discriminant variety approach [17], real comprehensive triangular decomposition [7], and so on. Symbolic methods for handling special biological systems also exist, for instance in [3,15]. In [5], we proposed a numerical approach for solving bi-parametric polynomial systems. This approach is essentially geometrical, which is based on curve tracing and projection of points rather than elimination. This approach is further generalized to computing stability boundary of dynamical systems defined by ODEs and applied directly to analyzing stability of biological systems [4].

Two important characteristics peculiar to biological models were not exploited in [4], namely the multiscale problem and non-negative requirement of variables and parameters. In this paper, we introduced heuristics such as variable rescaling and restricted curve tracing to address them. We re-examine two neurological disorder models studied in [9,20] but provide improved or new results by analyzing the bistability phenomenon in the models. To overcome the difficulty of traditional bifurcation analysis method for finding initial starting points, we reply on critical point techniques [14,22,28] in real algebraic geometry and homotopy continuation methods in numerical algebraic geometry [18,25] to find at least one witness point for each connected component of fold and Hopf bifurcation curves. As illustrated by the two models in Sect. 3, with respect to symbolic methods, our method has the advantage of producing smaller boundary since it only traces positive branches of the bifurcation in real space rather than computing the Zariski closure of the projection of the bifurcation curve in complex space.

## 2    Methodology

In this section, we first briefly review the theory introduced in [4,5] for computing the fold and Hopf bifurcation boundaries of dynamical systems. Then we present new strategies tuned for biological systems to enhance the algorithm in [4].

### 2.1    Basic Theory

Throughout this section, we consider continuous dynamical systems defined by autonomous ODEs of the form $\dot{x} = F(x, u)$, where $x = (x_1, \ldots, x_m)$

are unknowns, $u = (u_1, u_2)$ are parameters independent of time $t$, and $F = (F_1, \ldots, F_m)$ are rational functions of $\mathbb{R}(u, x)$, called the *vector field* of the system.

Let $\mathcal{J}$ be the Jacobian matrix of $F$ with respect to $x$. The following system defines the fold bifurcation [16], which is a bifurcation where the equilibrium has zero eigenvalue:

$$\{F = 0, \mathcal{J}v = 0, \alpha v - 1 = 0\}, \tag{1}$$

where $v = v_1, \ldots, v_m$ is a vector of auxiliary variables and $\alpha$ is a random vector of $\mathbb{R}^m$ to avoid $v = 0$. This system has $2m + 1$ equations and $2m + 2$ variables. "Generically" it defines a one-dimensional curve, called *fold bifurcation curve*. To avoid $v$ approaching to infinity or a large number, sometimes it is better to replace $\alpha v - 1 = 0$ by $vv - 1 = 0$ in Eq. (1).

Another defining system for fold bifurcation is $\{F = 0, \det(\mathcal{J}(F)) = 0\}$, which is suitable for symbolic solvers. But for our method, Eq. (1) is usually preferred for better numerical stability and taking advantage of sparsity.

Let $P(x, u)$ be the vector of the numerators of $F(x, u)$. Assume that the denominators of $F$ never vanish (which is usually automatically satisfied for biological systems due to the natural requirement that all the variables should take non-negative values). It is shown in [4] that one can safely replace $F$ by $P$ in the above two systems.

Next we derive the defining system for Hopf bifurcation [16], which is a bifurcation where the equilibrium has a pair of purely imaginary eigenvalues. Suppose that $\mathcal{J}$ has a pair of purely imaginary eigenvalues $\lambda = \pm \omega i$. Let $(\mu + i\nu)$ be the corresponding eigenvectors. Then we have $\mathcal{J}(\mu + i\nu) = (\omega i)(\mu + i\nu)$, which implies that $\mathcal{J}\mu = -\omega\nu$ and $\mathcal{J}\nu = \omega\mu$ hold. Thus a defining system for Hopf bifurcation can be defined as below

$$\begin{cases} F = 0 \\ \mathcal{J}\mu = -\omega\nu \\ \mathcal{J}\nu = \omega\mu \\ \alpha\mu - \beta\nu = 1 \\ \beta\mu + \alpha\nu = 0, \end{cases} \tag{2}$$

where both $\mu$ and $\nu$ are additional vectors of $m$ variables, $\omega$ is an additional scalar variable, and $\alpha$ and $\beta$ are random real vectors of $\mathbb{R}^m$ to avoid $\mu + i\nu = 0$. This system has $3m+2$ equations and $3m+3$ variables. "Generically", it defines a one-dimensional curve, called *Hopf bifurcation curve*. Note that in Eq. (2), when $\omega \neq 0$, it defines exactly the Hopf bifurcation. When $\omega = 0$, it defines exactly the fold bifurcation [4]. Another typical defining system for Hopf bifurcation is: $\{F = 0, \Delta_{m-1}(F) = 0\}$, where $\Delta_{m-1}(F)$ is the $(m-1)$-th Hurwitz determinant of $\mathcal{J}(F)$. This defining system is usually used for symbolic solvers as it does not introduce extra variables. Note that this defining system may contain points where $\mathcal{J}(F)$ has eigenvalues of opposite signs.

Let $\pi : \mathbb{R}^{m+2} \to \mathbb{R}^2$ be the projection defined by $\pi(x_1, \ldots, x_m, u_1, u_2) = (u_1, u_2)$. Let $R$ be a bounding box of the parametric space $(u_1, u_2)$. Let $P_F$ be the vector of numerators of $F$. Let $F_H$ be the right hand side of Eq. (2), which defines Hopf bifurcation. Let $P_H$ be the vector of numerators of $F_H$. Let $B_H := \pi(V_{\mathbb{R}}(P_H))$. Then $B_H$ is the fold and Hopf bifurcation boundaries of the vector field $F$. Note that when parameters take values crossing $B_H$, the sign of the real part of some eigenvalue of an equilibrium may change. So is stability of the equilibrium. We call $B_H$ the *stability boundary of F*.

Assuming that $P_F, P_H$ and $R$ satisfy the following assumptions:

- $(A_1)$ The set $V_{\mathbb{R}}(P_F) \cap \pi^{-1}(R)$ is compact.
- $(A_2)$ We have $\dim V_{\mathbb{R}}(P_H) = 1$.
- $(A_3)$ At each regular point of $V_{\mathbb{R}}(P_H)$, the Jacobian of $P_H$ has full rank.

Then the following property holds.

**Proposition 1** [4]. *Let $B_H$ be the stability boundary of the vector field $F(x, u)$ restricted to some bounding box $R$. Then $R \setminus B_H$ is divided into finitely many connected components, called cells, such that in each cell, the number of equilibria of $F(x, u)$ does not change. Moreover, in every given cell, each equilibrium is a smooth function of u with stability unchanged.*

**Remark 1.** *As argued in bifurcation analysis, "generically", the assumptions $A_2$ and $A_3$ are satisfied. It is argued in [4] that we can force $A_1$ to be satisfied for biological systems modeled by dynamical system $\dot{x} = F(x, u)$, since the variables $x$ usually denote concentrations of biological substances, which are non-negative and bounded by conservation laws. More precisely, in addition to $B_H$, one should also computes boundaries corresponding to constraints $0 \leq x_i \leq b_i$, namely $B_i := \pi(V_{\mathbb{R}}(P_F, x_i))$, $i = 1, \ldots, m$ and $B_{i+m} := \pi(V_{\mathbb{R}}(P_F, x_i - b_i))$, $i = 1, \ldots, m$.*

### 2.2  Computing Stability Boundary of Biological Systems

Next we present a numeric algorithm for computing the stability boundary of biological systems modeled by the dynamical system $F(x, u)$. It is specially tuned for biological systems and improves the algorithm in our earlier work [4]. See Remark 2 for details.

Let RealWitnessPoint be the routine introduced in [27] for computing a set of witness points $W$ of a real variety $V_{\mathbb{R}}(P_H)$ satisfying Assumption $(A_3)$. Recall that a set of witness points $W$ of a real variety $V$ is a finite subset of $V$ such that $W$ has non-empty intersection with every connected component of $V$. The basic idea of this routine is to introduce a random hyperplane $L$. Then "roughly speaking" the witness points of $V_{\mathbb{R}}(P_H)$ either belong to $V_{\mathbb{R}}(P_H) \cap L$ or are the critical points (points attaining local minima) of the distance from the connected components of $V_{\mathbb{R}}(P_H)$ to $L$.

**Algorithm** StabilityBoundary
Input: a bi-parametric biological system defined by $\dot{x} = F(x_1, \ldots, x_m, u_1, u_2)$, where $F$ is a vector of $m$ rational functions with real coefficients; a bounding box $R$ in the first quadrant of the $(u_1, u_2)$-plane.
Output: an approximation of the stability boundary of the vector field $F$ restricted to $R$.

Steps:

1. Let $F_H$ be the left hand side of Eq. (2) and let $P_H$ be the vector of numerators of $F_H$.
2. Choose a random point $u$ in $R$ and compute $S := P_H(u)^{-1}(0) \cap \mathbb{R}^m$ by homotopy continuation method [18].
3. Based on solutions in $S$, rescale each $x_i$ and each $u_i$ to the range $[0, \ell]$, where $\ell$ is a small integer between 1 and 10.
4. Rescale $R$ accordingly to $R'$. Choose a big integer $K \gg \ell$ and set $R_2 := R' \times [0, K]^m$.
5. Set $W := \emptyset$.
6. Compute the intersection of $V_{\mathbb{R}}(P_H)$ with $\partial R_2$ by a homotopy continuation method and add the points into $W$.
7. Compute RealWitnessPoint$(P_H)$ and add the points into $W$.
8. For $p \in W$, starting from $p$, follow both directions of the tangent line of $V_{\mathbb{R}}(P_H)$ at $p$, trace the curve $P_H$ by a prediction-projection method, until a closed curve is found or a boundary of $R_2$ is met.
9. Return the projections of the traced points in $R$.

**Remark 2.** Comparing with our algorithm in [4] for general dynamical systems, this algorithm is specially tuned for biological systems. In particular, in Steps 2 and 3, we apply a rescaling strategy considering the factor that variables and parameters in biological systems usually have different scales. In Step 4, we take into account the fact that the variables of biological systems are non-negative. Introducing the box $R_2$ has two advantages. First it will force the following curve tracing to be restricted in a positive box and thus avoid computing projections of bifurcation points with negative coordinates, which has no biological meanings. Secondly it will compute an approximation of the projection of positive infinity points. This is one key reason why our numeric geometrical method produces smaller boundaries than symbolic algebraic methods, which first computes Zariski closure of bifurcation boundaries in complex space. Another key reason is we use a defining system encoding exactly the fold and Hopf bifurcation curve and trace the curve in real space. It is usually infeasible for symbolic method to use this defining system due to the fact that almost twice more auxiliary variables are introduced.

**Remark 3.** *If only one parameter effectively appears in $F(x, u)$. Then computing the stability boundary boils down to computing zero-dimensional systems defined by Eq. (2) and $\{F, \partial R_2\}$.*

## 3    Two Examples

In this section, we analyze the bistability of two neurological disorder models in detail by means of the algorithm presented in last section. In particular, the strategies of rescaling and tracing only positive branches are illustrated.

## 3.1   Alzheimer's Disease Model

In [9], the authors proposed a model for studying Alzheimer's disease. Alzheimer's disease (AD) is a progressive neurodegenerative disorder characterized by progressive and irreversible cognitive decline. The pathogenesis of AD is only partially understood and there is no cure. The bistability is the property of the coexistence between a stable steady-state characterized by low levels of $Ca^{2+}$ and $A\beta$ (corresponding to a healthy situation) and another stable steady-state where the levels of both compounds are high (corresponding to a pathological situation). The study would like to reveal the fact that appropriate perturbations of various kinetic parameters can lead to a switch from the healthy to the pathological state.

The model is described by the following ODEs:

$$\begin{aligned}
\frac{da}{dt} &= V_1 + \frac{V_a c^n}{K_a{}^n + c^n} - k_1 a \\
\frac{dc}{dt} &= V_2 + k_b a^m - k_2 c,
\end{aligned} \tag{3}$$

where the suggested values of the parameters in [9] is $V_1 = \frac{13}{2000}$, $V_2 = 4$, $V_a = 1/20$, $K_a = 120$, $k_b = 1/5$, $k_1 = \frac{1}{100}$, $k_2 = 1/10$, $m = 4$, $n = 2$. Let $F$ be the right hand-side of the above ODE. We notice that it is impossible for the above system to have Hopf bifurcation since the trace of the Jacobian matrix $\mathcal{J}_F$ is negative no matter what values the parameters take. Thus to compute the stability boundary, it is enough to compute the fold bifurcation boundary.

First we free two parameters $V_1$ and $V_2$ while the other parameters take the suggested values. Now Eq. (1) defines a one-dimensional fold curve. Before solving Eq. (1), we first randomly set $V_1 = 0.7926710114$, $V_2 = 0.5581108504$. Solving $F$, we get one solution $a = 5.778935983$, $c = 2286.410222$. Since the value of $c$ is pretty high, we rescale $c = 1000c'$. Similarly we rescale $V_1 = V_1'/100$ and $V_2 = 10V_2'$. To take into account the infinity boundary and non-negative boundary, we plot the curve defined by Eq. (1) in the box $(V_1', V_2', a, c') \in [0, 2] \times [0, 2] \times [0, K] \times [0, K]$, where $K$ is a big number, say $10^4$. Finally, we rescale the values back and obtain the following fold boundary depicted in left subfigure of Fig. 1, which is also the border curve of $F$ in parameter space $(V_1, V_2)$. The number of (asymptotically) stable equilibria is also displayed. Similarly, we obtain stability boundary in parameter space $(V_a, k_b)$, depicted by the right subfigure of Fig. 1.

Finally, we free one parameter $K_a$ while using default values for other parameters, then Eq. (1) is a zero-dimensional system. After rescaling the variable $c$ as before, we solve Eq. (1) by homotopy continuation method restricted to the box $[0, \infty] \times [0, 10000]^2$, we get two boundary points $K_a = 109.6069757$, $501.1488977$ and the system is bistable if $K_a$ is between them, which corrects the value $K_a \in (105, 520)$ given in [9]. Indeed, there is only one (non-negative) equilibrium in $(105, 109.6069757)$ or $(501.1488977, 520)$. Bifurcation diagrams depending on $K_a$ is illustrated in Fig. 2.

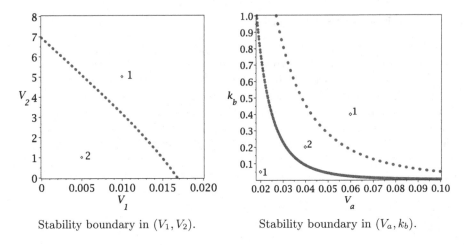

Stability boundary in $(V_1, V_2)$.          Stability boundary in $(V_a, k_b)$.

**Fig. 1.** Stability analysis of system (3).

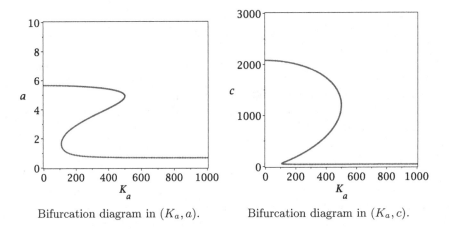

Bifurcation diagram in $(K_a, a)$.          Bifurcation diagram in $(K_a, c)$.

**Fig. 2.** Bifurcation diagrams of system 3.

### 3.2    The Protein Kinase $M\zeta$ Network

In [20], the author proposed a model for the protein kinase $M\zeta$ network. Protein kinase $M\zeta$ has drawn increasing attention as a molecule maintaining neuronal memory for an extremely long period of time. It can enhance excitatory postsynaptic currents and lead to the long-term potention of synapses. It is crucial for various modalities of learning, including spatial memory and fear conditioning. Bistable positive feedback loops of enzymatic reactions may provide a basis for cellular memory [9,20].

The model is described by the following ODEs:

$$\begin{aligned}
\frac{dP}{dt} &= \frac{j_1\,R(1-P)-P}{T_1} \\
\frac{dF}{dt} &= \frac{(Pj_3+j_2)(1-F)-F}{T_2} \\
\frac{dR}{dt} &= \frac{j_4\,F(P+s)(1-R)-R}{T_3},
\end{aligned} \tag{4}$$

where the default values of the parameters are: $T_1 = 1500, T_2 = 0.5, T_3 = 60, j_1 = 80, j_2 = 0.05, j3 = 0.5, j4 = 0.16, s = 0.003$.

First we free two parameters $j_2$ and $j_3$ while the other parameters take the suggested values. Now Eq. (2) defines a one-dimensional fold curve. First we rescale $j_1 = 100j_1'$ and $j_4 = j_4/10$. Since solving the right hand of Eq. (4) at random parameter values does not reveal equilibria with large coordinates, it is unnecessary to rescale the variables. To take into account the infinity boundary and non-negative boundary, we plot the curve defined by Eq. (2) in the box $(j_1', j_4, P, F, R) \in [0,3] \times [0,3] \times [0,K]^3$, where where $K$ is a big number, say $10^4$. Finally, we rescale the values back and obtain the following stability boundary depicted in left subfigure of Fig. 3. The number of (asymptotically) stable equilibria is also displayed. The right subfigure plots the stability boundary obtained by a symbolic approach by computing the discriminant variety [17] of the parametric system $\{F = 0, \Delta_{m-1}(F) \neq 0\}$, where the two redundant boundaries are projections of equilibria with negative coordinates. Similarly, we obtain stability boundary in parameter space $(j_2, j_3)$, which is exactly the same as Fig. 4C in [20].

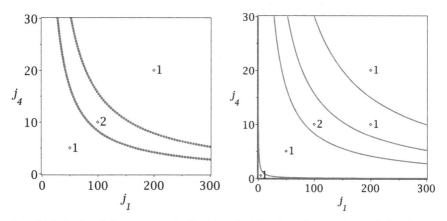

True (biological) stability boundary in $(j_1, j_4)$.    Stability boundary got by symbolic solver.

**Fig. 3.** Stability analysis of system 3.

Finally, we free one parameter $s$ while using default values for other parameters, then Eq. (2) is a zero-dimensional system. After rescaling the variable $c$ as before, we solve Eq. (2) by homotopy continuation method restricted to the box $[0, \infty] \times [0, 10000]^2$, we get one bifurcation point $s = 0.00984547072$ and the system is bistable if $s \in (0, 0.00984547072)$. Note that a direct use of symbolic approaches [17, 29] by solving the parametric system $\{F = 0, \Delta_{m-1}(F) \neq 0\}$ returns four points $0.009845470722, 0.4396431008, 2.100000000, 2.197170170$, while that last three does not have biological meanings as they are projections of equilibria with negative coordinates. A bifurcation diagram depending on $s$ is illustrated in Fig. 4. It is interesting to see that initially the system is bistable for $s < 0.00984547072$, where a stable equilibrium with high concentrations coexists with another stable one with low concentrations. As the stimulation $s$ increases, the system finally turns monotone and only one equilibrium with high concentrations is left and its concentration will never turn low even one reduces the value of $s$. This phenomenon is called irreversibility and the system behaves like maintaining memory.

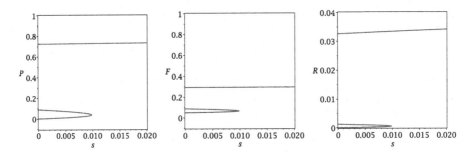

**Fig. 4.** Bifurcation diagrams of system 4.

## 4    Conclusion

In this paper, we presented a geometrical method for computing the stability boundary of biological systems, which may produce less redundant boundaries than symbolic methods. Its effectiveness was illustrated by revealing the bistability property of two neurological disorder models, which could be useful for a better understanding of molecular mechanisms of Alzheimer's disease and neuronal memory.

**Acknowledgements.** This work is partially supported by the projects NSFC (11471307, 11671377, 61572024), and the Key Research Program of Frontier Sciences of CAS (QYZDB-SSW-SYS026).

## References

1. Bard Ermentrout, G., Terman, D.H.: Mathematical Foundations of Neuroscience. Springer, Heidelberg (2010). https://doi.org/10.1007/978-0-387-87708-2

2. Bower, J.M. (ed.): 20 Years of Computational Neuroscience. Springer Series in Computational Neuroscience, vol. 9. Springer, Heidelberg (2013). https://doi.org/10.1007/978-1-4614-1424-7

3. Bradford, R.J., et al.: A case study on the parametric occurrence of multiple steady states. In: ISSAC 2017, pp. 45–52 (2017)

4. Chen, C., Wu, W.: A numerical method for analyzing the stability of bi-parametric biological systems. In: SYNASC 2016, pp. 91–98 (2016)

5. Chen, C., Wu, W.: A numerical method for computing border curves of bi-parametric real polynomial systems and applications. In: Gerdt, V.P., Koepf, W., Seiler, W.M., Vorozhtsov, E.V. (eds.) CASC 2016. LNCS, vol. 9890, pp. 156–171. Springer, Cham (2016). https://doi.org/10.1007/978-3-319-45641-6_11

6. Chen, C., Moreno Maza, M.: Quantifier elimination by cylindrical algebraic decomposition based on regular chains. J. Symb. Comput. **7**(5), 74–93 (2016)

7. Chen, C., Maza, M.M.: Semi-algebraic description of the equilibria of dynamical systems. In: Gerdt, V.P., Koepf, W., Mayr, E.W., Vorozhtsov, E.V. (eds.) CASC 2011. LNCS, vol. 6885, pp. 101–125. Springer, Heidelberg (2011). https://doi.org/10.1007/978-3-642-23568-9_9

8. Collins, G.E.: Quantifier elimination for real closed fields by cylindrical algebraic decompostion. In: Brakhage, H. (ed.) GI-Fachtagung 1975. LNCS, vol. 33, pp. 134–183. Springer, Heidelberg (1975). https://doi.org/10.1007/3-540-07407-4_17

9. De Caluwé, J., Dupont, G.: The progression towards Alzheimer's disease described as a bistable switch arising from the positive loop between amyloids and $Ca^{2+}$. J. Theor. Biol. **331**, 12–18 (2013)

10. Garfinkel, A., Shevtsov, J., Guo, Y.: Modeling Life: The Mathematics of Biological Systems. Springer, Heidelberg (2017). https://doi.org/10.1007/978-3-319-59731-7

11. Gerstner, W., Kistler, W.M., Naud, R., Paninski, L.: Neuronal Dynamics: From Single Neurons to Networks and Models of Cognition. Cambridge University Press, Cambridge (2014)

12. Govaerts, W.: Numerical Methods for Bifurcations of Dynamical Equilibria. Society for Industrial and Applied Mathematics, Philadelphia (2000)

13. Hassabis, D., Kumaran, D., Summerfield, C., Botvinick, M.: Neuroscience-inspired artificial intelligence. Neuron **95**(2), 245–258 (2017)

14. Hauenstein, J.D.: Numerically computing real points on algebraic sets. Acta Applicandae Mathematicae **125**(1), 105–119 (2012)

15. Hong, H., Tang, X., Xia, B.: Special algorithm for stability analysis of multistable biological regulatory systems. J. Symb. Comput. **70**, 112–135 (2015)

16. Kuznetsov, Y.A.: Elements of Applied Bifurcation Theory. Springer, Heidelberg (1995). https://doi.org/10.1007/978-1-4757-2421-9

17. Lazard, D., Rouillier, F.: Solving parametric polynomial systems. J. Symb. Comput. **42**(6), 636–667 (2007)

18. Li, T.Y.: Numerical solution of multivariate polynomial systems by homotopy continuation methods. Acta Numerica **6**, 399–436 (1997)

19. Niu, W., Wang, D.: Algebraic approaches to stability analysis of biological systems. Math. Comput. Sci. **1**(3), 507–539 (2008)

20. Ogasawara, H., Kawato, M.: The protein kinase Mζ network as a bistable switch to store neuronal memory. BMC Syst. Biol. **4**(1), 181 (2010)

21. Érdi, P., Bhattacharya, B.S., Cochran, A.L. (eds.): Computational Neurology and Psychiatry. SSB, vol. 6. Springer, Cham (2017). https://doi.org/10.1007/978-3-319-49959-8

22. Rouillier, F., Roy, M.F., Safey El Din, M.: Finding at least one point in each connected component of a real algebraic set defined by a single equation. J. Complex. **16**(4), 716–750 (2000)
23. Sacktor, T.C.: Memory maintenance by PKM$\zeta$ – an evolutionary perspective. Mol. Brain **5**(1), 31 (2012)
24. Schwartz, R.: Biological Modeling and Simulation. The MIT Press, Cambridge (2008)
25. Sommese, A., Wampler, C.: The Numerical Solution of Systems of Polynomials Arising in Engineering and Science. World Scientific Press, Singapore (2005)
26. Wang, D.M., Xia, B.: Stability analysis of biological systems with real solution classification. In: Kauers, M. (ed.) ISSAC 2005, pp. 354–361 (2005)
27. Wu, W., Reid, G.: Finding points on real solution components and applications to differential polynomial systems. ISSAC **2013**, 339–346 (2013)
28. Wu, W., Reid, G., Feng, Y.: Computing real witness points of positive dimensional polynomial systems. Theor. Comput. Sci. **681**, 217–231 (2017)
29. Yang, L., Xia, B.: Real solution classifications of a class of parametric semi-algebraic systems. In: A3L 2005, pp. 281–289 (2005)

# Early Ending in Homotopy Path-Tracking for Real Roots

Yu Wang[1(✉)], Wenyuan Wu[2(✉)], and Bican Xia[1(✉)]

[1] LMAM and School of Mathematical Sciences, Peking University, Beijing, China
yuxiaowang@pku.edu.cn, xbc@math.pku.edu.cn
[2] Chongqing Institute of Green and Intelligent Technology,
Chinese Academy of Sciences, Chongqing, China
wuwenyuan@cigit.ac.cn

**Abstract.** For computing only the isolated real solutions to a given polynomial system, a heuristic test is proposed to decide whether one homotopy path will converge to a real root, which is based on the asymptotic behavior of an angle defined by two points on the homotopy path. The data that the test requires is easily obtained from the points along the curve-following procedure in homotopy methods. The homotopy path-tracking may be sped up if we start the test before the endgames, since most divergent paths and paths heading to complex roots can be stopped tracking earlier and unnecessary endgames are avoided. Experiments show that the test works pretty well on tested examples.

## 1 Introduction

The homotopy continuation method was developed in 1970s [1,2] and has been greatly expanded and developed by many reseachers (see for example [3–8] for an overview of this area). An extensive description of polyhedral homotopy methods for sparse systems was given in [5] (see also [9–11]). A picture of the so-called *numerical algebraic geometry* up to the end of 2004 was described in [6]. The book [12] provided major developments of homotopy continuation methods up to 2013. Nowadays, homotopy continuation method has become one of the most reliable and efficient classes of numerical methods for finding the isolated solutions to a polynomial system and *numerical algebraic geometry*, based on homotopy continuation method, has been a blossoming area. There are many famous software packages implementing different algorithms in homotopy methods, including Bertini [12], Hom4PS-2.0 [13], NAG4M2 [14], PHCpack [15], etc.

Classical homotopy methods compute solutions in complex spaces, while in diverse applications, it is quite common that only real solutions have physical meaning. Computing real roots of an algebraic system is a difficult and fundamental problem in real algebraic geometry. In the field of symbolic computation,

The work is partly supported by the projects NSFC Grants 11471307, 61732001, 61532019 and CAS Grant QYZDB-SSW-SYS026.

J. Fleuriot et al. (Eds.): AISC 2018, LNAI 11110, pp. 181–194, 2018.
https://doi.org/10.1007/978-3-319-99957-9_12

there are some famous algorithms dealing with this problem. The cylindrical algebraic decomposition algorithm [16] is the first effective complete algorithm which has been implemented and used successfully to solve many real problems. However, in the worst case, its complexity is of doubly exponential in the number of variables. Based on the ideas of Seidenberg [17] and others, some algorithms for computing at least one point on each connected component of an real algebraic set were proposed through developing the formulation of critical points and the notion of polar varieties, see for example [18–21] and references therein. The idea behind is studying an objective function (or map) that reaches at least one local extremum on each connected component of a real algebraic set. For example, the function of square of the Euclidean distance to a randomly chosen point was used in [22,23]. These symbolic methods all have the complexity that limits their application in large problems. On the other hand, some homotopy based algorithms for real solving have been proposed in [24–32]. For example, a detailed description of the bifurcation phenomenon of the real homotopy paths was presented in [24]. In [30], a numerical homotopy method to find the extremum of Euclidean distance to a point as the objective function was presented. In [25,26], a combination of numerical algebraic geometry and sum of square programming was provided to verify the completeness of the real solution set of a real algebraic set. Algorithms for finding real roots based on homotopy methods and *interval analysis* was provided in [31]. The Euclidean distance to a plane was proposed as a linear objective function in [33].

Those homotopy based algorithms track all the paths defined by a given homotopy and select real roots from complex roots, or track carefully the real paths and deal with the complicate phenomenon of bifurcation. Motivated by the idea of truncation (see [34]), this paper proposes an early-stop test in path tracking procedure. By this test, we could identify the paths heading to isolated non-real solutions earlier and stop tracking them to save computation. One byproduct is that, if we start the test before the endgames of path tracking, we may even avoid the time-consuming procedure of tracking divergent paths. Compared to those truncation methods, our identification method is not rudely decided by a quantity of empirical data, but based on the asymptotic behavior of an angle defined by two points on the homotopy path which will be shown in Sect. 3. Experiments show that the test works pretty well.

The rest of this paper is organized as follows. Section 2 describes some preliminary concepts and symbols of homotopy continuation methods. Section 3 introduces an angle defined by two points on one homotopy path, and analyzes its asymptotic behaviors. It naturally leads to an EST (Early Stop Test) algorithm. The experimental performance of the EST is given in Sect. 4.

## 2    Preliminary

### 2.1    Homotopy Continuation Methods

Homotopy continuation methods aim to numerically compute the isolated solutions of a zero-dimensional square polynomial system $F(x)$ (as for positive dimensional systems, the idea of "witness sets" is developed, see [3]), mostly including two steps. Firstly, we look into the structure of the system $F(x)$ to construct a start system $G(x)$, and define the homotopy

$$H(x,t) = \gamma \cdot (1-t) \cdot G(x) + t \cdot F(x) = 0, t \in [0,1]$$

with $\gamma \in \mathbb{C}$ a random number. Such a homotopy in general (by Implicit Function Theorem) defines curves (one-dimensional manifold) of solutions. Secondly, we numerically trace the paths that originate at the solutions of the start system, $G(x)$, towards the solutions of the target system $F(x)$. The coefficient-parameter homotopy [35] detailed and ensured many aspects of the above approach. The *good properties* we expect for the system $G(x)$ are (borrowed from [36]):

1. (triviality) The solutions for $t = 0$ are trivial to find.
2. (smoothness) No singularities along the solution paths occur (the **gamma-trick**).
3. (accessibility) An isolated solution of multiplicity $m$ is reached by exactly $m$ paths.

The choice of the start system $G(x)$ is crucial in the first step. The root count of the start system is the number of paths tracked in the second step, and determines the efficiency of the homotopy. In the early applications [37–40], $G(x)$ is chosen from dense polynomials, where the number of paths equals the product of the degrees in the system. Multi-homogeneous homotopies were introduced and applied in [41, 42]. Linear-product start systems were developed in [43, 44]. Product structures exploiting approach was introduced in [45]. In the middle of 90's, because of Bernshtein's Theorem [46], the polyhedral homotopy was developed [9] to solve sparse systems efficiently. A combination of linear-product homotopy and polyhedral homotopy was introduced in [32] for systems derived from optimization problems.

## 2.2 Trackable Paths

In homotopy continuation methods, the notion of path tracking is fundamental, the following definition of trackable solution path is adapted from [47].

**Definition 1.** *Let $H(x,t) : \mathbb{C}^n \times \mathbb{C} \to \mathbb{C}^n$ be polynomial in $x$ and complex analytic in $t$, and let $x^*$ be nonsingular isolated solution of $H(x,0) = 0$. We say $x^*$ is* trackable *for $t \in [0,1)$ from 0 to 1 using $H(x,t)$ if there is a smooth map $\xi_{x^*} : [0,1) \to \mathbb{C}^n$ such that $\xi_{x^*}(0) = x^*$, and for $t \in [0,1)$, $\xi_{x^*}(t)$ is a nonsingular isolated solution of $H(x,t) = 0$. The solution path started at $x^*$ is said to be* convergent *if $\lim_{t\to 1} \xi_{x^*}(t) \in \mathbb{C}^n$, and the limit is called the* endpoint *of the path.*

When the start system $G(x)$ is appropriately chosen, each solution path originating at the isolated solutions of the system $G(x)$ is trackable. See [35, 47].

## 2.3  Path Tracking

Assuming that the solution path is $x(t)$ and $x(0) = x^*$, then we have $H(x(t), t) = 0$ for all $t$. By applying the operator $\frac{\partial}{\partial t}$ on the homotopy, via the chain rule, we have

$$0 \equiv \frac{dH(x(t), t)}{dt} = H_x(x(t), t)\frac{dx(t)}{dt} + H_t(x(t), t).$$

then

$$\begin{cases} \frac{dx(t)}{dt} = -H_x(x(t), t)^{-1} H_t(x(t), t) \\ x(0) = x^* \end{cases}.$$

This is an ordinary differential equation for $x(t)$, with initial value $x(0) = x^*$. We seek the value of the roots $x(1)$. When tracking the paths, the Predictor-Corrector method is widely used. There are three commonly used Predictors: the Secant Predictor, the Euler Predictor, and the Hermite Predictor. The Secant Predictor avoids solving a linear system at each step, but is less accurate. The Euler Predictor uses the tangent vectors at points along the path. And the Hermite Predictor uses the tangent vectors at two points along the path for cubic interpolation. See Fig. 1 for a comparison. When the new point is "Predicted", it comes to Newton's methods for "correction" step. For simplicity, we only go into details about the Euler Predictor-Corrector method as follows:

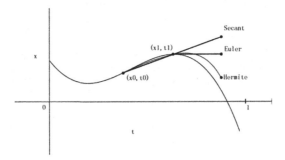

**Fig. 1.** Predictors: Secant, Euler and Hermite

**Step 1: Euler Prediction**

For a given step size $\delta > 0$, let $t_1 = t_0 + \delta < 1$ and

$$\widetilde{x}(t_1) = x(t_0) - \delta \cdot H_x(x(t_0), t_0)^{-1} H_t(x(t_0), t_0).$$

**Step 2: Newton's Correction**

For fixed $t_1$, $H(x, t_1) = 0$ becomes a system of $n$ equations in $n$ unknowns, and we have an approximate solution $\widetilde{x}(t_1)$. So, Newton's iteration can be employed, i.e.

$$x^{(k+1)} = x^{(k)} - H_x(x^{(k)}, t_1)^{-1} H(x^{(k)}, t_1), k = 0, 1, \ldots$$

with $x^{(0)} = \widetilde{x}(t_1)$. Eventually, a more accurate approximation value of $x(t_1)$ can be obtained. The following algorithm sketches the procedure of Euler Predictor-Corrector path tracing method:

---

**Algorithm 1.** Euler Prediction-Newton Correction

---

    **input** : $t_0 = 0$, $(x_0, t_0) \in \mathbb{C}^n \times \mathbb{R}$ and $\delta > 0$ such that $H(x_0, t_0) = 0$ and
        $t_0 + \delta < 1$
    **output**: $x_0$

1 **repeat**
2     $x^{(0)} := x_0 - \delta \cdot H_x(x_0, t_0)^{-1} H_t(x_0, t_0)$;
3     **repeat**
4        $x^{(1)} := x^{(0)} - H_x(x^{(0)}, t_0 + \delta)^{-1} H(x^{(0)}, t_0 + \delta)$;
5        $x^{(0)} := x^{(1)}$;
6     **until** *Convergence*;
7     $x_0 := x^{(0)}$;
8     $t_0 := t_0 + \delta$;
9 **until** *Path tracking is stopped*;
10 **return** $x_0$;

---

*Remark 1.* It should be mentioned that, in homotopy continuation methods, the paths tracked may be divergent (some coordinates will become arbitrarily large) at the end, i.e. when $t = 1$. For divergent paths, the Newton Corrector method will be invalid when $t$ is close to 1. For identifying those divergent paths, a procedure called end-game should be added [48,49]. But for simplicity, we omit it in Algorithm 1. For divergent path tracking, there will be no return at Line 10 in Algorithm 1.

## 3   Heuristic Early-Ending Homotopy Procedure

In this section, we give a description of our idea. First we give a theorem which characterizes the asymptotic behavior of the angle, defined by two points on a trackable path. Then, we propose a strategy for early ending in homotopy continuation methods for real roots. We make the assumption that the start system $G(x)$ has the abovementioned "good properties" (thus all the solution paths are trackable).

**Theorem 1.** *Let $x(t) : [0, 1) \to \mathbb{C}^n$ be one of the trackable paths of homotopy $H(x, t) = \gamma \cdot (1 - t) \cdot G(x) + t \cdot F(x) = 0, t \in [0, 1]$, $u \in [0, 1)$, $h > 0$ and $u + h \in [0, 1)$. Define $\alpha(u, h) := \langle (x'(u), 1), (x(u + h) - x(u), h) \rangle$, ( i.e. the angle between vector $(x'(u), 1)$ and vector $(x(u + h) - x(u), h)$). Then the following expansion holds:*

$$\alpha(u, h) = \kappa(u) \cdot h + O(h^3).$$

*Proof.* Because of the smoothness of the trackable path $x(t)$, we have the following expansion from Taylor's formula

$$x(u + h) = x(u) + x'(u) \cdot h + \frac{1}{2} \cdot x''(u) \cdot h^2 + \frac{1}{6} \cdot x'''(u) \cdot h^3 + O(h^4)$$

Hence

$$\frac{x(u + h) - x(u)}{h} = x'(u) + \frac{1}{2} \cdot x''(u) \cdot h + \frac{1}{6} \cdot x'''(u) \cdot h^2 + O(h^3)$$

$$\alpha(u, h) = \left\langle (x'(u), 1), (x'(u) + \frac{1}{2} \cdot x''(u) \cdot h + \frac{1}{6} \cdot x'''(u) \cdot h^2 + O(h^3), 1) \right\rangle$$

Then

$$\cos(\alpha(u, h)) = \frac{1 + x'(u)^T [x'(u) + \frac{1}{2} \cdot x''(u) \cdot h + \frac{1}{6} \cdot x'''(u) \cdot h^2 + O(h^3)]}{\|(x'(u), 1)\| \, \|(x'(u) + \frac{1}{2} \cdot x''(u) \cdot h + \frac{1}{6} \cdot x'''(u) \cdot h^2 + O(h^3), 1)\|}$$

$$= 1 - \kappa_1 \cdot h^2 + O(h^3)$$

$$(1)$$

where

$$\kappa_1 = \frac{1}{8} \cdot \frac{x''(u)^2 + x'(u)^2 \cdot x''(u)^2 - (x'(u) \cdot x''(u)^T)^2}{(1 + x'(u)^2)^2} > 0$$

Thus we have

$$\alpha(u, h) = \pm \arcsin(\sqrt{1 - \cos^2(\alpha(u, h))}) = \kappa h + O(h^3)$$

where $\kappa = \sqrt{2\kappa_1}$.    $\square$

*Remark 2.* 1. For the simplicity, in the proof of Theorem 1, we omit the Taylor expansion of functions: $\arcsin(x)$ at $x = 0$, $\sqrt{x}$ at $x = 1$, and $\frac{1}{x}$ at $x = 1$.
2. The reason we using the Taylor expansion of function $\arcsin(x)$ but not $\arccos(x)$, is that $\arccos(x)$ cannot be expanded as Taylor series at $x = 1$.
3. From the proof of Theorem 1, the constant $\kappa$ is independent of $h$, and depends smoothly on $u$.

For many polynomial systems derived from applications in science, engineering, and economics, the real roots are the only ones of interest. In the method of homtopy continuation, $H(x, t) = \gamma \cdot (1 - t) \cdot G(x) + t \cdot F(x) = 0$, the number of paths tracked is larger than the number of the isolated roots of the target system, especially for sparse systems. The extra paths are divergent as $t$ close to 1. The tracking of divergent paths is time-consuming in computation [48, 49]. And, most of the time, the number of the isolated roots (complex roots) is greater than the number of real roots. So, when interested in computing only the real roots, we propose a heuristic test for detecting the paths that are not convergent to real roots and stop tracking them earlier to save time. We look into how the size of the imaginary part of the points on the path is changing as $t$ changes.

If we denote $\tilde{x}$ to be the vector of the imaginary part of a complex vector $x$, i.e. $\tilde{x} = \frac{x - \bar{x}}{2 \cdot i}$, and by the same procedure of the proof of Theorem 1, we obtain:

**Theorem 2.** *Let $x(t) : [0,1) \to \mathbb{C}^n$ be one of the trackable paths of homotopy $H(x,t) = \gamma \cdot (1-t) \cdot G(x) + t \cdot F(x) = 0, t \in [0,1], u \in [0,1), h > 0$ and $u+h \in [0,1)$. Define $\widetilde{\alpha(u,h)} := \left\langle (\widetilde{x'(u)}, 1), (\frac{\widetilde{x(u+h)} - \widetilde{x(u)}}{h}, 1) \right\rangle$. Then the following expansion holds:*

$$\widetilde{\alpha(u,h)} = \widetilde{\kappa(u)} \cdot h + O(h^3).$$

*where $\widetilde{\kappa(u)}$ is independent of $h$ and depends only on $u$.*

Based on Theorem 2, and since real roots have zero imaginary part, the steps of a heuristic test, called *Early Stop Test* (EST for short), can be as follows:

1. Choose a reasonable $0 < u < 1$ to start the test, and let $h = 1 - u$.
2. Compute an approximate value of $\widetilde{\kappa(u)}$, $\widetilde{\kappa(u)} \approx \frac{\widetilde{\alpha(u,-\delta)}}{-\delta}$ and let $\widetilde{\alpha} = \widetilde{\kappa(u)} \cdot h$.
3. Compute the angle $\alpha_1$ between vector $(\widetilde{x'(u)}, 1)$ and vector $(-\widetilde{x(u)}, h)$.
4. If $\alpha_1$ is less than $\Delta$ and the absolute error $|\widetilde{\alpha} - \alpha_1|$ is in a reasonable tolerance $\Delta_1$, then the test passes. If $\alpha_1$ is greater than $\Delta$ and the relative error $\left| \frac{\widetilde{\alpha} - \alpha_1}{\alpha_1} \right|$ is in another reasonable tolerance $\Delta_2$, then the test passes.

*Remark 3.* We give detailed explanations about these four steps respectively:

1. For different systems and different homotopies, the reasonability of choice of $u$ to start test differs. We think it is reasonable that $u$ should not be too far from 1. And the test should be started before the endgame, because some of the divergent paths may be stopped tracking due to the test.
2. By the definition of $\widetilde{\alpha(u,-\delta)}$, it is the angle between the vectors $(\widetilde{x'(u)}, 1)$ and $(\frac{\widetilde{x(u-\delta)} - \widetilde{x(u)}}{-\delta}, 1)$. These two vectors are easily obtained if we reserve the consecutive points, $x_{old}$ and $x_0$, and the two consecutive tangent vectors, $x'_{old}$ and $x'_0$, along the path tracking process. The value of these vectors are computed in Euler Prediction and Newton Correction, so it causes no extra cost of time. Then $\widetilde{\alpha}$ is the approximation of the angle between vectors $(\widetilde{x'(u)}, 1)$ and $(\frac{\widetilde{x(1)} - \widetilde{x(u)}}{h}, 1)$ based on Theorem 2, i.e. $\widetilde{\alpha} \approx \widetilde{\alpha(u,h)}$.
3. If the tracked path is convergent to an isolated real root, the equation $(\widetilde{x(1)}, 1) = (0, 1)$ holds. Then we have

$$\alpha_1 = \left\langle (\widetilde{x'(u)}, 1), (-\widetilde{x(u)}, h) \right\rangle = \left\langle (\widetilde{x'(u)}, 1), (\widetilde{x(1)} - \widetilde{x(u)}, h) \right\rangle = \widetilde{\alpha(u,h)}.$$

We give an illustrative description of the angle $\alpha_1$ in Fig. 2.
4. By the explanation of 2 and 3, if the tracked path is convergent to a real isolated root, we have $\widetilde{\alpha} \approx \widetilde{\alpha(u,h)} = \alpha_1$. So, when $\alpha_1$ is small, less than $\Delta$, due to the numerical stability, it is better to use the absolute error $|\widetilde{\alpha} - \alpha_1|$ as a criterion. while, when $\alpha_1$ is not really small, it is reasonable to use the relative error $\left| \frac{\widetilde{\alpha} - \alpha_1}{\alpha_1} \right|$ as a criterion.

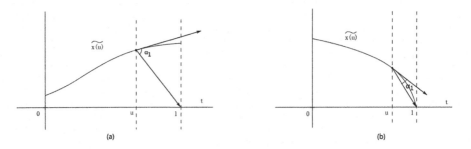

**Fig. 2.** (*a*) illustrates $\alpha_1$ of complex roots, (*b*) illustrates $\alpha_1$ of real roots.

---

**Algorithm 2.** EST (Early Stop Test)

---

**input** : step size $\delta > 0$, error tolerance $\Delta, \Delta_1, \Delta_2 > 0$ and test starts at
$\qquad\quad 0 < u < 1$

**output:** False or True

1   $\widetilde{\kappa(u)} :\approx \frac{\widetilde{\alpha(u, -\delta)}}{-\delta}$ ;

2   $\widetilde{\alpha} := \widetilde{\kappa(u)} \cdot h$;

3   $\alpha_1 := \arccos((x'(u), 1) \cdot (\widetilde{x(u)}, h)^T)$;

4 **if** $\alpha_1 \leqslant \Delta$ **then**

5      **if** $|\widetilde{\alpha} - \alpha_1| \leqslant \Delta_1$ **then**

6        |   **return** True;

7      **end**

8      **else**

9        |   **return** False;

10     **end**

11 **end**

12 **else**

13      **if** $\left| \frac{\widetilde{\alpha} - \alpha_1}{\alpha_1} \right| \leqslant \Delta_2$ **then**

14        |   **return** True;

15     **end**

16      **else**

17        |   **return** False;

18     **end**

19 **end**

---

The following algorithm combines the Euler Predictor-Newton Corrector path tracking method and the Early Stop Test.

---

**Algorithm 3.** ESTPT (Early Stop Test Path Tracking)

---

> **input :**
> $t_0 = 0, (x_0, t_0) \in \mathbb{R}^{n+1}$ such that $H(x_0, t_0) = 0$ ;
> $(x_{old}, t_{old}) = (x_0, t_0)$;
> step size $\delta > 0$, $t_0 + \delta < 1$;
> error tolerance $\Delta, \Delta_1, \Delta_2 > 0$ and
> $0 < u < 1$;
> PASS:= True, TESTED:=False;
>
> **output:** $x_0$ or Null

1  **repeat**
2      $x^{(0)} := x_0 - \delta \cdot H_x(x_0, t_0)^{-1} H_t(x_0, t_0).$ ;
3      **repeat**
4          $x^{(1)} := x^{(0)} - H_x(x^{(0)}, t_0 + \delta)^{-1} H(x^{(0)}, t_0 + \delta)$ ;
5          $x^{(0)} := x^{(1)}$ ;
6      **until** *Convergence*;
7      $x_{old} := x_0$;
8      $t_{old} := t_0$;
9      $x_0 := x^{(0)}$ ;
10     $t_0 := t_0 + \delta$ ;
11     **if** $t_0 > u$ && TESTED==False **then**
12         PASS:= EST($\delta$, $\Delta$, $\Delta_1$, $\Delta_2$, $u$);
13     **end**
14     TESTED:= True;
15 **until** Path tracking is stopped or PASS==False;
16 **if** PASS == False **then**
17     **return** Null;
18 **end**
19 **else**
20     **return** $x_0$;
21 **end**

---

## 4 Experiments

We have checked many benchmarks and randomly generated polynomial systems with algorithm ESTPT. Because we cannot get the source code of path tracking of those famous homotopy software packages (such as Bertini, Hom4PS2.0, HOMPACK and PHCpack), it is hard to make a suitable comparison with them. We implement our algorithm in the path tracking procedure of program LPH (available at http://arcnl.org/PDF/LHP.zip) and compare the early ending path tracking procedure ESTPT to the traditional path tracking procedure. In this section, all the homotopies are generated by total degree homotopy and all the experiments are computed on a PC with Intel Core i5 processor (2.5GHz CPU, 4 Cores, and 6 GB RAM) in the Windows environment.

In Table 1, we provide the comparison of ESTPT to traditional path tracking on benchmarks which are available at [50]. For those benchmarks, we do the EST at $u = 0.98$. N1 is the total degree of the system, and N2 is the number of paths pass the EST test. T1 and T2 are the time (in millisecond) of traditional path tracking and ESTPT respectively. R1 is the number of real roots, and R2 is the number of real roots detected by ESTPT. "C" is the number of complex roots of the system. "TR" is "Time Ratio", i.e. the ratio of T2 to T1. "PR" is "Pass Ratio", i.e. the ratio of N2 to N1. "Precision" is the ratio of number of real roots found by ESTPT to the number of paths that pass the EST test. "Recall" is the percentage of real roots found by ESTPT.

In Table 1, for some benchmarks, N2 is greater than "C", because some of the divergent paths passed the EST test. Those divergent paths show their divergent character after where we do the EST test. Actually, if we do the EST test at $u = 0.99$, almost all of these divergent paths will not pass the EST test.

In Table 2, we provide the comparison on randomly generated square polynomial systems. "n" is the number of variables in the system. "degree" is the maximal degree of the polynomials in the system. "Terms" is the maximal number of monomials in the polynomials. For those randomly generated systems, we do the EST test at $u = 0.95$.

In Table 2, when the system is very sparse, N2 is much smaller than N1 and "C" . It is because that there are many divergent paths and paths heading to complex roots, and most of them are early ended by EST. So we save a lot of time for the endgames. However, when the system is (relatively) dense, because the number of isolated roots are close to the total degree, there are few divergent paths, we follow each path till we do the EST test. So, the time ratio is close to $u$ which is the place we do the EST test.

**Table 1.** Quantities compared on benchmarks

| Equations | T1 | N1 | R1 | T2 | N2 | C | R2 | TR | PR | Precision | Recall |
|---|---|---|---|---|---|---|---|---|---|---|---|
| des18_3 | 20763 | 324 | 5 | 20124 | 253 | 46 | 5 | 0.969 | 0.780864 | 0.0197628 | 1 |
| eco7 | 5663 | 486 | 8 | 5226 | 301 | 32 | 8 | 0.922 | 0.619342 | 0.0265781 | 1 |
| cyclic5 | 281 | 120 | 10 | 234 | 28 | 70 | 9 | 0.832 | 0.233333 | 0.321429 | 0.9 |
| eco8 | 17831 | 1458 | 8 | 16412 | 954 | 64 | 8 | 0.920 | 0.654321 | 0.008386 | 1 |
| geneig | 873 | 243 | 10 | 624 | 62 | 10 | 10 | 0.714 | 0.255144 | 0.16129 | 1 |
| kinema | 468 | 64 | 8 | 468 | 37 | 40 | 7 | 1 | 0.578 | 0.189189 | 0.875 |
| reimer4 | 202 | 120 | 8 | 156 | 37 | 36 | 8 | 0.772 | 0.308333 | 0.216216 | 1 |
| virasoro | 842 | 256 | 224 | 827 | 224 | 256 | 224 | 0.982 | 0.875 | 1 | 1 |
| boon | 3651 | 1024 | 8 | 2200 | 64 | 8 | 8 | 0.602 | 0.0625 | 0.125 | 1 |

**Table 2.** Randomly generated examples

| n | Degree | Terms | N1 | N2 | C | R1 | R2 | TR | PR | Precision | Recall |
|---|--------|-------|-----|-----|-------|-----|----|-------|--------|-----------|--------|
| 4 | 3 | 3 | 54 | 6 | 8 | 4 | 4 | 0.715 | 0.111 | 0.667 | 1 |
| 5 | 4 | 8 | 1024 | 67 | 598 | 14 | 14 | 0.823 | 0.065 | 0.209 | 1 |
| 6 | 3 | 10 | 729 | 24 | 704 | 15 | 15 | 0.843 | 0.033 | 0.625 | 1 |
| 7 | 4 | 15 | 16384 | 209 | 14767 | 104 | 98 | 0.839 | 0.0128 | 0.469 | 0.94 |
| 6 | 5 | 40 | 15625 | 135 | 15395 | 86 | 82 | 0.925 | 0.0086 | 0.607 | 0.95 |
| 8 | 3 | 50 | 6561 | 72 | 6488 | 30 | 29 | 0.967 | 0.011 | 0.403 | 0.97 |
| 9 | 2 | Dense | 512 | 18 | 512 | 14 | 14 | 0.959 | 0.035 | 0.778 | 1 |

# 5    Conclusions

For isolated real roots identification in homotopy continuation methods, we present a heuristic early-stop test in path tracking procedure. The test is mainly based on Theorem 1, which shows the asymptotic behavior of an angle defined by two points on one homotopy path. The EST test may stop tracking most of the divergent paths and paths heading to complex roots. The parameter $u$ we choose influences the performance of algorithm ESTPT. There is a trade-off we have to make. If $u$ is too far from 1, the approximation $\tilde{\alpha}$ of the angle $\alpha_1$ will be bad, and none of the paths will pass the test ESTPT. On the other hand, if $u$ is too close to 1, it will save little time for the path tracking procedure. We do the test with the same $u$ for all the examples in Table 1 (and Table 2 respectively).

It may happen that, paths tracking to real roots may not pass the EST test, and divergent paths pass the EST test. We mention some possible future work. Some other data (such as the absolute value of the imaginary part of $x$) in homotopy procedure could be added into this heuristic test. And, most benchmarks are actually sparse systems. There are many other homotopies (such as polyhedral homotopy) for sparse systems while the ESTPT in Sect. 4 is based on total degree homotopy. It would be interesting to test the performance of ESTPT on those homtopies. We may do the EST test in the ESTPT procedure more than once based on some data in homotopy procedure.

# References

1. Garcia, C.B., Zangwill, W.I.: Finding all solutions to polynomial systems and other systems of equations. Math. Program. **16**(1), 159–176 (1979)
2. Drexler, F.J.: Eine methode zur berechnung sämtlicher lösungen von polynomgleichungssystemen. Numerische Mathematik **29**(1), 45–58 (1977)
3. Sommese, A.J., Verschelde, J., Wampler, C.W.: Numerical algebraic geometry. In: The Mathematics of Numerical Analysis. Lectures in Applied Mathematics, vol. 32, pp. 749–763 AMS (1996)
4. Allgower, E.L., Georg, K.: Introduction to numerical continuation methods. Reprint of the 1979 Original. Society for Industrial and Applied Mathematics (2003)

5. Li, T.Y.: Numerical solution of polynomial systems by homotopy continuation methods. In: Handbook of Numerical Analysis, vol. 11, pp. 209–304. Elsevier (2003)
6. Sommese, A.J., Wampler, C.W.: The Numerical Solution of Systems of Polynomials Arising in Engineering and Science. World Scientific, Singapore (2005)
7. Morgan, A.: Solving Polynominal Systems Using Continuation for Engineering and Scientific Problems. Society for Industrial and Applied Mathematics, Philadelphia (2009)
8. Hauenstein, J.D., Sommese, A.J.: What is numerical algebraic geometry? J. Symb. Comput. **79**, 499–507 (2017). SI: Numerical Algebraic Geometry
9. Huber, B., Sturmfels, B.: A polyhedral method for solving sparse polynomial systems. Math. Comput. **64**(212), 1541–1555 (1995)
10. Huber, B., Sturmfels, B.: Bernstein's theorem in affine space. Discret. Comput. Geom. **17**(2), 137–141 (1997)
11. Verschelde, J., Verlinden, P., Cools, R.: Homotopies exploiting newton polytopes for solving sparse polynomial systems. SIAM J. Numer. Anal. **31**(3), 915–930 (1994)
12. Bates, D.J., Haunstein, J.D., Sommese, A.J., Wampler, C.W.: Numerically Solving Polynomial Systems with Bertini. Society for Industrial and Applied Mathematics, Philadelphia (2013)
13. Lee, T.L., Li, T.Y., Tsai, C.H.: HOM4PS-2.0: a software package for solving polynomial systems by the polyhedral homotopy continuation method. Computing **83**(2), 109 (2008)
14. Leykin, A.: Numerical algebraic geometry for macaulay2. https://msp.org/jsag/2011/3-1/p02.xhtml
15. Verschelde, J.: Algorithm 795: PHCpack: a general-purpose solver for polynomial systems by homotopy continuation. ACM Trans. Math. Softw. **25**(2), 251–276 (1999)
16. Collins, G.E.: Quantifier elimination for real closed fields by cylindrical algebraic decompostion. In: Brakhage, H. (ed.) GI-Fachtagung 1975. LNCS, vol. 33, pp. 134–183. Springer, Heidelberg (1975). https://doi.org/10.1007/3-540-07407-4_17
17. Seidenberg, A.: A new decision method for elementary algebra. Ann. Math. **60**(2), 365–374 (1954)
18. Safey El Din, M., Schost, E.: Polar varieties and computation of one point in each connected component of a smooth real algebraic set. In: Proceedings of the 2003 International Symposium on Symbolic and Algebraic Computation, ISSAC 2003, pp. 224–231. ACM, New York (2003)
19. Safey El Din, M., Spaenlehauer, P.J.: Critical point computations on smooth varieties: degree and complexity bounds. In: Proceedings of the ACM on International Symposium on Symbolic and Algebraic Computation, ISSAC 2016, pp. 183–190. ACM, New York (2016)
20. Bank, B., Giusti, M., Heintz, J., Pardo, L.M.: Generalized polar varieties and an efficient real elimination. Kybernetika **40**(5), 519–550 (2004)
21. Bank, B., Giusti, M., Heintz, J., Pardo, L.: Generalized polar varieties: geometry and algorithms. J. Complex. **21**(4), 377–412 (2005)
22. Rouillier, F., Roy, M.F., Safey El Din, M.: Finding at least one point in each connected component of a real algebraic set defined by a single equation. J. Complex. **16**(4), 716–750 (2000)
23. El Din, M.S., Schost, É.: Properness defects of projections and computation of at leastone point in each connected component of a real algebraic set. Discret. Comput. Geom. **32**(3), 417–430 (2004)

24. Li, T.Y., Wang, X.: Solving real polynomial systems with real homotopies. Math. Comput. **60**(202), 669–680 (1993)
25. Brake, D.A., Hauenstein, J.D., Liddell, A.C.: Numerically validating the completeness of the real solution set of a system of polynomial equations, February 2016
26. Cifuentes, D., Parrilo, P.A.: Sampling algebraic varieties for sum of squares programs. **27**, November 2015
27. Lu, Y., Bates, D.J., Sommese, A.J., Wampler, C.W.: Finding all real points of a complex curve. Technical report, In Algebra, Geometry and Their Interactions (2006)
28. Bates, D.J., Sottile, F.: Khovanskii-Rolle continuation for real solutions. Found. Comput. Math. **11**(5), 563–587 (2011)
29. Besana, G.M., Rocco, S., Hauenstein, J.D., Sommese, A.J., Wampler, C.W.: Cell decomposition of almost smooth real algebraic surfaces. Numer. Algorithms **63**(4), 645–678 (2013)
30. Hauenstein, J.D.: Numerically computing real points on algebraic sets. Acta Applicandae Mathematicae **125**(1), 105–119 (2013)
31. Shen, F., Wu, W., Xia, B.: Real root isolation of polynomial equations based on hybrid computation. In: Feng, R., Lee, W., Sato, Y. (eds.) ASCM 2009, pp. 375–396. Springer, Heidelberg (2014). https://doi.org/10.1007/978-3-662-43799-5_26
32. Wang, Y., Wu, W., Xia, B.: A special homotopy continuation method for a class of polynomial systems. In: Gerdt, V.P., Koepf, W., Seiler, W.M., Vorozhtsov, E.V. (eds.) CASC 2017. LNCS, vol. 10490, pp. 362–376. Springer, Cham (2017). https://doi.org/10.1007/978-3-319-66320-3_26
33. Wu, W., Reid, G.: Finding points on real solution components and applications to differential polynomial systems. In: Proceedings of the 38th International Symposium on Symbolic and Algebraic Computation, ISSAC 2013, pp. 339–346. ACM, New York (2013)
34. Hauenstein, J.D., Regan, M.H.: Adaptive strategies for solving parameterized systems using homotopy continuation. Appl. Math. Comput. **332**, 19–34 (2018)
35. Morgan, A.P., Sommese, A.J.: Coefficient-parameter polynomial continuation. Appl. Math. Comput. **29**(2), 123–160 (1989)
36. Li, T.Y.: Numerical solution of multivariate polynomial systems by homotopy continuation methods. **6**, 399–436, January 1997
37. Chow, S.N., Mallet-Paret, J., Yorke, J.A.: A homotopy method for locating all zeros of a system of polynomials. **730**, January 1979
38. Morgan, A.P.: A method for computing all solutions to systems of polynomials equations. ACM Trans. Math. Softw. **9**(1), 1–17 (1983)
39. Wright, A.H.: Finding all solutions to a system of polynomial equations. Math. Comput. **44**(169), 125–133 (1985)
40. Zulehner, W.: A simple homotopy method for determining all isolated solutions to polynomial systems. Math. Comput. **50**(181), 167–177 (1988)
41. Morgan, A., Sommese, A.: A homotopy for solving general polynomial systems that respects m-homogeneous structures. Appl. Math. Comput. **24**(2), 101–113 (1987)
42. Wampler, C., P. Morgan, A., Sommese, A.: Complete solution of the nine-point path synthesis problem for four-bar linkages. **114**, March 1992
43. Verschelde, J., Haegemans, A.: The GBQ-algorithm for constructing start systems of homotopies for polynomial systems. SIAM J. Numer. Anal. **30**(2), 583–594 (1993)
44. Verschelde, J., Cools, R.: Symbolic homotopy construction. **4**, 169–183, September 1993

45. Morgan, A.P., Sommese, A., Wampler, C.: A product-decomposition theorem for bounding Bezout numbers, March 2018
46. Bernshtein, D.N.: The number of roots of a system of equations. Funct. Anal. Appl. **9**(3), 183–185 (1975)
47. Hauenstein, J.D., Sommese, A.J., Wampler, C.W.: Regeneration homotopies for solving systems of polynomials. Math. Comp. **80**(273), 345–377 (2011)
48. Huber, B., Verschelde, J.: Polyhedral end games for polynomial continuation. Numer. Algorithms **18**(1), 91–108 (1998)
49. Sosonkina, M., Watson, L.T., Stewart, D.: Note on the end game in homotopy zero curve tracking. **22**, 281–287, September 1996
50. Gerdt, V., Blinkov, Y., Yanovich, D.: GINV Project. http://invo.jinr.ru/ginv/

# Autocorrelation via Runs

Ilias S. Kotsireas[1] and Jing Yang[2(✉)]

[1] Wilfrid Laurier University, Waterloo, ON N2L 3C5, Canada
ikotsire@wlu.ca
[2] SMS-HCIC, Guangxi University for Nationalities,
Nanning 530006, China
yangjing0930@gmail.com

**Abstract.** A problem of interest in the realm of autocorrelation of (binary) finite sequences is to find sequences of length $n$ with given (pre-defined) autocorrelation profiles. This amounts to solving a system of $\lfloor n/2 \rfloor$ quadratic equations over the boolean cube $\{-1, +1\}^n$. We establish and discuss a computational approach to this autocorrelation problems, using the concept of runs. An algorithm is given to solve this problem and its application is illustrated with non-trivial examples.

**Keywords:** D-optimal design · Periodic autocorrelation function
Supplementary difference set · Hadamard matrix

## 1 Introduction

In this work we are interested in finite (binary) sequences $A = [a_0, \ldots, a_{n-1}]$ of length $n$, with elements from $\{-1, +1\}$. Throughout the paper, we will use $A$ to denote sequences of this form. The periodic autocorrelation function (PAF) of the sequence $A$ is defined as

$$PAF(A, k) = \sum_{i=0}^{n-1} a_i a_{i+k}, \text{ for } k = 0, 1, \ldots, \lfloor n/2 \rfloor$$

(where $i + k$ is taken modulo $n$, when needed). The periodic autocorrelation function is a concept of central importance, as it can be used to define several classes of combinatorial matrices in a unified manner (See [8] for more details).

Given a sequence $A$ of length $n$, the PAF profile of $A$ is the following list of $\lfloor n/2 \rfloor$ numbers:

$$[PAF(A, 1), \ldots, PAF(A, \lfloor n/2 \rfloor)]$$

Note that $PAF(A, 0)$ is not used in the above definition. Since for binary $\{-1, +1\}$ sequences of length $n$, we have that $PAF(A, 0) = n$. It is also clear that any rotation or reverse rotation of $A$ has the same PAF profile as $A$, since the PAF values remain unchanged under such rotations of the sequence. Therefore, without loss of generality, we assume that $A$ *always starts with $+1$ and ends with $-1$*.

© Springer Nature Switzerland AG 2018
J. Fleuriot et al. (Eds.): AISC 2018, LNAI 11110, pp. 195–205, 2018.
https://doi.org/10.1007/978-3-319-99957-9_13

In the realm of autocorrelation, we are typically interested in $\{-1, +1\}$ sequences with a constant element sum, i.e. such sequences satisfy the linear equation/constraint $a_1 + \cdots + a_n = s$, for a fixed constant $s$. This linear equation/constraint implies that there is a fixed number of "$+1$" and "$-1$" elements in the sequence. In particular, there are exactly $\dfrac{n+s}{2}$ "$+1$" elements and there are exactly $\dfrac{n-s}{2}$ "$-1$" elements.

We are interested in the following problem:

**PAF Profile Problem (PPP)**

For a specific length $n$ and an element sum $s$, given a specific sequence of numbers, $P = [p_1, \ldots, p_{\lfloor n/2 \rfloor}]$, is there a binary $\{-1, +1\}$ sequence of length $n$, with a constant sum of elements $s$, whose PAF profile is equal to $P$?

Note that it is possible that the answer to the PPP is negative. In case the answer to the PPP is positive, we are interested in exhibiting one sequence of length $n$ that materializes the given PAF profile $P$. The question of finding all sequences of length $n$ that materialize the given PAF profile $P$ is more general and we will not be concerned with it in the current work.

**Example**

We illustrate the PPP with $n = 9$, $s = -1$ and $P = [-3, 1, -3, 1]$. A solution to the PPP is the sequence $A = [+1, +1, -1, +1, -1, +1, -1, -1, -1]$. Note that in fact we have solved the following system of one linear and four quadratic equations in the boolean cube $\{-1, +1\}^9$:

$$
\begin{aligned}
a_1 + a_2 + \cdots + a_8 + a_9 &= -1 \\
a_1 a_2 + a_2 a_3 + \cdots + a_8 a_9 + a_9 a_1 &= -3 \\
a_1 a_3 + a_2 a_4 + \cdots + a_8 a_1 + a_9 a_2 &= 1 \\
a_1 a_4 + a_2 a_5 + \cdots + a_8 a_2 + a_9 a_3 &= -3 \\
a_1 a_5 + a_2 a_6 + \cdots + a_8 a_3 + a_9 a_4 &= 1
\end{aligned}
$$

Another way to describe a $(+1, -1)$-sequence is with runs. A *run* of a $(+1, -1)$-sequence is defined as a fragment of the sequence such that: (a) its elements are all $+1$'s or all $-1$'s; (b) the last element before the fragment and the first element after the fragment are both with the opposite sign to the elements in the fragment. For example, given $A = [1, 1, -1, 1, 1, -1, -1]$, there are four runs in $A$, i.e., $[1, 1], [-1], [1, 1], [-1, -1]$.

The basic idea in our work is to convert the problem of solving a PPP into that of finding all the runs of $A$, which is equivalent to finding a $t$-partition of $n$ (i.e., $n = n_1 + \cdots + n_t$ where $n_i > 0$ for $i = 1, \ldots, t$) such that a sequence $A$ satisfying the given PPP can be constructed from the partition by the rules that the first run starts with $+1$ and the $i$th run has exactly $n_i$ $+1$'s (when $i$ is odd) or $-1$'s (when $i$ is even). This idea is based on the observation that every sequence of length $n$ corresponds to an ordered partition of $n$. For example, given $A = [+1, +1, -1, +1, -1, +1, -1, -1, -1]$, one can get a 6-partition of 9 which is $(2, 1, 1, 1, 1, 3)$; vice versa, given a 6-partition $(2, 1, 1, 1, 1, 3)$ of 9, one can get a sequence $[+1, +1, -1, +1, -1, +1, -1, -1, -1]$ of length 9. The two processes are called *encoding* (from sequences to partitions) and *decoding* (from partitions to sequences), respectively. We explore some nice combinatorial properties related to the PAF and runs, which are very helpful to discard candidate sequences that cannot possibly materialize the given PAF profile and thus can significantly improve the efficiency of the designed algorithm for solving PPP. Computational results show that the proposed algorithm can solve very difficult PAF profile problems. It should be pointed out that the algorithm presented in this paper can be used for solving any PPP without any constraints on $n$. To the best of our knowledge, there is no prior work for this problem.

This paper is organized as follows. In Sect. 2, some nice properties are proved that can be used to explore some features of the partition corresponding to the sequence $A$. Based on these properties, we design an algorithm with a series of filters, to discard candidate sequences that cannot possibly materialize the given PAF profile in Sect. 3 and it is followed by an application for solving non-trivial problems in D-optimal designs in Sect. 4.

## 2    Combinatorial Properties

In this section we state and prove several combinatorial properties related to the PAF as well as runs. These properties will be used in the algorithm, to implement a series of filters, to discard candidate sequences that cannot possibly materialize the given PAF profile.

For proving these properties, we introduce a new sequence $C = [c_0, \ldots, c_{n-1}]$ associated to $A$ with $c_i = \frac{a_i+1}{2}$. Obviously, $c_i = 0$ when $a_i = -1$; $c_i = 1$ when $a_i = 1$. Let $P_k := PAF(A, k)$. Then we have

$$s' := \sum_{i=0}^{n-1} c_i = \sum_{i=0}^{n-1} \frac{a_i + 1}{2} = \frac{1}{2}(s + n),$$

$$Q_k := \sum_{i=0}^{n-1} c_i c_{i+k} = \frac{1}{4} \sum_{i=0}^{n-1} (a_i + 1)(a_{i+k} + 1) = \frac{P_k + 2s + n}{4}.$$

Under the assumption that $A$ always starts with $+1$ and ends with $-1$, the numbers of runs in $A$ is even because the number of runs consisting of $+1$'s is equal to that of runs consisting of $-1$'s. We denote these three numbers of runs by $r_A$, $r_A^{(+1)}$ and $r_A^{(-1)}$, respectively. Then $r_A = 2r_A^{(+1)}$. Similarly, we can define $r_C$, $r_C^{(1)}$ and $r_C^{(0)}$ for the sequence $C$. It is obvious that

$$r_A = r_C, \quad r_C^{(1)} = r_A^{(+1)} \quad \text{and} \quad r_C^{(0)} = r_A^{(-1)}.$$

We will prove some nice combinatorial properties related to the PAF and runs with the help of $C$.

**Proposition 1.** *For a sequence $A = [a_0, \ldots, a_{n-1}]$ of length $n$, we have that $PAF(A, k) \equiv n \mod 4$, for $k = 1, \ldots, n - 1$.*

See [4] for a proof of the above proposition.

**Proposition 2.** *The number of runs in the sequence $A$ is $(P_0 - P_1)/2$.*

*Proof.* With the above notations, we first prove $r_C^{(1)} = Q_0 - Q_1$.

Suppose that $A$ has $2m$ runs. Thus so does $C$. Assume the $j$-th run of $C$ starts with $c_{\tau_j+1}$ and ending with $c_{\tau_{j+1}}$ where $\tau_1 := 0$. Then the runs consisting of 1's are those which start with the $(\tau_{2j-1} + 1)$th elements in $C$ and end with the $\tau_{2j}$th elements in $C$ for $j = 1, \ldots, m$. Therefore, we have

$$Q_0 = \sum_{i=0}^{n-1} c_i^2 = \sum_{i=\tau_1+1}^{\tau_2} 1 \times 1 + \sum_{i=\tau_3+1}^{\tau_4} 1 \times 1 + \cdots + \sum_{i=\tau_{2m-1}+1}^{\tau_{2m}} 1 \times 1, \tag{1}$$

$$Q_1 = \sum_{i=0}^{n-1} c_i c_{i+1} = \left( \sum_{i=\tau_1+1}^{\tau_2-1} c_i c_{i+1} + c_{\tau_2} c_{\tau_2+1} \right) + \left( \sum_{i=\tau_3+1}^{\tau_4-1} c_i c_{i+1} + c_{\tau_4} c_{\tau_4+1} \right) + \cdots$$

$$+ \left( \sum_{i=\tau_{2m-1}+1}^{\tau_{2m}-1} c_i c_{i+1} + c_{\tau_{2m}} c_{\tau_{2m}+1} \right)$$

$$= \left( \sum_{i=\tau_1+1}^{\tau_2-1} 1 \times 1 + 1 \times 0 \right) + \left( \sum_{i=\tau_3+1}^{\tau_4-1} 1 \times 1 + 1 \times 0 \right) + \cdots$$

$$+ \left( \sum_{i=\tau_{2m-1}+1}^{\tau_{2m}-1} 1 \times 1 + 1 \times 0 \right). \tag{2}$$

It follows that $Q_0 - Q_1 = \#\{\tau_1, \tau_3, \ldots, \tau_{2m-1}\} = r_C^{(1)}$. Hence

$$r_A = 2r_A^{(+1)} = 2r_C^{(1)} = 2(Q_0 - Q_1)$$

$$= 2 \left( \frac{P_0 + 2s + n}{4} - \frac{P_1 + 2s + n}{4} \right) = \frac{P_0 - P_1}{2}.$$

$\square$

*Remark 1.* The evenness of $r_A$ can be verified via Propositions 1 and 2.

Proposition 2 implies that if the given PPP has a solution, the decoded partition of $n$ must have $\frac{p_0-p_1}{2}$ parts. If $\frac{p_0-p_1}{2}$ is a fraction or an odd number, we immediately conclude that the given PPP has no solution.

If the length of a run is 1, we call it a 1-run. The number of 1-runs in a sequence $A$ can be obtained from $PAF(A, i)$ where $i = 0, 1, 2$ using the following proposition.

**Proposition 3.** *The number of 1-runs in $A$ is $(P_0 + P_2 - 2P_1)/4$.*

*Proof.* Let $C$ be a sequence associated to $A$ as defined above. Then the number of 1-runs in $A$ is equal to that in $C$. We only need to show that the number of 1-runs in $C$ is $Q_0 + Q_2 - 2Q_1$. which can be proved by induction on the number of 1-runs, denoted by $N_1$.

1. $N_1 = 0$.

    With the settings in the proof of Proposition 2,

$$Q_2 = \sum_{i=0}^{n-1} c_i c_{i+1}$$

$$= \left( \sum_{i=\tau_1+1}^{\tau_2-2} c_i c_{i+1} + c_{\tau_2-1} c_{\tau_2+1} + c_{\tau_2} c_{\tau_2+2} \right)$$

$$+ \left( \sum_{i=\tau_3+1}^{\tau_4-2} c_i c_{i+1} + c_{\tau_4-1} c_{\tau_4+1} + c_{\tau_4} c_{\tau_4+2} \right) + \cdots$$

$$+ \left( \sum_{i=\tau_{2m-1}+1}^{\tau_{2m}-2} c_i c_{i+1} + c_{\tau_{2m}-1} c_{\tau_{2m}+1} + c_{\tau_{2m}} c_{\tau_{2m}+2} \right)$$

$$= \left( \sum_{i=\tau_1+1}^{\tau_2-2} 1 \times 1 + 1 \times 0 + 1 \times 0 \right) + \left( \sum_{i=\tau_3+1}^{\tau_4-2} 1 \times 1 + 1 \times 0 + 1 \times 0 \right) +$$

$$\cdots + \left( \sum_{i=\tau_{2m-1}+1}^{\tau_{2m}-2} 1 \times 1 + 1 \times 0 + 1 \times 0 \right). \tag{3}$$

The substitution of (1)–(3) into $Q_0 + Q_2 - 2Q_1$ yields 0, which is equal to $N_1$.

2. $N_1 = 1$.

    Assume $C = [\underbrace{1, 1, \ldots, 1}_{n-1}, 0]$, i.e., $c_i = 1$ when $i < n-1$ and $c_{n-1} = 0$. Then

$$Q_0 = \sum_{i=0}^{n-2} c_i^2 + c_{n-1}^2 = \sum_{i=0}^{n-2} 1 \times 1 + 0 \times 0 = n - 1,$$

$$Q_1 = \sum_{i=0}^{n-3} c_i c_{i+1} + c_{n-2} c_{n-1} + c_{n-1} c_0 = \sum_{i=0}^{n-3} 1 \times 1 + 1 \times 0 + 0 \times 1 = n - 2,$$

$$Q_2 = \sum_{i=0}^{n-4} c_i c_{i+2} + c_{n-3} c_{n-1} + c_{n-2} c_0 + c_{n-1} c_1$$

$$= \sum_{i=0}^{n-4} 1 \times 1 + 1 \times 0 + 1 \times 1 + 0 \times 1 = n - 2.$$

Thus $Q_0 + Q_2 - 2Q_1 = n - 1 + (n-2) - 2(n-2) = 1$, which is equal to $N_1$. The case for $C = [1, \underbrace{0, \ldots, 0}_{n-1}]$ can be proved similarly.

3. Suppose the conclusion holds for $N_1 < k$. We now prove that it holds for $N_1 = k$. Let $C$ be a sequence with $k$ 1-runs.

    (a) If $C$ has a fragment of the form $1, 1, 0, 1, 1$ and 0 in this fragment is the $j$th element of $C$, we construct $C'$ by replacing the $j$th element of $C$ with

1 and keeping the other $n-1$ elements unchanged. Then the number of 1-runs in $C'$ is $k-1$. Let $Q'_i = PAF(C', i)$. By induction, we have $Q'_0 + Q'_2 - 2Q'_1 = k-1$.

Noting that $c_{j-2}c_j = c_{j-1}c_j = c_j^2 = c_jc_{j+1} = c_jc_{j+2} = 0$, we have

$$Q'_0 = \sum_{i \neq j} c'^2_i + c'^2_j = \sum_{i \neq j} c^2_i + 1 = Q_0 + 1,$$

$$Q'_1 = \sum_{i \neq j-1,j} c'_i c'_{i+1} + c'_{j-1}c'_j + c'_j c'_{j+1}$$

$$= \sum_{i \neq j-1,j} c_i c_{i+1} + 1 \times 1 + 1 \times 1 = Q_1 + 2,$$

$$Q'_2 = \sum_{i \neq j-2,j} c'_i c'_{i+2} + c'_{j-2}c'_j + c'_j c'_{j+2}$$

$$= \sum_{i \neq j-2,j} c_i c_{i+2} + 1 \times 1 + 1 \times 1 = Q_2 + 2.$$

Thus

$$Q_0 + Q_2 - 2Q_1 = (Q'_0 - 1) + (Q'_2 - 2) - 2(Q'_1 - 2) = Q'_0 + Q'_2 - 2Q'_1 + 1 = k.$$

(b) If $C$ has a fragment of the form $0,0,1,0,0$ and 1 in this fragment is the $j$th element of $C$, we construct $C'$ by replacing the $j$th element of $C$ with 0 and keeping the other $n-1$ elements unchanged. With similar reasoning, we get $Q_0 = Q'_0 + 1, Q_1 = Q'_1$ and $Q_2 = Q'_2$, which leads to $Q_0 + Q_2 - 2Q_1 = Q'_0 + Q'_2 - 2Q'_1 + 1 = k$.

(c) If $C$ has a fragment of the form $1,0,1,0$ and the first 0 in this fragment is the $j$th element of $C$, we construct $C'$ by replacing the $j$th element of $C$ with 1 and keeping the other $n-1$ elements unchanged. Then the number of 1-runs in $C'$ is $k - 2 - (1 - c_{j-2})$. By induction, we have $Q'_0 + Q'_2 - 2Q'_1 = k - 2 - (1 - c_{j-2})$. Observe that

$$c_{j-2}c_j = c_{j-1}c_j = c_j^2 = c_jc_{j+1} = c_jc_{j+2} = 0,$$

$$c'_{j-2}c'_j = c_{j-2}, \quad c'_{j-1}c'_j = c'^2_j = c'_j c'_{j+1} = 1, \quad c'_j c'_{j+2} = 0.$$

With similar reasoning, we get $Q'_0 = Q_0 + 1$, $Q'_1 = Q_1 + 2$ and $Q'_2 = Q_2 + c_{j-2}$. It follows that

$$Q_0 + Q_2 - 2Q_1 = Q'_0 + Q'_2 - 2Q'_1 + 3 - c_{j-2} = k.$$

Recall that $Q_i = (P_i + 2s + n)/4$. One may immediately get $N_1 = Q_0 + Q_2 - 2Q_1 = (P_0 + P_2 - 2P_1)/4$. $\qquad\square$

Proposition 3 can help us discard the partitions of $n$ where the number of ones is not equal to $(P_0 + P_2 - 2P_1)/4$. Furthermore, it can be checked that the constraint $r_A = (n - p_1)/2 \wedge N_1 = (n + p_2 - 2p_1)/4$ with $p_1, p_2$ as shown in the PAF Profile Problem is equivalent to $PAF(A, 1) = p_1 \wedge PAF(A, 2) = p_2$.

# 3    Description of the Algorithm

Now we summarize the ideas discussed above into the following algorithm.

**Algorithm 1.** (PPP_solving)

---

Input:     $n$, a natural number; $s$, an integer;
           $P = [p_1, \ldots, p_{\lfloor n/2 \rfloor}]$, a sequence of integers.

Output: $A$, a solution to the PPP determined by $n$, $s$ and $P$ if it exists; $[]$
           otherwise.

---

1.    $r_A \leftarrow (n - p_1)/2$, $N_1 \leftarrow (n + p_2 - 2p_1)/4$, $n_{+1} \leftarrow (n + s)/2$, $\mathcal{Q} \leftarrow \{\}$.

2.    If one of the following occurs, return $[]$.

   (a) $r_A$ is fractional/odd/negative;

   (b) $n_{+1}$ is fractional/negative;

   (c) $N_1$ is fractional/negative.

3.    Compute the set of all the $r_A$-partitions of $n$, denoted by $\mathcal{P}$.

4.    Discard $\alpha$ from $\mathcal{P}$ if the occurrence of 1 in $\alpha$ is not equal to $N_1$.

5.    While $\mathcal{P} \neq \emptyset$

   5.1  Pop $\alpha$ out of $\mathcal{P}$.

   5.2  Choose $r_A/2$ elements $e_1, \ldots, e_{r_A/2}$ from $\alpha$ such that $\sum_{i=1}^{r_A/2} e_i = n_{+1}$
        and let its supplementary sequence be $[e'_1, \ldots, e'_{r_A/2}]$. Collect all such
        possibilities and form a set of sequences with the designated form $[e_1, e'_1,$
        $e_2, e'_2, \ldots, e_{r_A/2}, e'_{r_A/2}]$, denoted by $\mathcal{C}$.

   5.3  For each $\beta = [\beta_1, \ldots, \beta_{r_A}] \in \mathcal{C}$
      5.3.1 Compute the sets of all the permutations of $[\beta_1, \beta_3, \ldots, \beta_{r_A-1}]$ and
            $[\beta_2, \beta_4, \ldots, \beta_{r_A}]$, denoted by $\mathcal{P}_\beta^{(+1)}$ and $\mathcal{P}_\beta^{(-1)}$, respectively.

      5.3.2 $\mathcal{P}_\beta^{(+1)} \leftarrow$ Removing_repeated_rotation($\mathcal{P}_\beta^{(+1)}$).*

      5.3.3 $\mathcal{P}_\beta \leftarrow \{[\gamma_1, \gamma'_1, \gamma_2, \gamma'_2, \ldots, \gamma_{r_A/2}, \gamma'_{r_A/2}] : \gamma \in \mathcal{P}_\beta^{(+1)}, \gamma' \in \mathcal{P}_\beta^{(-1)}\}$.
      5.3.4 $\mathcal{Q} \leftarrow \mathcal{Q} \cup \mathcal{P}_\beta$.**

6.    While $\mathcal{Q} \neq \emptyset$

   6.1 Pop $\alpha$ out of $\mathcal{Q}$.

   6.2 Construct $A = [\underbrace{+1, \ldots, +1}_{\alpha_1}, \underbrace{-1, \ldots, -1}_{\alpha_2}, \ldots, \underbrace{-1, \ldots, -1}_{\alpha_{r_A}}]$ from $\alpha$.

   6.3 If $A$ satisfies $PAF(A, k) = p_k$ for $k = 3, \ldots, \lfloor n/2 \rfloor$, then return $A$.

7.    Return $[]$.

---

*Removing_repeated_rotation is an algorithm which removes the rotations or reverse rotations of any sequence $\alpha \in \mathcal{P}^{(+1)}$, only leaving the representative $\alpha$. It can help to avoid considerable redundant computation. For example, $[1, 1, 2, 3]$, $[2, 3, 1, 1]$ and $[2, 1, 1, 3]$ are viewed as identical partitions because we can obtain the other two after rotating or reversely rotating one sequence. If one of them is the solution to a PPP, so are the other two.

** One may also remove the repeated cycles or reverse cycles from $\mathcal{P}_\beta$. However, it should be pointed out that only those cycles obtained from rotations or reverse rotations by even positions can be discarded.

The above algorithm is illustrated with the following example.

*Example 1.* Consider the PPP given in Sect. 1 where with $n = 9$, $s = -1$ and $P = [-3, 1, -3, 1]$. We solve the system by using Algorithm 1.

Step 1. By calculation, $r_A = 6$, $n_{+1} = 4$, $N_1 = 4$ and $\mathcal{Q}$ is initialized to $\{\}$.

Step 2. Since $r_A$, $N_1$, and $n_{+1}$ are all positive integers and $r_A$ is an even number, we go to Step 3.

Step 3. Compute all the 6-partitions of 9 and get

$$\mathcal{P} = \{[1, 1, 1, 2, 2, 2], [1, 1, 1, 1, 2, 3], [1, 1, 1, 1, 1, 4]\}.$$

Step 4. When $\alpha = [1, 1, 1, 2, 2, 2]$, the number of occurrences of 1 in $\alpha$ is 3, not equal to $N_1$; thus $[1, 1, 1, 2, 2, 2]$ should be removed from $\mathcal{P}$. Similarly, $[1, 1, 1, 1, 1, 4]$ should also be removed from $\mathcal{P}$. So $\mathcal{P}$ is updated with $\{[1, 1, 1, 1, 2, 3]\}$.

Step 5. Since $\mathcal{P} \neq \emptyset$, pop $\alpha = [1, 1, 1, 1, 2, 3]$ out of $\mathcal{P}$ (which becomes an empty set). Then the groups with 3 elements from $\alpha$ are $[1, 1, 1]$, $[1, 1, 2]$, $[1, 1, 3]$ and $[1, 2, 3]$. Only $[1, 1, 2]$ satisfies $1 + 1 + 2 = n_{+1} = 4$ and its supplementary sequence is $[1, 1, 3]$. Thus $\mathcal{C} = \{[1, 1, 1, 1, 2, 3]\}$.

Next let $\beta = [1, 1, 1, 1, 2, 3]$ and compute the permutations of $[1, 1, 2]$ and $[1, 1, 3]$, which yields

$$\mathcal{P}_\beta^{(+1)} = \{[1, 1, 2], [1, 2, 1], [2, 1, 1]\}, \quad \mathcal{P}_\beta^{(-1)} = \{[1, 1, 3], [1, 3, 1], [3, 1, 1]\}.$$

After removing the repeated cycles in $\mathcal{P}_\beta^{(+1)}$, we get $\mathcal{P}_\beta^{(+1)} = \{[1, 1, 2]\}$. Then $\mathcal{P}_\beta$ is assigned with

$$\{[1, 1, 1, 1, 2, 3], [1, 1, 1, 3, 2, 1], [1, 3, 1, 1, 2, 1]\}.$$

Add $\mathcal{P}_\beta$ to $\mathcal{Q}$ and we get

$$\mathcal{Q} = \{[1, 1, 1, 1, 2, 3], [1, 1, 1, 3, 2, 1], [1, 3, 1, 1, 2, 1]\}.$$

Step 6. Check whether the sequence $A$ constructed from every $\alpha \in \mathcal{Q}$ satisfies the conditions $PAF(A, 3) = -3$ and $PAF(A, 4) = 1$. Eventually, we get a solution to the given PPP, which is $[+1, -1, +1, -1, +1, +1, -1, -1, -1]$. Its decoded partition is $[1, 1, 1, 1, 2, 3]$.

# 4    Application in D-Optimal Designs

The Maximal Determinant problem asks for the largest possible determinant (in its absolute value) of an $\nu \times \nu$ matrix whose entries are chosen from the set $\{-1, +1\}$. It was first posed by Hadamard [4,7] and has many applications in the areas of experimental design and coding theory. However, it is quite difficult when $\nu$ is big. In the worst case, the search space could have $2^{\nu^2}$ possibilities with $\nu^2$ to be the number of unknowns. The Maximal Determinant problem is still an open problem in the full generality. Significant progress on the problem comes from Ehlich [6]. He proved that if $R_1$ and $R_2$ are circulant $(-1, 1)$-matrices of order $n$ such that $R_1 R_2^t + R_2 R_2^t = 2(n-1)I_n + 2J_n$ where $I_n$ is the $n \times n$ identity matrix and $J_n$ is the $n \times n$ matrix with all entries 1, then the matrix

$$H = \begin{pmatrix} R_1 & R_2 \\ -R_2^t & R_1^t \end{pmatrix}$$

has the maximal determinant. $R_1$ and $R_2$ can be constructed from cyclic supplementary difference sets which can be compressed into periodic autocorrelation functions. The method of compression for periodic autocorrelation is introduced in [5]. We also refer to [1–3,9] for the motivation of our work.

Given two PAF sequences $PAF_A$ and $PAF_B$ of length $n$, if $PAF_A + PAF_B = [\alpha, \ldots, \alpha]$, i.e., the sum of PAF values are constant, then $PAF_A$ and $PAF_B$ are called *complementary sequences*. For instance, given $\alpha = 2$ and two PPPs with $n = 41$, $s_A = 9$, $s_B = -9$ and

$$PAF_A = [1, 5, -3, 5, 1, 1, 1, 1, -3, 5, 1, 5, 1, -3, 1, 5, 9, -3, -7, -3],$$
$$PAF_B = [1, -3, 5, -3, 1, 1, 1, 1, 5, -3, 1, -3, 1, 5, 1, -3, -7, 5, 9, 5]$$

which are complementary sequences. Solving the PPPs with the above algorithm, we get the following two sequences: [1]

$$\begin{aligned} A = [\,&+1, -1, +1, -1, +1, -1, +1, +1, -1, -1, +1, +1, +1, -1, -1, \\ &-1, -1, +1, -1, +1, +1, +1, +1, +1, +1, +1, -1, +1, +1, +1, \\ &-1, +1, +1, -1, +1, +1, +1, +1, -1, -1, -1\,], \\ B = [\,&+1, -1, -1, +1, -1, +1, +1, -1, -1, -1, +1, -1, -1, -1, -1, \\ &-1, -1, -1, +1, +1, -1, +1, -1, -1, +1, +1, -1, -1, +1, +1, \\ &+1, -1, +1, -1, -1, -1, -1, -1, +1, +1, -1\,]. \end{aligned}$$

Next we construct two circulant matrices $R_1, R_2$ from $A$ and $B$ (as shown below). More explicitly, let $A$ (or $B$) be the first row of $R_1$ (or $R_2$) and every row (except the first) of $A$ (or $B$) is a right cyclic shift by one of the previous row. Then the D-optimal matrix $H$ can be constructed as

$$H = \begin{pmatrix} R_1 & R_2 \\ -R_2^t & R_1^t \end{pmatrix}.$$

---

[1] The experiments are performed in Maple 2017 on a Windows PC with an Intel(R) Core(TM) i7-6700U CPU @3.40 GHz and 8 GB RAM. The program terminates with success after 7750 s and 14156 s, respectively.

Thus the determinant of $H$ is $2^{161}\, 3^4\, 5^{40}$.

**Acknowledgments.** This work was made possible by the facilities of the SMS International at GXUN, Nanning, P.R. China. ISK's work is supported by NSERC grants. JY's work is supported by the Special Fund for Guangxi Bagui Scholars (WBS 2014-01) and the Startup Foundation for Advanced Talents in Guangxi University for Nationalities (2015MDQD018). ISK would like to thank JY for providing excellent working conditions at the SMS International and warm hospitality in Nanning.

# References

1. Brent, R.P., Orrick, W., Osborn, J.H., Zimmermann, P.: Maximal determinants and saturated D-optimal design of orders 19 and 37. arXiv: 1112.4160v1 (2015)
2. Brent, R.P., Osborn, J.H.: General lower bounds on maximal determinants of binary matrices. Electron. J. Comb. **20**(2), 12 (2013). Paper 15
3. Brent, R.P., Osborn, J.H.: On minors of maximal determinant matrices. J. Integer Seq. **16**(4), 30 (2013). Article 13.4.2
4. Craigen, R., Kharaghani, H.: Hadamard matrices and Hadamard designs. In: Colbourn, C.J., Dinitz, J.H. (eds.) Handbook of Combinatorial Designs. Discrete Mathematics and Its Applications, 2nd edn, pp. 273–280. Chapman & Hall/CRC, Boca Raton (2007)
5. Dzokovic, D.Z., Kotsireas, I.S.: Compression of periodic complementary sequences and applications. Des. Codes Cryptogr. **74**(2), 365–377 (2015)
6. Ehlich, H.: Determinantenabsschätzungen für binäre matrizen. Mathematische Z. **83**, 517–531 (1964)
7. Hadamard, J.: Résolution dùne question relative aux déterminants. Bull. des Sci. Mathématiques **17**, 240–246 (1893)
8. Kotsireas, I.S.: Algorithms and metaheuristics for combinatorial matrices. In: Pardalos, P., Du, D.Z., Graham, R. (eds.) Handbook of Combinatorial Optimization. Springer, New York (2013). https://doi.org/10.1007/978-1-4419-7997-1_13
9. Orrick, W.P.: On the enumeration of some D-optimal designs. J. Stat. Plan. Inference **138**(1), 286–293 (2008)

# Intelligent Documents and Collective Intelligence

# LAText: A Linear Algebra Textbook System

Xiaoyu Chen[1,2]($\boxtimes$), Haotian Shuai[3], Dongming Wang[1,2,3], and Jing Yang[3]

[1] Beijing Advanced Innovation Center for Big Data and Brain Computing,
Beihang University, Beijing 100191, China
`chenxiaoyu@buaa.edu.cn`
[2] LMIB-SKLSDE, School of Mathematics and Systems Science, Beihang University,
Beijing 100191, China
[3] SMS-HCIC, College of Software and Information Security,
Guangxi University for Nationalities, Nanning 530006, China

**Abstract.** Mathematical textbooks play a key role in disseminating systematized mathematical knowledge of study. Most textbooks are published in printed or online electronic format without machine-understandable semantics. In this paper we present an intelligent system for managing linear algebra knowledge in the form of textbook with open access to users. Fine-grained data schemas are designed to represent structural semantics of knowledge contents and implemented by using a graph database with an interface of authoring and browsing knowledge contents and structures. A vector-based retrieving method is implemented to rank knowledge objects with respect to query. We report the results of our investigations on semantic representation of mathematical knowledge with experimental implementations for the development of such textbooks.

**Keywords:** Intelligent textbook · Knowledge retrieval
Mathematical knowledge management · Semantic representation

## 1 Motivation

Mathematical textbooks have been used to systematically introduce mathematical knowledge of study to students or learners and played a fundamental role in education and knowledge dissemination. Traditional textbooks are published usually in printed or online electronic formats, such as PDF and HTML, as whole documents. Most textbooks are stored and managed in libraries accessible only via meta information including titles, authors, publishers, and keywords. To enhance their efficiency and adaptivity, a number of learning systems have been developed to manage learning objects that encapsulate pieces of mathematical knowledge in a database. Based on teaching objectives and student requirements, the systems could assemble necessary learning objects into a visual dependent graph or a document accordingly. This makes it possible for students to learn

© Springer Nature Switzerland AG 2018
J. Fleuriot et al. (Eds.): AISC 2018, LNAI 11110, pp. 209–214, 2018.
https://doi.org/10.1007/978-3-319-99957-9_14

knowledge contents in a more flexible way. However, the textbook contents are not yet machine-understandable to a desired extent, so that computer programs still have difficulties in effectively dealing with the underlying knowledge to exactly respond users' requirements of query, browsing, and question answering, etc.

Many efforts have been dedicated to tackling the difficulties of managing textbook knowledge effectively. For example, electronic geometry textbooks were studied from the point of view of dynamic software specially designed for interactively managing geometric knowledge that can be processed and interfaced with software tools developed for geometric reasoning, diagram visualization, multi-versioned textbook generation and consistency checking [3, 5]. An intelligent and dynamic geometry book for the future was proposed and sketched with adaptive, collaborative, visual, and intelligent features to bring together a whole new generation of mathematical tools with impact in all levels of education [13]. Interactive mathematical documents were introduced and realized within the MathDox system that supports recording context information of both mathematical and personal data for users [6]. Moreover, an intelligent textbook, named Inquire Biology, was designed to answer users' questions by applying knowledge representation and question-answering technology to electronic textbooks [2]. Intelligent tutoring systems have been intensively studied and developed to provide immediate and customized instruction or feedback to learners by synthesizing domain knowledge, problem solving strategies, pedagogical approaches, student capabilities and profiles, etc. into interactive exercises [11].

Based on previous explorations, we started a new project named LAText (for Linear Algebra Textbook system). The main objective is to investigate the methodologies of creating intelligent textbook systems capable of not only managing mathematical knowledge with semantics in the form of textbook but also making use of the semantic representations of knowledge for fine-grained management and retrieval, automated computation and reasoning, effective interaction and tutoring. This is an ongoing project and we will present our general view and design principles on building such textbooks and report our progress and experiments on the implementation of a system.

## 2    General View on Intelligent Textbooks

We view intelligent textbooks as software platforms for managing both domain knowledge with semantics and software toolkits of various kinds to provide users with facilities of dynamic authoring and rendering, fine-grained retrieving and browsing, and automated problem solving. Their features can be characterized from the following three aspects.

**Modularization and Structurization.** In order to be effectively manageable and processable, systematic knowledge contents need to be classified and encapsulated into manipulable semantic units, called *knowledge objects*. Each knowledge object organizes contents with the same semantics but in different representation formats for specific applications of rendering, computing, and reasoning, etc. According to different types of semantic relations, knowledge objects are linked with each other in a multi-scale way so that knowledge retrieving, browsing, and problem solving can be efficiently performed consequently. Some standard formats, such as OMDoc [8] and OntoMath [7], have been developed for the modularization and structurization of mathematical documents.

**Formalization and Interoperation.** It is important to make knowledge contents capable of interoperating with external software tools and other knowledge resources to efficiently realize specific intelligent settings, such as multi-platform displaying and automated problem solving. Transformations between different knowledge representation formats are required with semantics unchanged. Task-oriented formalization approaches are provided for representing the semantics of knowledge objects in a machine-processable way and the formalized contents can thus play a standard and intermediate role in corresponding transformation tasks. In addition, semantic features can also be extracted from the formalized representations for accurate knowledge object matching and retrieving. Some standard formats, such as OpenMath [12], have been designed and implemented for the formalization of mathematical expressions and interfacing with selected computer algebra systems and dynamic geometry software. Some systems, such as Theorema [15] and Mizar [1], support mathematical theorem proving, verification, and theory exploration through formalizing a certain amount of mathematical knowledge.

**Crowdsourcing and Personalization.** The textbook is freely accessible on the web to teachers, students, and other interested users. Structured and formalized knowledge contents are allowed to be created, revised, and reviewed via collective intelligence in contrast to traditional textbooks usually authored by a few domain experts with their own intelligence. Effective mechanisms are implemented to automatically or interactively justify the assessments of revisions and update the textbook with its acceptable version adaptively in real time [4]. According to specified teaching or learning objectives and tasks, the platform can automatically retrieve and interactively assemble target knowledge contents into usable courseware and documents for teachers. Furthermore, every student has access to a personalized dynamic textbook under comprehensive consideration of the student's learning capability and weakness and tutoring pedagogy. To personalize learning processes of students, intensive efforts have been made on adaptive learning and intelligent tutoring methods and systems, such as ActiveMath [10] and Knewton [14].

## 3   Design Principles and Elements of LAText

Based on our general view, we have investigated concrete methods for the construction of an intelligent textbook system by taking linear algebra as case study. The following principles have been used for the design of the LAText system.

**Architecture and Modules.** The system manages both a fine-grained knowledge base and a knowledge processing toolkit to respond users' requests through a user interface with dialog box on the web (see Fig. 1). The knowledge base stores knowledge objects encapsulating multi-format contents together with specific relations among the objects. The knowledge toolkit includes a search engine for retrieving knowledge objects via free text, mathematical expressions, and semantic relations, a convertor for equivalently transforming knowledge contents from one format to another, and interfaces with external computing and reasoning tools.

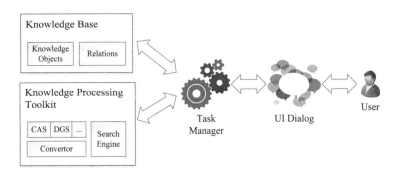

**Fig. 1.** Framework of the LaTeXt system

**Knowledge Management.** It is difficult to select an appropriate granularity of knowledge objects satisfying all requirements. We deal with the problem in a scalable way. Firstly, knowledge objects are created, encapsulating contents of definitions, axioms, theorems, proofs, problems, solutions, and examples, etc. that are usually marked in traditional mathematical textbooks. Then related objects are linked with each other through tagged edges that indicate meanings of the relations, such as definition of Determinant $\rightarrow_{\text{isContextOf}}$ Cramer's Rule, symmetric matrix $\rightarrow_{\text{inherit}}$ square matrix, and orthonormalizing a set of vectors $\rightarrow_{\text{hasAlgorithm}}$ Gram–Schmidt process. Finally, knowledge objects together with their relations are viewed as a tagged relationship graph, called *knowledge development graph*, depicting the dependent structure of domain knowledge. For the purposes of accurate retrieving, computing, and reasoning, selected subgraphs of the knowledge development graph can further derive more detailed tagged graphs by analyzing and processing the encapsulated contents. On the whole, the knowledge base can be viewed as consisting of multiple tagged relationship graphs for different views and applications of the system. To formally represent the structural semantics of knowledge contents, we have designed an ontology defining concepts of knowledge objects, properties and relations of the concepts, and required constraints.

## 4    Implementation Issues

We have implemented a preliminary version of L&Text [9] built with Node.js server, AngularJS front-end web application framework, and Ionic client, to be efficiently deployed on both desktop and mobile platforms. A graph database system Neo4j is used as a knowledge base to store and manage structural textbook contents. Currently, representation formats for mathematical expressions in L&TEX, MathML, and OpenMath are allowed. A rich text editor CKEditor with MathJax plugin is embedded into the client for users to edit knowledge objects with mathematical contents. Traditional readable documents in HTML with hybrid links are automatically converted and generated from the stored knowledge contents (see Fig. 2).

Textbook content adding, revising, and removing are allowed dynamically without breaking the non-redundancy and completeness of the textbook. To avoid creating a knowledge object already existing in the knowledge base, we adopt the following interactive mechanism. A classic text-based search engine is implemented based on vector space model with TF-IDF scheme. The search engine retrieves from the knowledge base a ranked list $\mathbb{L}$ of knowledge objects $\mathcal{O}_1, \mathcal{O}_2, \ldots, \mathcal{O}_n$ whose contents are similar with the content of a knowledge object $\mathcal{O}$ that a user is adding or revising. Then the user determines whether there exists an $\mathcal{O}_i$ in $\mathbb{L}$ which has the same semantics as $\mathcal{O}$. A path searching tool is also implemented for retrieving related objects in the view of tagged relationship graphs.

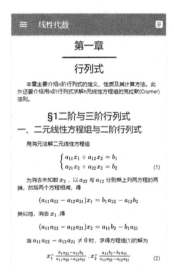

**Fig. 2.** Rendering L&Text

## 5    Conclusion and Future Work

In this paper, we present our general view on intelligent mathematical textbooks from three aspects: structurization, interoperation, and personalization. As an ongoing project of case study, a linear algebra textbook system L&Text has been designed and a preliminary implementation of the system is reported. Current version of the system provides users with basic manipulations of adding, removing, revising, browsing linear algebraic knowledge objects, and retrieving similar ones. Traditional textbook-like documents with hybrid links can be generated and rendered. To realize accurate knowledge retrieval and automated problem solving, we will focus our future work on the design of suitable formal languages for encoding knowledge object contents that require efficient transformations with both natural languages and internal representation formats of software tools developed for algebraic computation and reasoning.

Furthermore, the creation of textbooks via collective intelligence and the personalization of textbooks adaptive to learners will be studied in a late stage.

**Acknowledgements.** This work was supported by NSFC (61702025 and 11771034), SKLSDE-2017ZX-11, Special Fund for Guangxi Bagui Scholar Project, and Scientific Research Foundation of the Education Department of Guangxi Zhuang Autonomous Region (2017KY0173).

# References

1. Bancerek, G., et al.: Mizar: state-of-the-art and beyond. In: Kerber, M., Carette, J., Kaliszyk, C., Rabe, F., Sorge, V. (eds.) CICM 2015. LNCS (LNAI), vol. 9150, pp. 261–279. Springer, Cham (2015). https://doi.org/10.1007/978-3-319-20615-8_17
2. Chaudhri, V., et al.: Inquire biology: a textbook that answers questions. AI Mag. **34**(3), 55–72 (2013)
3. Chen, X.: Electronic geometry textbook: a geometric textbook knowledge management system. In: Autexier, S., et al. (eds.) CICM 2010. LNCS (LNAI), vol. 6167, pp. 278–292. Springer, Heidelberg (2010). https://doi.org/10.1007/978-3-642-14128-7_24
4. Chen, X., Li, W., Luo, J., Wang, D.: Open geometry textbook: a case study of knowledge acquisition via collective intelligence. In: Jeuring, J., et al. (eds.) CICM 2012. LNCS (LNAI), vol. 7362, pp. 432–437. Springer, Heidelberg (2012). https://doi.org/10.1007/978-3-642-31374-5_31
5. Chen, X., Wang, D.: Towards an electronic geometry textbook. In: Botana, F., Recio, T. (eds.) ADG 2006. LNCS (LNAI), vol. 4869, pp. 1–23. Springer, Heidelberg (2007). https://doi.org/10.1007/978-3-540-77356-6_1
6. Cohen, A., Cuypers, H., Verrijzer, R.: Mathematical context in interactive documents. Math. Comput. Sci. **3**(3), 331–347 (2010)
7. Elizarov, A., Kirillovich, A., Lipachev, E., Nevzorova, O.: Digital ecosystem OntoMath: mathematical knowledge analytics and management. In: Kalinichenko, L., Kuznetsov, S.O., Manolopoulos, Y. (eds.) DAMDID/RCDL 2016. CCIS, vol. 706, pp. 33–46. Springer, Cham (2017). https://doi.org/10.1007/978-3-319-57135-5_3
8. Kohlhase, M.: OMDoc – An Open Markup Format for Mathematical Documents [Version 1.2]. LNCS (LNAI), vol. 4180. Springer, Heidelberg (2006). https://doi.org/10.1007/11826095
9. LAText: A Linear Algebra Textbook System. http://www.latext.net/
10. Melis, E., Siekmann, J.: ActiveMath: an intelligent tutoring system for mathematics. In: Rutkowski, L., Siekmann, J.H., Tadeusiewicz, R., Zadeh, L.A. (eds.) ICAISC 2004. LNCS (LNAI), vol. 3070, pp. 91–101. Springer, Heidelberg (2004). https://doi.org/10.1007/978-3-540-24844-6_12
11. Nkambou, R., Azevedo, R., Vassileva, J. (eds.): ITS 2018. LNCS, vol. 10858. Springer, Cham (2018). https://doi.org/10.1007/978-3-319-91464-0
12. OpenMath Home. http://www.omdoc.org/
13. Quaresma, P.: Towards an intelligent and dynamic geometry book. Math. Comput. Sci. **11**(3–4), 427–437 (2017)
14. Wilson, K., Nichols, Z.: The Knewton platform: a general-purpose adaptive learning infrastructure. A Knewton White Paper (2015)
15. Windsteiger, W.: Theorema 2.0: a brief tutorial. In: Proceedings of SYNASC 2017, pp. 1–3. IEEE Explore (2017)

# Towards an Automated Geometer

Francisco Botana[1], Zoltán Kovács[2(✉)], and Tomás Recio[3]

[1] Department of Applied Mathematics I, University of Vigo,
Campus A Xunqueira, 36005 Pontevedra, Spain
fbotana@uvigo.es
[2] The Private University College of Education of the Diocese of Linz,
Salesianumweg 3, 4020 Linz, Austria
zoltan@geogebra.org
[3] Universidad de Cantabria, Avenida de los Castros, s/n,
39071 Santander, Spain
tomas.recio@unican.es

**Abstract.** We report on preliminary work towards the automated finding of theorems in elementary geometry. The resulting system is being currently implemented on top of *GeoGebra*, a dynamic geometry system with millions of users at high schools and universities. Our system exploits *GeoGebra*'s recently added new functionalities concerning automated reasoning tools in geometry. We emphasize that the method for finding geometric properties that are present on a user-provided construction is purely symbolic, thus giving such properties rigorous mathematical certainty. We describe some generalities about the system we are developing, which are illustrated through an example.

**Keywords:** Automated discovery · Automated theorem proving
Computer algebra · *GeoGebra*

## 1 Introduction

Half a century ago Lenat's *AM* [1] introduced a rule based system able to successfully discover (or rediscover) non-trivial mathematical results in number theory. It tried to replicate a human approach to "doing mathematics". Our aim, somehow similar, but in the realm of automated discovery in geometry, has been inspired by the strategy reported in [2, p. 44]. Roughly, it consists of using automatic reasoning tools for checking mechanically produced statements involving elements of a geometric construction, both in the case where those elements are actually present in the construction or when the elements are automatically generated by the program from the given ones. For example, given a triangle and a point on its plane, the system will develop some elementary operations between the point and the vertices (or sides). These operations (drawing perpendicular lines, adding midpoints of sides and lines from vertices to midpoints, ...) can be also imposed by the user or suggested somehow by the system. Then, the

ⓒ Springer Nature Switzerland AG 2018
J. Fleuriot et al. (Eds.): AISC 2018, LNAI 11110, pp. 215–220, 2018.
https://doi.org/10.1007/978-3-319-99957-9_15

program will verify the truth/failure of different statements concerning collinearity, parallelism, ... of the different elements, given or generated, in the input construction.

This line of work concerning automated discovery (i.e. finding statements holding in a given figure) in geometry was initiated, to the best of our knowledge, in [3]. There, the authors developed a system with a generator of constructions where a systematic search is performed to find new conjectures which are then proved through Wu's algebraic method. A related proposal, able to discover all properties of a construction given a set of rules, was reported in [4]. Finally, a report on discovering properties from scanned images has been described in [5]. Some strategies are used to generate conjectures involving the image translation to a geometric figure, and algebraic computations return their truth or falsity.

Our system is being built on top of *GeoGebra*, exploiting recent abilities on automatic reasoning tools in geometry [6]. The *Automated Geometer, AG*, (also meaning *Amateur Geometer*) intends to be a *GeoGebra* module where pure automatic discovery is performed. It includes a generator of further geometric elements from those of a given construction, and a set of rules for producing conjectures on the whole set of elements. But the ultimate *AG* aim is not just performing a systematic exploration of the space of possible conjectures, but mimicking human thought when doing elementary geometry.

*GeoGebra*, from its first versions, incorporates a *Relation* tool that returns the results of some basic checks (for instance, incidence, parallelism, perpendicularity, equal length, ...) between a pair of selected elements. This command is not exclusive to *GeoGebra*: it also existed in previous dynamic software like *Cabri*, but until the inclusion of automated reasoning tools all these approaches in widespread environments were based on numerical checking. A paradigmatic example of this numerical checking is *OK Geometry* [7]: this tool detects relevant facts in a construction by slightly moving its free points, checking which relations among them remain then invariant, and filtering the results through a library of well-known properties.

The *GeoGebra AG* module does not perform numerical checking. Rather, all facts are symbolically managed by means of the `Prove` command [6]. The module runs on modern browsers, thus providing universal accessibility, and it is controlled by the Javascript API [8]. Currently, *AG* is able to accept a user defined construction (that could be also the result of loading a preexistent one) and it searches for meaningful relations between the construction elements. All possible relations are listed on a combinatorial basis, and those classified as *generally true* by the prover algorithm are returned. So, *AG* only outputs certified true properties in constructions. Furthermore, since it is built on top of *GeoGebra*, it can reach an audience of millions of mathematics students.

From the technical perspective we highlight that the *AG* module currently runs in a web browser and it implicitly uses a precompiled version of the *Giac* [9] computer algebra system as a piece of JavaScript or WebAssembly [10] code. To our knowledge, these kinds of technologies ensure the users the quickest performance to obtain results on heavy computations in a popular, user-friendly way. For instance, running our *Prove* algorithm, that is based on an algebraic geometry approach to automatic proving [11], requires Gröbner bases computations and variable elimination, two time and space consuming operations that are efficiently solved in our framework.

## 2  *AG* at Work

The *AG* module can be freely tested at http://htmlpreview.github.io/?https://github.com/kovzol/ag/blob/master/automated-geometer.html, and its development is shown at https://github.com/kovzol/ag. Figure 1 shows part of the default screen of the web application. There, a simple construction is displayed, involving three user-defined points $A, B, C$, their midpoints $D, E, F$ and a fourth midpoint $G$ between $D$ and $E$. The user is requested to select, among a list of possible choices, the type of generic properties to be tested. Currently, the choices are: collinearity or equality of distances between three points, and perpendicularity or parallelism of segments defined by two points.

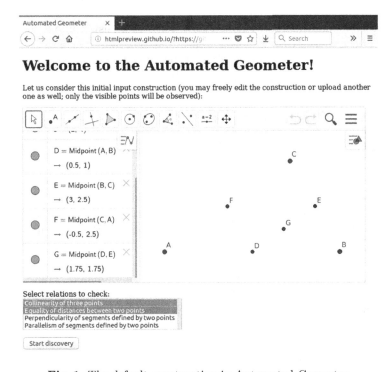

**Fig. 1.** The default construction in Automated Geometer.

Soon, after starting the discovery process and assuming that collinearity and distance equality between three points have been selected, $AG$ returns a list of sixteen theorems:

$$D \in AB, F \in AC, E \in BC, G \in BF, G \in DE, AD = BD,$$
$$AD = EF, AF = CF, AF = DE, BD = EF, BE = CE,$$
$$BE = DF, BG = FG, CE = DF, CF = DE, DG = EG.$$

It could be argued that some of these theorems are, in fact, mere reformulations of the constructions constraints. For instance, it is true that $D \in AB$ since $D$ is the midpoint of $A$ and $B$. An identical argument can also be applied to $AD = BD$. Only theorems like $AD = EF, AF = DE, BD = EF, BE = DF, CE = DF, CF = DE$ are of different nature, although, for the expert reader, they could seem also quite obvious. Yet, we remark we are just describing a prototype and a basic toy example. Anyway, deciding, automatically, the relevance of the obtained mathematical properties is an interesting, on-going, research question [12]. On the other hand let us remark that $AG$ is being designed not only as an academic tool, but as a geometry discovery product for the masses (i.e. for being used in touristic walks, as a helper mechanism to appreciate better the geometry of some monuments [13]), so we do not have yet a final answer about the level of difficulty of the results that we would expect our program to achieve. A balance between cognitive load and the richness of discovered facts will be investigated with the help of didactics experts and field tests.

## 3   Future $AG$ Improvements

A major improvement will consist of connecting the LocusEquation GeoGebra command [14] with $AG$. Currently, this command can accept two parameters, a Boolean condition and a free point in the construction, returning as an implicit curve the equation that the free point must satisfy in order to verify the Boolean condition. A well-known example of this situation is the Simson-Wallace theorem about the collinearity of the projections from a point on the sides of a triangle (Fig. 2, the projections are collinear if the point $D$ lies on the circumcircle).

The actual version of LocusEquation gives a conic as the necessary path of $D$ for the sought collinearity. Under the current $AG$ development, points $D, E, F$ are not, in general, aligned, so the theorem $E \in FG$ will not be returned. $AG$ aims to complement user intelligence (when he/she states the boolean condition) by a heuristic search on the construction elements, that is, using in a systematic way the LocusEquation command with simple conditions and relevant points of the construction. The outputs of the command will also be filtered trying to identify meaningful relations in the construction.

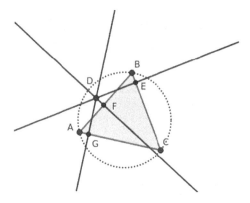

**Fig. 2.** The projections from a circumcircle point to the sides are collinear.

**Acknowledgement.** Authors partially supported by the grant MTM2017-88796-P from the Spanish MINECO and the ERDF (European Regional Development Fund).

# References

1. Lenat, D.B.: Automated theory formation in mathematics. Contemp. Math. **29**, 287–314 (1984)
2. de Guzmán, M.: La experiencia de descubrir en geometría. Nivola (2002)
3. Bagai, R., Shanbhogue, V., Żytkow, J.M., Chou, S.C.: Automatic theorem generation in plane geometry. In: Komorowski, J., Raś, Z.W. (eds.) ISMIS 1993. LNCS, vol. 689, pp. 415–424. Springer, Heidelberg (1993). https://doi.org/10.1007/3-540-56804-2_39
4. Chou, S.C., Gao, X.S., Zhang, J.Z.: A deductive database approach to automated geometry theorem proving and discovering. J. Autom. Reason. **25**, 219–246 (2000)
5. Chen, X., Song, D., Wang, D.: Automated generation of geometric theorems from images of diagrams. Ann. Math. Artif. Intell. **74**, 333–358 (2015)
6. Botana, F., Hohenwarter, M., Janicic, P., Kovács, Z., Petrovic, I., Recio, T., Weitzhofer, S.: Automated theorem proving in GeoGebra: current achievements. J. Autom. Reason. **55**, 39–59 (2015)
7. Magajna, Z.: OK Geometry. http://z-maga.si/index?action=article&id=40. Accessed 3 May 2018
8. The GeoGebra Team: Reference: GeoGebra Apps API. https://wiki.geogebra.org/en/Reference:GeoGebra_Apps_API. Accessed 10 May 2018
9. Kovács, Z., Parisse, B.: Giac and GeoGebra - improved Gröbner basis computations. In: Gutierrez, J., Schicho, J., Weimann, M. (eds.) Computer Algebra and Polynomials. LNCS, vol. 8942, pp. 126–138. Springer, Cham (2015). https://doi.org/10.1007/978-3-319-15081-9_7
10. Bright, P.: The web is getting its bytecode: WebAssembly. Condé Nast (2015)
11. Recio, T., Vélez, M.P.: Automatic discovery of theorems in elementary geometry. J. Autom. Reason. **23**, 63–82 (1999)
12. Gao, H., Goto, Y., Cheng, J.: A set of metrics for measuring interestingness of theorems in automated theorem finding by forward reasoning: a case study in NBG

set theory. In: He, X. (ed.) IScIDE 2015. LNCS, vol. 9243, pp. 508–517. Springer, Cham (2015). https://doi.org/10.1007/978-3-319-23862-3_50

13. Botana, F., Kovács, Z., Martínez-Sevilla, A., Recio, T.: Automatically augmented reality with GeoGebra (to appear)

14. Abánades, M., Botana, F., Kovács, Z., Recio, T., Sólyom-Gecse, C.: Development of automatic reasoning tools in GeoGebra. ACM Commun. Comput. Algebr. **50**, 85–88 (2016)

# Automatic Deduction in an AI Geometry Book

Pedro Quaresma$^{(\boxtimes)}$ (iD)

CISUC/Department of Mathematics, University of Coimbra, Coimbra, Portugal
pedro@mat.uc.pt

**Abstract.** The pursuit of an *AI Geometry Book* should involve the study of how currently developing methodologies and technologies of geometry knowledge representation, management, deduction and discovery can be incorporated effectively into a computational application, a "book" of the future.

In the geometry book of the future statements and proofs should be en-lighted by dynamic geometry sketches and diagrams, and the correctness of the proofs should be ensured by computer checking. The book will be intelligent, the reader should be able to ask closed or open questions, and can also ask for proof hints. The book should also provide interactive exercises with automatic correction.

To fulfil such a goal the development of an open library of geometry automated theorem provers with a carefully design application interface protocol, must be considered. This would allow to link computer platforms for geometry with theorem provers, providing the automatic deduction capabilities for the *AI Geometry Book*.

## 1 Introduction

The geometry book of the future should be intelligent, correct, visual, adaptive, and collaborative in its production and use. Although the previous efforts and achievements help set aside many obstacles on the way to an intelligent geometry book, all the approaches, methods, techniques, basic tools, and systems, are still developed, applied, and conceived in separated, and relatively small, circles [13, 19].

When considering the application of automated reasoning in a learning context two, somehow opposing, goals are to be considered: efficiency and readability of the proofs. Whenever the help of the computer is considered, the user wants a fast and friendly answer [6].

Overall, the existing methods for automated theorem proving and discovering in geometry are very efficient and, in some cases, providing an output that can be used in a learning context. However, finding a method/implementation capable of both (fast and friendly) is only possible for a small number of cases. There is still a lot of room for further improvements [3].

Automated deduction methods can be more tightly integrated within interactive theorem provers and dynamic geometry [8,9]. Automated deductions methods are currently unable to produce guidance and explanations for educational

© Springer Nature Switzerland AG 2018
J. Fleuriot et al. (Eds.): AISC 2018, LNAI 11110, pp. 221–226, 2018.
https://doi.org/10.1007/978-3-319-99957-9_16

purposes. New techniques are needed involving synthetic geometry, formalisation techniques and artificial intelligence [2,4,5,16].

Computer checking of geometric proofs has been addressed in the literature but: current formal proofs are not readable, natural and visual languages should be connected with formal deductions; several foundations of geometry and automated theorem provers are available, but all connections between them have not been obtained; the editing of a pedagogical corpus from formalisation have not been studied deeply enough [4,7,10,20,21].

Building on the work already developed in the systems: *OpenGeoProver* (*OpenGeoProver*) an open source project implementation of GATPs [1,2]; the *Thousand of Geometric problems for geometric Theorem Provers* (*TGTP*) a problem repository and test bench [12]; and the standardised format for constructions and conjectures in geometry, I2GATP [14], we intend to extend and integrate these projects, providing deductive services to other applications.

The immediate program include the development/implementation of new methods to incorporate into the *OpenGeoProver*; finalising the standard format I2GATP to be able to define a programming protocol between *OpenGeoProver* and third-party programs; extend the platform *TGTP* in such a way that it can become the base for a competition between GATPs developed by the scientific community, to boost the improvement of methods/implementations.

*Overview of the Paper.* This paper is organised as follows: In Sect. 2 the current status and future developments of *OpenGeoProver* are described. In Sect. 3, the *TGTP* platform is described and the goal for a future competition between GATPs are set. In Sect. 4 the common format I2GATP is described and future developments are foreseen. In Sect. 5 the integration issues are discussed. In Sect. 6 some final remarks are drawn.

## 2   Open Geometry Prover

The *OpenGeoProver*[1] is an open source project, aiming to implement various geometry automated theorem provers (GATPs). It can be used as a stand-alone tool but can also be integrated into other geometry tools, such as dynamic geometry software, e.g., work is being made to integrate *OpenGeoProver* with *GeoGebra* [2]. In its current state, *OpenGeoProver* implements two algebraic methods, the Wu's method and the Gröbner basis method, as well as one semi-synthetic method, the *area method*. Some initial work on the implementation of the *full-angle method* was already done [1].

As future developments we have: the implementation of new methods, finishing the work done with *full-angle method*, implementation of the deductive databases method, among others; the design of an application programming interface (API), allowing the different implemented GATPs to be easily used by other programs; write an extensive documentation, with a special care in the description of the API.

---

[1] https://github.com/ivan-z-petrovic/open-geo-prover/.

The *OpenGeoProver* will constitute an excellent mean to provide automated reasoning resources to the *AI Geometry Book*.

## 3   The *TGTP* Platform

Thousands of Geometric problems for geometric Theorem Provers (*TGTP*) is a Web-based library of problems in geometry, a comprehensive common library of problems with a significant size and unambiguous reference mechanism, easily accessible to all researchers in the automated reasoning in geometry community. It share the database of problems with the *GeoThms*, a Web-based framework for exploring geometrical knowledge that integrates Dynamic Geometry Software (DGS), Automatic Theorem Provers (ATP), and a repository of geometrical constructions, figures and proofs [5,12].

One of the main motivations in building *TGTP* is to support the testing and evaluation of geometric automated theorem proving (GATP) systems. Providing a common library of problems for the testing of GATPs it will allow the pursue of the development of better methods/implementations.

Along the lines of the competition CASC [18], the conception and operationalization of a competition among GATPs is part of the future developments for the *TGTP*, it will allow the development of better methods/implementations, but will also push in the direction of a better interface between the geometric information and the programs that can use the information.

For the *AI Geometry Book* a repository such as *TGTP* is of extreme importance, it will constitute a source of valuable information to enrich the "book".

## 4   The I2GATP Common Format

The I2GATP format is an extension of the I2G (*Intergeo*) common format aimed to support conjectures and proofs produced by geometric automatic theorem provers. The goal in building such a format is to provide a communication channel between different tools from the field of geometry, allowing linking such tools, as well as allowing the use of geometric knowledge kept in different repositories [14,17].

The I2GATP format is a combination of four XSD files: `information.xsd`, with all the meta-information regarding the conjecture; `intergeo.xsd`, the *Intergeo* I2G format, for the description of geometric constructions; `conjecture.xsd`, the translation of `geocons.dtd` to the XSD format [11]; `proofInfo.xsd`, the meta-information regarding the proof generated by a given GATP on a given computing platform..

The I2GATP container is an extension of the I2G container. In addition to the information in the I2G container, all the information regarding the geometric conjecture and all the proofs attempts are kept in the I2GATP container.

The I2GATP library is an open source project[2], to support the writing of filters from/to the I2GATP container and different geometric computation tools.

---

[2] https://github.com/GeoTiles/libI2GATP.

The I2GATP library will enable to specify an API, opening the possibility of geometric information interchange and the use of the GATPs in *OpenGeoProver* (and others, supporting the I2GATP format) and also the information contained in *TGTP*.

## 5    Deductive Tools in an AI Geometry Book

The integration that can be seen in systems like *GeoThms*, *GeoGebra*, *Cinderella*, *JGEX*, *GCLC* and *GeoProof* [2,4,5,10,16,21], should be used as an inspiration to the *AI Geometry Book*.

The *AI Geometry Book* should be able to interface with repositories of geometric information and with DGSs and GATPs. Using the I2GATP common format as interface we could include deductive services via the *OpenGeoProver* and information via the *TGTP* repository.

Another two development that the author and other researcher are working on are: a taxonomy for geometric problems allowing to adjust the queries to each and every user [15]; a semantic search mechanism that will allow to search for geometric information in a geometric fashion.[3]

## 6    Conclusions and Future Work

The pursuit of an *AI Geometry Book*—where the "book" is understood as a computation platform that will extend the concept of (non-digital) object of study, a repository of (non-digital) knowledge, to the digital world—needs the use of many AI techniques. AI is needed for the automatic deduction, for the adaptation to the user's profiles, for the search of information.

Some of this issue are already, partially, solved, e.g. there are already many efficient GATPs, but the integration of all the "players" in such a way that a *AI Geometry Book* can be built is still to be done. As described above we are working on those issues, trying to integrate *OpenGeoProver* with *TGTP* through the I2GATP common format, and also on some issued related with the usability of such integrated platform, i.e. the adaptability and the semantic geometric search.

**Acknowledgements.** Financed by national funding via the Foundation for Science and Technology and by the European Regional Development Fund (FEDER), through the COMPETE 2020 – Operational Program for Competitiveness and Internationalisation (POCI).

---

[3] Geometric Figure Mining via Conceptual Graphs, Yannis Haralambous and Pedro Quaresma, submitted to ADG2018.

# References

1. Baeta, N., Quaresma, P.: The full angle method on the OpenGeoProver. In: Lange, C., et al. (eds.) MathUI, OpenMath, PLMMS and ThEdu Workshops and Work in Progress at the Conference on Intelligent Computer Mathematics, No. 1010 in CEUR Workshop Proceedings, Aachen (2013). http://ceur-ws.org/Vol-1010/paper-08.pdf
2. Botana, F., et al.: Automated theorem proving in Geogebra: current achievements. J. Autom. Reason. **55**(1), 39–59 (2015). https://doi.org/10.1007/s10817-015-9326-4
3. Chou, S.C., Gao, X.S.: Automated reasoning in geometry. In: Robinson, J.A., Voronkov, A. (eds.) Handbook of Automated Reasoning, pp. 707–749. Elsevier Science Publishers B.V., San Diego (2001)
4. Janičić, P.: GCLC—a tool for constructive euclidean geometry and more than that. In: Iglesias, A., Takayama, N. (eds.) ICMS 2006. LNCS, vol. 4151, pp. 58–73. Springer, Heidelberg (2006). https://doi.org/10.1007/11832225_6
5. Janičić, P., Quaresma, P.: System description: GCLCprover + geoThms. In: Furbach, U., Shankar, N. (eds.) IJCAR 2006. LNCS (LNAI), vol. 4130, pp. 145–150. Springer, Heidelberg (2006). https://doi.org/10.1007/11814771_13
6. Jiang, J., Zhang, J.: A review and prospect of readable machine proofs for geometry theorems. J. Syst. Sci. Complex. **25**(4), 802–820 (2012). https://doi.org/10.1007/s11424-012-2048-3
7. Matsuda, N., Vanlehn, K.: Gramy: a geometry theorem prover capable of construction. J. Autom. Reason. **32**, 3–33 (2004)
8. Moraes, T.G., Santoro, F.M., Borges, M.R.: Tabulæ: educational groupware for learning geometry. In: Fifth IEEE International Conference on Advanced Learning Technologies, ICALT 2005, pp. 750–754, July 2005. https://doi.org/10.1109/ICALT.2005.251
9. Moriyón, R., Saiz, F., Mora, M.: GeoThink: an environment for guided collaborative learning of geometry. In: Sánchez, J. (ed) Nuevas Ideas en Informática Educativa, Santiago de Chile, vol. 4, pp. 200–208 (2008)
10. Narboux, J.: A graphical user interface for formal proofs in geometry. J. Autom. Reason. **39**, 161–180 (2007). https://doi.org/10.1007/s10817-007-9071-4
11. Quaresma, P., Janičić, P., Tomašević, J., Vujošević-Janičić, M., Tošić, D.: XML-bases format for descriptions of geometric constructions and proofs. In: Communicating Mathematics in the Digital Era, pp. 183–197. A. K. Peters Ltd., Wellesley (2008)
12. Quaresma, P.: Thousands of geometric problems for geometric theorem provers (TGTP). In: Schreck, P., Narboux, J., Richter-Gebert, J. (eds.) ADG 2010. LNCS (LNAI), vol. 6877, pp. 169–181. Springer, Heidelberg (2011). https://doi.org/10.1007/978-3-642-25070-5_10
13. Quaresma, P.: Towards an intelligent and dynamic geometry book. Math. Comput. Sci. **11**(3), 427–437 (2017). https://doi.org/10.1007/s11786-017-0302-8
14. Quaresma, P., Baeta, N.: Current status of the I2GATP common format. In: Botana, F., Quaresma, P. (eds.) ADG 2014. LNCS (LNAI), vol. 9201, pp. 119–128. Springer, Cham (2015). https://doi.org/10.1007/978-3-319-21362-0_8
15. Quaresma, P., Santos, V., Graziani, P., Baeta, N.: Taxonomies of geometric problems. J. Symb. Comput. (2018, Submitted)
16. Richter-Gebert, J., Kortenkamp, U.: The Interactive Geometry Software Cinderella. Springer, Heidelberg (1999)

17. Santiago, E., Hendriks, M., Kreis, Y., Kortenkamp, U., Marquès, D.: I2G Common File Format Final Version, Technical report D3.10, The Intergeo Consortium (2010). http://i2geo.net/xwiki/bin/view/I2GFormat/
18. Sutcliffe, G.: The 5th IJCAR automated theorem proving system competition - CASC-J5. AI Commun. **24**(1), 75–89 (2011). http://dl.acm.org/citation.cfm?id=1937696.1937700
19. Wang, D., Chen, X., An, W., Jiang, L., Song, D.: OpenGeo: an open geometric knowledge base. In: Hong, H., Yap, C. (eds.) ICMS 2014. LNCS, vol. 8592, pp. 240–245. Springer, Heidelberg (2014). https://doi.org/10.1007/978-3-662-44199-2_38
20. Wang, K., Su, Z.: Automated geometry theorem proving for human-readable proofs. In: Proceedings of the 24th International Conference on Artificial Intelligence, IJCAI 2015, pp. 1193–1199. AAAI Press (2015). http://dl.acm.org/citation.cfm?id=2832249.2832414
21. Ye, Z., Chou, S.-C., Gao, X.-S.: An introduction to Java geometry expert. In: Sturm, T., Zengler, C. (eds.) ADG 2008. LNCS (LNAI), vol. 6301, pp. 189–195. Springer, Heidelberg (2011). https://doi.org/10.1007/978-3-642-21046-4_10

# A Chinese New Word Detection Approach Based on Independence Testing

Dongchen Jiang[1]([email protected]), Xiaoyu Chen[2,3]([email protected]), and Xin Yang[1]

[1] School of Information Science and Technology,
Beijing Forestry University, Beijing 100083, China
jiangdongchen@bjfu.edu.cn
[2] Beijing Advanced Innovation Center for Big Data and Brain Computing,
Beihang University, Beijing 100191, China
[3] School of Mathematics and Systems Science,
Beihang University, Beijing 100191, China
chenxiaoyu@buaa.edu.cn

**Abstract.** New word detection is of great significance for Chinese text information processing, which directly affects the capabilities of word segmentation, information retrieval and automatic translation. Focusing on the problem of Chinese new word detection, this paper proposes an independence-testing-based detection approach with no need of prior information. The paper analyzes statistical characteristics of new words in Chinese texts, uses statistical hypothesis testing to infer the correlations between adjacent semantic units, and proposes an iterative algorithm to detect new words gradually. Our algorithm is evaluated on both large-scale corpus and short news texts. Experimental results show that this approach can effectively detect new words from all kinds of news.

**Keywords:** New word detection · Hypothesis testing
Test of independence · Semantic unit

## 1 Introduction

Words, the basic unit of a language, are important in information processing. In the fields of information retrieval, automatic translation, part-of-speech tagging and text semantic analysis, words are the basic symbolic units with particular meanings for processing. However, unlike English and other western languages, Chinese is based on characters without white spaces to mark word boundaries. Moreover, there are not unified definitions or rules to identify Chinese words. In order to process Chinese texts, it always needs a dictionary for word segmentation.

This work was supported by Fundamental Research Funds for Central Universities (No. BLX2015-17) and National Nature Science Foundation of China (No. 61702025).

J. Fleuriot et al. (Eds.): AISC 2018, LNAI 11110, pp. 227–236, 2018.
https://doi.org/10.1007/978-3-319-99957-9_17

However, the form of Chinese words is flexible and diverse. New words can be generated from existing words and characters through derivation, compounding, abbreviation, etc. The occurrence of new words makes it difficult to handle word segmentation with a fixed dictionary. Especially with the rapid development of the Internet, unknown names and places, new companies and expressions and other kinds of new words emerge frequently. The out-of-vocabulary problem becomes the most important factor that affects the accuracy of Chinese word segmentation [1]. Therefore, effective methods of new word detection are very important for Chinese language processing.

As words are not segmented by special symbols in Chinese and there are no morphological rules for Chinese word identification, it is not feasible to detect new words by syntax analysis or morphological analysis. Currently, frequently used methods for new word detection are based on statistics, semantical rules, or both. Candidates of new words are extracted according to basic statistical features, and they are filtered according to more complex statistical features or semantical rules. These methods are effective in practice, but they usually require prior knowledge or large-scale training corpus which leads to unexpected correlations between the detection results and the prior information.

To eliminate the dependencies on prior knowledge and training corpus, this paper studies the problem of Chinese new word detection from the perspective of statistical hypothesis testing, and proposes a new word detection approach with no need of prior information. The main contributions of this paper are as follows.

(1) Three criterias are proposed to describe the statistical characteristics of new words in Chinese texts;
(2) Statistical hypothesis testing techniques are used for Chinese new word detection, and an independence-testing-based detection approach is proposed.

## 2   Related Work

Currently, most effective methods of Chinese new word detection adopt a two-step approach. In the first step, all possible candidates are extracted from the target text. As the result may include many garbage strings, in the second step different strategies and methods are used to filter out garbage strings and improve the accuracy.

In the candidate extraction step, the most frequently used method is providing some kind of frequency threshold for character strings. Once the frequency of a string exceeds the threshold, the string will be extracted as a candidate. Zou et al. assumed that new words need to be repeated in a certain number of texts in a period of time. Thus, the total frequency of a string and the number of texts that contain the string are used as thresholds for candidate extraction [2]. Luo and Song proposed the concept of suffix array to handle candidate extraction. Strings that have the same prefix or suffix are indexed in right or left suffix array, and the ones that appear in both suffix arrays are treated as candidates

[3]. Li et al. set a fixed frequency threshold for n-gram candidate extraction [4]. Zhang et al. used hierarchical pruning to improve the n-gram based method, which reduces the number of garbage strings [5].

The methods for candidate-word filtering are mainly based on statistics or semantic rules. In statistics-based methods, different statistical features of new words are used to describe the internal association and external boundary of new words. He et al. proposed the concepts of inside word probability and double character coupling for filtering candidate words [6]. Luo and Zhao et al. used mutual information, left/right entropy and average entropy of left/right neighbor to filter candidate words [3,7]. Statistical models where various lexicons or statistical features are used, such as predication by partial matching (PPM) and conditional random fields (CRF), were also applied to new word detection [8,9].

Besides statistical features, manually constructed word-formation rules are also used for candidate words filtering. Zou et al. defined a set of rules by regular expressions for candidate-word filtering [2]. Cui et al. trained three garbage lexicons and one suffix lexicon using a large-scale corpus to remove garbage strings [10]. Zhang and Lin et al. integrated statistical features, semantical rules and other tactics to filter candidate words [5,11].

These methods are effective in new word detection. However, in most of the methods, the parameters of statistical features and semantic rules are obtained from training large-scale corpus or from prior knowledge. As there are dependent relations between the results and the prior knowledge or training corpus, once the prior knowledge is not proper for the target text or the training corpus has a different type with the target text, the accuracy of the results will be affected. To avoid the dependencies on prior information, this paper will study the problem of Chinese new word detection based on statistical characteristics of the target text.

## 3    Statistics-Based Modelling

### 3.1    Statistical Characteristics of New Words

Generally, people identify Chinese words according to their personal experiences and the meanings of the context. But for machines, context understanding is still a challenge in natural language processing, and the flexibility of Chinese makes the results of rule-based detection incomplete.

For a given Chinese text, we believe that a new word in the text can be identified if it satisfies the following characteristics: (1) the characters in the word are interrelated; (2) the word shows a certain independence and flexibility, i.e. it can be used as some specific component of a sentence and can be connected with different words or characters; (3) the word appears in the text at a certain frequency so that people can recognize it from the context. Based on the above analysis, the statistical characteristics of new words can be described by following three criterias.

1. The Chinese characters that compose a new word show strong correlations in a given text.

2. The occurrence of a new word and the occurrence of its adjacent characters (or words) are independent.
3. The frequency of a new word reaches a certain threshold in in a given text.

In order to obtain all new words of a given text automatically, all three criterias above need to be translated into mathematical methods, which can be processed on computer. As the correlations in Criteria 1 and the independency in Criteria 2 are both concepts in statistics, we can use hypothesis testing to infer whether Criteria 1 and 2 are satisfied. For example, for any two Chinese characters $X$ and $Y$, the hypothesis can be set as for any two adjacent characters, the event of $X$ being the first character and the event of $Y$ being the second character are independent. According to statistics knowledge, this hypothesis can be tested by observation and computation. If the test result shows that the hypothesis is accepted, then the two characters come together at random. If the result shows the rejection of the hypothesis, then there is a correlation between the occurrences of the two characters. Moreover, it is possible that the two characters constitute a word (or part of a word). Therefore, both the internal correlations between characters in a word and the external independencies between different words or characters can be inferred by hypothesis testing. In addition, for the last criteria, it only needs to set a basic frequency threshold for filtering.

## 3.2  Basic Concepts and Modelling

The idea of the independence-testing-based new word detection is to combine all interrelated characters into the candidates of new words. In this paper, a Chinese character string that represent a relatively complete meaning is called a *semantic unit*. Without additional information, all Chinese characters are semantic units. For a given text, the string of two adjacent semantic units is called a *semantic pair*. In a semantic pair, the former semantic unit is called the *pre-unit* of the semantic pair, and the latter one is called the *post-unit* of the semantic pair.

For a given text $T$, let the number of all semantic pairs be $n$ and all of these semantic pairs constitute a sample of hypothesis testing. For a semantic unit $u$, the number of all the semantic pairs which have $u$ as pre-units is denoted as $n_{u+}$. Then the probability of $u$ being a pre-unit of a semantic pair in $T$ can be estimated by $p_{u+} = n_{u+}/n$. Similarly, $n_{+u}$ denotes the number of all the semantic pairs which have $u$ as post-units and $p_{+u}$ is the relevant probability. Then, for any semantic units $u$ and $v$, the independence hypothesis can be stated as follows:

$H_0$: for any semantic pair in $T$, the event of $u$ being its pre-unit and the event of $v$ being its post-unit are independent.

Based on this hypothesis, the frequency $n_{u,v}$ of the semantic pair $uv$ in $T$ can be estimated by $np_{u+}p_{+v}$. Similarly, if we use $\tilde{u}$ to denote any semantic unit that is not $u$, then $n_{\tilde{u},v}$ can be estimated by $np_{\tilde{u}+}p_{+v}$ where $p_{u+} + p_{\tilde{u}+} = 1$.

With above analysis, the statistic $Q_{u,v}^2$ can be constructed to characterize the total frequency errors of semantic pairs associated with $u$ and $v$.

$$Q_{u,v}^2 = \frac{(n_{u,v} - np_{u+}p_{+v})^2}{np_{u+}p_{+v}} + \frac{(n_{u,\tilde{v}} - np_{u+}p_{+\tilde{v}})^2}{np_{u+}p_{+\tilde{v}}} + \frac{(n_{\tilde{u},v} - np_{\tilde{u}+}p_{+v})^2}{np_{\tilde{u}+}p_{+v}} + \frac{(n_{\tilde{u},\tilde{v}} - np_{\tilde{u}+}p_{+\tilde{v}})^2}{np_{\tilde{u}+}p_{+\tilde{v}}} \tag{1}$$

If we assume that in Eq. (1) each error indicating the difference between the actual frequency and the relevant estimated frequency fits a normal distribution, then $Q_{u,v}^2$ fits the chi-squared distribution with 1 degree of freedom, i.e. $Q_{u,v}^2 \sim \chi(1)^2$.

For any semantic pair $uv$ in $T$, the $2 \times 2$ contingency table for $uv$ can be constructed accordingly, i.e.

**Table 1.** $2 \times 2$ Contingency table of $uv$

| Pre-unit | Post-unit | | |
|---|---|---|---|
| | $+v$ | $+\tilde{v}$ | Row total |
| $u+$ | $a$ | $b$ | $a+b$ |
| $\tilde{u}+$ | $c$ | $d$ | $c+d$ |
| Column total | $a+c$ | $b+d$ | $n$ |

In Table 1, $a$ is the frequency of the semantic pair $uv$ in $T$, $b$, $c$ and $d$ are the frequencies of $u\tilde{v}$, $\tilde{u}v$ and $\tilde{u}\tilde{v}$, respectively, and $n = a+b+c+d$ holds. With this table, Eq. (1) can be simplified, and we have

$$Q_{u,v}^2 = \frac{n(ad - bc)^2}{(a+b)(c+d)(a+c)(b+d)}. \tag{2}$$

Accordingly, the correlation between the occurrences of $u$ and $v$ in one semantic pair can be inferred by independence testing: given a significance level $\alpha$, if $Q_{u,v}^2$ is in the critical region, $H_0$ is rejected, which means that the occurrences of $u$ and $v$ in one semantic pair is correlated; otherwise, $H_0$ should be accepted, which indicates that $u$ and $v$ occur in the semantic pair $uv$ independently.

## 4    Algorithm Based on Independence Testing

### 4.1    Problem Analysis

Independence testing can be used to infer the correlation between the occurrences of two semantic units in one semantic pair. However, even if the testing result is rejection of the hypothesis, it is still problematic to identify the semantic pair of the two units as one new word.

Firstly, the rejection of $H_0$ means that the event of $u$ being a pre-unit of one semantic pair and the event of $v$ being the post-unit of the same semantic pair are correlated. However, the correlation may not be determined by the co-occurrence of $u$ and $v$. It may be determined by the co-occurrence of $u$ and $\tilde{v}$ or the co-occurrence of $\tilde{u}$ and $v$. Therefore, to show that the correlation is indeed determined by the co-occurrence of $u$ and $v$, it is still necessary to demonstrate $n_{u,v} \geq \delta n p_{u+} p_{+v}$ where the coefficient $\delta$ is significant larger than 1.

Secondly, two interrelated semantic pairs often share one semantic unit, i.e. the post-unit of one semantic pair is the pre-unit of the other semantic pair. In this case, improper word identification tactic may result in inaccuracy or irrationality.

If we identify one of these overlapped interrelated semantic pairs as a word, the correlation of the other semantic pair is affected, which may cause inaccurate word identification. Take the name " 特朗普 " (Trump) as an example, the consecutive semantic units " 特朗普 " contains two overlapped semantic pairs " 特朗 " and " 朗普 ", but neither of them is a word. Thus, it is inappropriate to select one of the overlapped semantic pairs as a new word.

However, if we identify all the consecutive semantic units that constitute several overlapped semantic pairs as one word, it may also cause irrationality. For example, " 人民 " (people) and " 民办 " (civilian-run) are two interrelated semantic pairs sharing the common character " 民 ". And it is unwise to identify " 人民办 " as one new word in the sentence " 为人民办实事 " (do practical work for the people).

The main reason for this problem is that the test of independence is used to infer the correlation between two semantic units and it is insufficient to analyze the correlations between multiple consecutive semantic units. Therefore, it needs other tactic to detect words formed by multiple semantic units.

## 4.2   Algorithm Description

In order to solve the problem of interference between overlapped interrelated semantic pairs, we apply an iterative approach to merge the most interrelated semantic pair as a new semantic unit for each iteration and gain new words gradually.

More specifically, for a text $T$, the statistic $Q^2$ of each semantic pair can be calculated under the independence hypothesis $H_0$ according to Eq. (2). For any semantic pair whose $Q^2$ is in the critical region, the larger $Q^2$ is the stronger the correlation between its internal semantic units is. Therefore, we can select the internal related semantic pair with the largest $Q^2$ as a semantic unit. Based on this idea, we merge the most related semantic units and obtain new words gradually. The independence-testing-based Chinese word detection algorithm is proposed as follows:

In the algorithm, $U$ and $V$ are used to record the adjacent semantic units which have the largest $Q^2$. $freq$ is used to record the frequency of semantic pair $UV$. For each iteration, $UV$ will be merged into one semantic unit and

---

**Algorithm 1.** $IHT - WD$

---

**Input**: A string of Chinese text $T$
**Output**: A list *wordlist* of new detected words
1 **foreach** *character c of T* **do**
2      **if** *c is a Chinese character* **then**
3          insert $(c, pos_c)$ into *semanticUnitList*;

4 $Q^2 = \chi(1)^2_\alpha$, *freq = threshold*;
5 set $U$ and $V$ to be empty;
6 **while** $Q^2 \geq \chi(1)^2_\alpha$ **do**
7      **forall the** *semantic pair UV $\in$ semanticUnitList* **do**
8          set $UV$ as one semantic unit and update *semanticUnitList*;

9      update all frequency information of *semanticUnitList*;
10     set $Q^2 = 0$, *freq = threshold*;
11     **foreach** *semantic pair uv $\in$ semanticUnitList* **do**
12        **if** $n_{u,v} \geq threshold \wedge n_{u,v} \geq \delta n p_{u+} p_{+v}$ **then**
13          calculate $Q^2_{u,v}$ according to Equation (2);
14          **if** $Q^2_{u,v} > Q^2 \vee Q^2_{u,v} = Q^2 \wedge n_{u,v} > freq$ **then**
15             update $Q^2$, *freq*, $U$ and $V$ according to $Q^2_{u,v}$, $n_{u,v}$, $u$ and $v$;

16 **foreach** *semantic unit u $\in$ semanticUnitList* **do**
17      **if** *u is not a character* $\wedge n_u \geq threshold$ **then**
18          insert $u$ into *wordlist*;

19 **return** *wordlist*;

---

*semanticUnitList* will be updated accordingly (the two operations will not be performed in the first loop as $U$ and $V$ are empty at the beginning). After these operations, the algorithm will find the most interrelated semantic pair in *semanticUnitList*. If its error statistic is in the critical region and its actual frequency exceeds the frequency threshold *threshold*, the merging and the updating operations will be performed in next iteration.

In this algorithm, *threshold* is a frequency threshold which is determined by two factors. Firstly, for any text, there should be a basic frequency threshold *basic_freq* for all new words according to Criteria 1 of Sect. 3. *threshold* must not be smaller than *basic_freq*. Secondly, as the number $n$ of all semantic units may vary from text to text, if *threshold* is fixed, the noise will increase as the text grows. To avoid this, the threshold should grow as the length of the text increases. In this paper, the average word frequency is used for the threshold. As the number of words are unknown before processing, Heaps' law is used for estimation, i.e. if there are $N$ words in the text, the text contains $KN^\beta$ different words. In this paper, $N$ is estimated by half of the initial semantic pairs $n$. Then, the frequency threshold *threshold* can be calculated as follows:

$$threshold = max(basic\_freq, (\frac{n}{2})^{1-\beta}/K).$$

This iterative merging algorithm can solve the problems mentioned above. Take the name " 特朗普 " as an example: if " 特朗 " instead of " 朗普 " is merged in one iteration, then " 特朗 " will be merged with " 普 " in some following iteration. For the example of " 人民 " and " 民办 ", if " 人民 " is merged first, then " 人民办 " and " 民办 " will become two different semantic pairs. " 民 " and " 办 " still can be merged, but the merge of " 人民 " and " 办 " is rare in practice.

## 5    Experiments

The detection algorithm is estimated by precision. As the three criterias for new words in Sect. 3 also accord with the characteristics of some commonly used Chinese words, the detection results may contain existing words, new words and garbage strings. If we use $R$, $E$, $N$ and $G$ to denote the set of result strings, the set of existing words, the set of new words and the set of garbage strings, respectively, then word detection precision $P_w$ and new word detection precision $P_n$ can be used for evaluation of our approach.

$$P_w = \frac{|E| + |N|}{|R|}, \quad P_n = \frac{|N|}{|N| + |G|}.$$

In the experiments, parameters in the detection algorithm are set as follows: the significant level $\alpha$ is set to 0.5%; the basic frequency $basic\_freq$ is set to 4; in Heaps' law, we set $K = 1$, $\beta = 0.75$, which is suitable for news and other short texts; the coefficient $\delta$ for co-occurrence judgement is set to 4.

The algorithm is firstly evaluated by using the ICTCLAS testing corpus - People's Daily of Jan 1998, which is provided by Peking University. All news are input as one single text, and the output is a list of semantic units. The existing words and the new words are obtained in different methods. All words in the segmentation result of the ICTCLAS corpus are set as the existing words. After filtering out the existing words from the result, new words are manually annotated according to the following criteria: (1) a new word should be an existing entry in Wikipedia or Baidu encyclopedia; (2) a new word should represent a clear concept in real life; (3) a new word should be the abbreviation of an existing word. If a semantic unit meets one of the criteria, it is identified as a new word (Table 2).

**Table 2.** Result on ICTCLAS corpus

| Total | Existing | New | $P_w$ |
|-------|----------|-----|-------|
| 4471  | 3308     | 350 | 81.82% |

The algorithm has detected 4471 different semantic units from the ICTCLAS corpus, 3308 of them are existing words, and 350 of them are new words. The precision of word detection is 81.82%. As the word segmentation result of the ICTCLAS corpus has already included some of the new words, it is not appropriate to show the precision of new word detection by this experiment. To illustrate the effect of the algorithm on new word detection, we randomly download 100 pieces of news from www.gmw.cn, which cover politics, international news, economy, life, sport, education, etc (Table 3).

**Table 3.** Result on news from www.gmw.cn

| Total | Existing | New | $P_w$ | $P_n$ |
|-------|----------|-----|-------|-------|
| 1217  | 767      | 368 | 93.26% | 81.78% |

The algorithm obtains 1217 semantic units in total, which includes 767 existing words and 367 new words. The new word detection precision is 81.78% while the total word detection precision is 93.26%. Because there is no standard test set for Chinese new word detection, it is improper to compare our method with existing ones directly. But in terms of the new word detection precision, our approach is competitive. Some examples of new words are listed as follows:

- Entries in Baidu encyclopedia: 公办学校 (the public school), 智能芯片 (intelligent chip), 基本医疗保险 (basic medical insurance), 非洲裔美国人 (African American), etc.
- Semantic units representing a clear concept: 一带一路 (the Belt and Road), 安倍政府 (Abe administration), 星巴克 (Starbucks), 积极废人 (an active loser), etc.
- Abbreviations: 上合组织 (Shanghai Cooperation Organization), 外研社 (Foreign Language Teaching and Research Press), 世卫组织 (World Health Organization), etc.

## 6   Conclusion and Future Work

This paper presents three statistical criterias for new word detection and proposes an independence-testing-based approach for Chinese new word detection. Compared with the existing methods, our approach does not need prior knowledge or large scale training corpus and it is more suitable for detecting new words from news texts. The experiment on randomly selected news shows that the new word detection precision of the algorithm is over 80%, which is competitive compared with the existing methods. As the method only use some statistical characteristics of new words, it is recommended to combine this method with semantic-rule-based methods to improve the accuracy.

In the future, we will construct a test set of Chinese new word detection for comparison and calculate the recall rate of our approach. Furthermore, since both the coefficient $\delta$ and the frequency threshold *threshold* can be determined by statistical methods, the methods of setting these parameters in Algorithm 1 and the impact of these parameters will be studied in a late stage.

# References

1. Huang, C.N., Hai, Z.: Chinese word segmentation: a decade review. J. Chin. Inf. Process. **21**(3), 8–19 (2007)
2. Zou, G., Liu, Y., Liu, Q.: Internet-oriented Chinese new words detection. J. Chin. Inf. Process. **18**(6), 1–9 (2004)
3. Luo, Z., Song, R.: An integrated method for Chinese unknown word extraction. In: Proceedings of the 3rd SIGHAN Workshop on Chinese Language Processing, pp. 148–154. Association for Computational Linguistics (2004)
4. Li, D., Tu, W., Shi, L.: Chinese new word identification algorithm based on context-aware. Comput. Eng. Des. **33**(10), 4022–4027 (2012)
5. Zhang, H., Yong, L.I., Yan, Q.: Method of new Chinese words identification from large scale network corpora. Comput. Eng. Appl. **51**(5), 208–213 (2015)
6. He, M., Gong, C., Zhang, H., Cheng, X.: Method of new word identification based on lager-scale corpus. Comput. Eng. Appl. **43**(21), 157–159 (2007)
7. Zhao, X., Zhang, H.: New words identification based on iterative algorithm. Comput. Eng. **40**(7), 154–158 (2014)
8. Zeng, H.L., Zhou, C.L., Shi, X.D., et al.: New word detection algorithm for Chinese based on extraction of local context information. In: Proceedings of the 3rd International Conference on Intelligent System and Knowledge Engineering, pp. 797–801. IEEE Xplore (2008)
9. Peng, F., Feng, F., Mccallum, A.: Chinese segmentation and new word detection using conditional random fields. In: Proceedings of the 20th International Conference on Computational Linguistics, Geneva, Switzerland, pp. 562–568 (2004)
10. Cui, S.: New word detection based on large-scale corpus. J. Comput. Res. Dev. **43**(5), 927–932 (2006)
11. Zhang, H., Luan, J., Li, Y., Qi, X.: Method of new Chinese word detection based on statistical learning framework. Comput. Sci. **39**(2), 232–235 (2012)

# The Accessibility of Mathematical Formulas for the Visually Impaired in China

Wei Su[1], Chuan Cai[1(✉)], and Jinzhao Wu[2]

[1] School of Information Science and Engineering,
Lanzhou University, Lanzhou, China
{suwei,caichuan}@lzu.edu.cn
[2] School of Software, Guangxi University for Nationalities, Nanning, China

**Abstract.** Accessing mathematical information, including inputting and reading mathematical formulas on computer, still has big difficulty and barrier for people who are blind or partially sighted. The paper introduces new progress of assistive technology of mathematical documents and Web pages for Chinese visually impaired people. In the paper, we give two conversion models from MathML to Chinese mathematics Braille and a verbalization method of mathematical formulas in Chinese. The work described in the paper is part of "China Digital Platform of Braille" Web site and can be found in http://www.braille.org.cn.

**Keywords:** Braille · Mathematical formula · MathML
Information accessibility

## 1 Background

Information accessibility refers to that no matter if you are healthy or disabled, you would benefit from the information technology. Braille and speech are two main ways of getting information for the visually impaired [1]. Mathematics as the language of science underpins almost all applications of science, education and engineering. Mathematics is also at the heart of representing and reasoning about scientific and education data and knowledge. The accessibility of mathematics information refers to the inclusive practice of making mathematics accessible and usable by all the people of abilities and disabilities. However, accessing mathematical formulas, including inputting and reading mathematical formulas on computer, still has big difficulty and barrier for people who are blind or partially sighted. The main contributions of the paper is to introduce our new progress of assistive technology of mathematical documents and Web pages for visually impaired people in China. The Sect. 2 analyzes the reasons of verbalization ambiguity and discusses the method of mathematical verbalization in Chinese. Section 3 presents our proposed system of transforming various mathematical expression format to Chinese Braille. Finally, the conclusion is presented in Sect. 4.

© Springer Nature Switzerland AG 2018
J. Fleuriot et al. (Eds.): AISC 2018, LNAI 11110, pp. 237–242, 2018.
https://doi.org/10.1007/978-3-319-99957-9_18

## 2    Verbalization of Mathematical Formula in Chinese

Recently, with the rapid development of the TTS (Text To Speech) technology, the visually impaired can access Chinese Web pages by using Chinese screen reader such as Sunshine [2] and NVDA [3]. However, the current screen reader cannot process mathematical expressions due to their flexible and complex two-dimensional structures. Currently there is no specification on how to read mathematical expressions both in English and in Chinese. Plenty of institutions have set foot in the verbalization field of mathematical formulas many years ago [4–14]. MathPlayer [4], MathGenie [5] and MathSpeak [6], LAMBDA [7], and AudioMath [8] could speak mathematical notations, which use the indicator words such as "begin/end" to clearly point out where the notations begin or end, and use the abbreviation to replace the long math pronunciation. Fateman [9] has done some research on analyzing of the pronunciation rules of operators and operands, and designing some ad-hoc methods on eliminating the pronunciation ambiguity of mathematical formulas via adding "all, quantity" and other indicators. The major research field of these studies is on mathematical formulas pronunciation in English, and few start on other languages [10,11]. Mathematical formulas speech are different in orders and rules for different languages and countries [12]. The paper introduces a way of translating mathematical expressions into Chinese voice for the visually impaired of China.

In Lanzhou University (LZU), we are doing research on converting the mathematical formulas into unambiguous and concise Chinese text [12]. As shown in Fig. 1, the various forms of mathematical formulas including LaTeX, OpenMath, OMML, and Infix (Mathematica, Maple, and Maxima) are translated into MathML firstly, and then MathML formulas are converted into Chinese text, which can output to TTS system for producing voice.

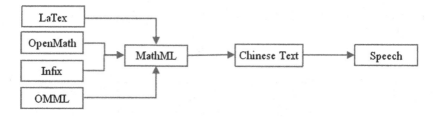

**Fig. 1.** Converting mathematical formulas to Chinese speech

A series of verbalization rules for mathematical formulas are created in our system to solve the ambiguity problem. When creating the rules we mainly consider three basic requirements:

- The verbalization rules should decrease and/or avoid ambiguous of different mathematical formulas as far as possible.
- The rules should ensure the verbalization of most formulas meet current usual habit.
- The verbalization rules should be simple to learn and easy to understand.

A mathematical formula may contain one or several operators and operands. A single operator or operand will not produce ambiguity for listeners. Most ambiguity problems are caused by the combination of operators and operands. According to their relative position of operands, most operators can be categorized into prefix, infix and postfix operator. In the paper we use "⊣" to denote prefix operator, "⊥" to denote infix operator, and "⊢" to denote postfix operator, and "$\Delta$" to denote operand or sub-expression. We can classify most elements of Content MathML into the three categories [12]. The combination of operators and operands can get eight different forms (See Table 1). The experiment in Lanzhou University in 2014 [12] shows that most ambiguity problems occur in the forms of "⊣ $\Delta$ ⊥", "⊣ $\Delta$ ⊢", "$\Delta$ ⊥ $\Delta$ ⊥ $\Delta$", "$\Delta$ ⊥ $\Delta$ ⊢", while there is almost no ambiguity problems in other four forms of, "⊣⊣ $\Delta$", "$\Delta$ ⊥⊣ $\Delta$", "$\Delta$ ⊢⊥ $\Delta$", and "$\Delta$ ⊢⊢". For example, the Chinese verbalization of "x plus y squared" (see the first line of Fig. 2), which is the form of "$\Delta$ ⊥ $\Delta$ ⊢", could be understood either as "$x + y^2$" or "$(x + y)^2$". Another example, the Chinese verbalization of "b plus c over a" ("$\Delta$ ⊥ $\Delta$ ⊥ $\Delta$", see the second line of Fig. 2) could be considered as formula of "b/a+c" or "(b+c)/a". However the Chinese verbalization of "x plus sin y" (the third line of Fig. 2) refers only the expression of "x+sin y". To avoid ambiguity problem, we add a start-tag or end-tag into the verbalization of mathematical expressions in order to constrain the effect scope of each operator. For example, in the Chinese verbalization of "x plus y all squared" (the fourth line of Fig. 2), an end-tag "all" is used to indicate the affecting score of "plus". The verbalization of the last line in Fig. 2 can also eliminate the ambiguity. The rules are easy to be accepted by the visually impaired because start and end tags are used in mathematical Braille too.

x加y的平方

x加 sin y

a分之b加c

x加y整体的平方

a分之b分式结束加c

**Fig. 2.** Chinese verbalization of four formulas

**Table 1.** Combination of operators and operands

| Type | Prefix | Infix | Postfix |
|---|---|---|---|
| Prefix | ⊣⊢ Δ | ⊣ Δ ⊥ Δ | ⊣ Δ ⊢ |
| Infix | Δ ⊥⊣ Δ | Δ ⊥ Δ ⊥ Δ | Δ ⊥ Δ ⊢ |
| Postfix | Null | Δ ⊢⊥ Δ | Δ ⊢⊢ |

# 3   Braille of Mathematical Formulas

Braille is another pervasive format for the visually impaired. Many countries have developed some software in order to enable the visually impaired to access mathematical information via Braille. The assistant tools and projects support mathematical formulas transformations to/from different Braille formats including Marburg, Nemeth, French, Netherland, BAUK, UEBC or Czech Braille. Table 1 lists the researches and software of in recent two decades [13–19]. Braille dots and rules for mathematics vary among different countries. Zhiwei and Youyang in LZU have introduced our assistant tool for translating mathematical formulae (Content MathML) into Chinese Braille in the paper [20] (Table 2).

**Table 2.** Research and software on transforming mathematical formulas to/from Braille

| Name | Math format | Braille format | Work |
|---|---|---|---|
| LaBraDoor | Latex | Marburg Braille, Austrian Braille | ↔ |
| BraMaNet | MathML Presentation, Latex | French Braille | → |
| Math2Braille | MathML | Netherlands Braille | → |
| MAVIS | Latex | Nemeth Braille | ↔ |
| DBT | MathM | Nemeth, UEBC, French Braille | → |
| Infty | Latex | UEBC, Japanese Braille | → |
| Insight | Latex | Nemeth Braille | ← |
| MathGenie | MathML | Nemeth Braille | → |
| Vickie | MathML, Latex | French Braille | → |
| Tiger | Text | Nemeth, UK and French Braille | → |
| Lambda | MathML | 8-dots Braille | → |
| BrailleCz | MathML | Czech Braille | → |
| Dooley | Content MathML | Nemeth Braille | ↔ |
| RoboBraille | MathML, Latex | Nemeth, Marburg Braille | → |
| Latex2nemeth | Latex | Nemeth Braille | → |

Note: →←↔ respectively denotes translation math formulas to, from, to/from Braille

In LZU, we developed two tools: SunMath$^C$ and SunMath$^P$, for translating MathML to Braille. MathML supports both a presentation encoding and a content encoding for different purposes. Content MathML is a semantic mathematical markup language, focused on their semantics. SunMath$^C$ is a XSLT-based tool, which can convert Content MathML to Chinese Braille. Content-based editing enables the user to enter/modify well-formed expressions that represent meaningful mathematical operations. Presentation MathML is very good for the display layout of mathematical formulas. SunMath$^P$, developed in Java, is Web service tool, which can convert Presentation MathML to Chinese Braille. In both tools, we create conversion modules to improve the accuracy of conversion and build several simplification rules to remove the redundant Braille cells.

## 4    Conclusion

The paper gives a Chinese verbalization method of mathematical expression and introduces two translation tools for Content MathML and Presentation MathML to Chinese Braille. In LZU, we are implementing a Web site of "China Digital Platform of Braille" (http://www.braille.org.cn). The method and tools mentioned in the paper have been applied in the "Mathematics Braille" module. We are also trying to provide audio and Braille feedback when users enter expressions in MathEdit [21]. In the next step, we will complete all the functions of translating mathematical expressions to speech and Braille.

**Acknowledgements.** We would like to thank the anonymous reviewers for their valuable comments. This work is supported by Science Foundation of Guangxi (AA17204096, AD16380076) and Innovation Project for Lanzhou Talent (2014-RC-3).

## References

1. Karshmer, A., et al.: Mathematics and accessibility: a survey. In: Proceedings of the 9th International Conference on Computers Helping People with Special Needs (2007)
2. Chen, H.: Research on Screen Reader Technology and Its Application to the Sunshine Screen Reader. Nankai University, Tianjin (2010)
3. NonVisual Desktop Access: NVDA screen-reader. http://www.nvda-project.org/. Accessed 18 May 2018
4. Soiffer, N.: A flexible design for accessible spoken math. In: Stephanidis, C. (ed.) UAHCI 2009. LNCS, vol. 5616, pp. 130–139. Springer, Heidelberg (2009). https://doi.org/10.1007/978-3-642-02713-0_14
5. Gillan, D.J., Barraza, P., Karshmer, A.I., Pazuchanics, S.: Cognitive analysis of equation reading: application to the development of the math genie. In: Miesenberger, K., Klaus, J., Zagler, W.L., Burger, D. (eds.) ICCHP 2004. LNCS, vol. 3118, pp. 630–637. Springer, Heidelberg (2004). https://doi.org/10.1007/978-3-540-27817-7_94
6. Sheikh, W., Schleppenbach, D., Leas, D.: MathSpeak: a non-ambiguous language for audio rendering of MathML. Int. J. Learn. Technol. **13**(1), 3–25 (2018)
7. Schweikhardt, W., Bernareggi, C., Jessel, N., Encelle, B., Gut, M.: LAMBDA: a European system to access mathematics with Braille and audio synthesis. In: Miesenberger, K., Klaus, J., Zagler, W.L., Karshmer, A.I. (eds.) ICCHP 2006. LNCS, vol. 4061, pp. 1223–1230. Springer, Heidelberg (2006). https://doi.org/10.1007/11788713_176

8. Subagya, S: Design of mathematics audiobooks for students with visual impairment at the secondary school. Eur. J. Spec. Educ. Res. (2017)
9. Fateman, R.: How can we speak math. J. Symb. Comput. **25**(2) (2013)
10. Bier, A., Sroczyski, Z.: Adaptive math-to-speech interface. In: Proceedings of the Mulitimedia, Interaction, Design and Innnovation. ACM (2015)
11. Boonprakong, N., et al.: Reading mathematical expression in Thai. In: Proceedings of the 11th International Convention on Rehabilitation Engineering and Assistive Technology, Singapore (2017)
12. Hou, H., Su, W., Cai, C., Li, L., Li, H.: Research for the MathML-based mathematical formulas pronunciation in Chinese. J. Syst. Sci. Complex. **34**(4), 401–412 (2014)
13. Coen, C.S., Zacchiroli, S.: Efficient ambiguous parsing of mathematical formulae. In: Asperti, A., Bancerek, G., Trybulec, A. (eds.) MKM 2004. LNCS, vol. 3119, pp. 347–362. Springer, Heidelberg (2004). https://doi.org/10.1007/978-3-540-27818-4_25
14. Batusic, M., Miesenberger, K., Stoger, B.: Labradoor: a contribution to making mathematics accessible for the blind. In: The 6th International Conference on Computers Helping People with Special Needs (ICCHP), pp. 307–315. Vienna (1998)
15. Cosma, V.P., Stevns, T., Christensen, L.B.: Braille math extension to RoboBraille. In: Miesenberger, K., Bühler, C., Penaz, P. (eds.) ICCHP 2016. LNCS, vol. 9758, pp. 15–18. Springer, Cham (2016). https://doi.org/10.1007/978-3-319-41264-1_2
16. Crombie, D., Lenoir, R., McKenzie, N., Barker, A.: math2braille: Opening access to mathematics. In: Miesenberger, K., Klaus, J., Zagler, W.L., Burger, D. (eds.) ICCHP 2004. LNCS, vol. 3118, pp. 670–677. Springer, Heidelberg (2004). https://doi.org/10.1007/978-3-540-27817-7_100
17. Armano, T., et al.: An overview on ICT for the accessibility of scientific texts by visually impaired students. Congresso Nazionale SIREM (2014)
18. Annamalai, N., Gopal, D., Gupta, G., Guo, H., Karshmer, A.: INSIGHT: a comprehensive system for converting Braille based mathematical documents to LaTeX. In: The 10th International Conference on Universal Access in Human-Computer Interaction, Mahwah, USA (2003)
19. Soiffer, N.: A study of speech versus Braille and large print of mathematical expressions. In: Miesenberger, K., Bühler, C., Penaz, P. (eds.) ICCHP 2016. LNCS, vol. 9758, pp. 59–66. Springer, Cham (2016). https://doi.org/10.1007/978-3-319-41264-1_8
20. Liu, Z., Su, W., Li, L., et al.: Automatic translation for Chinese mathematical Braille code, pp. 340–345. IEEE (2010)
21. Su, W., et al.: MathEdit, a browser-based visual mathematics expression editor. In: Proceedings of ATCM (2006)

# Specialty-Aware Task Assignment in Spatial Crowdsourcing

Tianshu Song[(⊠)], Feng Zhu, and Ke Xu

SKLSDE Laboratory, School of Computer Science and Engineering,
Beihang University, Beijing 100191, China
{songts,zfone,kexu}@buaa.edu.cn

**Abstract.** With the rapid development of mobile Internet, spatial crowdsourcing is gaining more and more attention from both academia and industry. In spatial crowdsourcing, spatial tasks are sent to workers based on their locations. A wide kind of tasks in spatial crowdsourcing are specialty-aware, which are complex and need to be completed by workers with different skills collaboratively. Existing studies on specialty-aware spatial crowdsourcing assume that each worker has a unified charge when performing different tasks, no matter how many skills of her/him are used to complete the task, which is not fair and practical. In this paper, we study the problem of specialty-aware task assignment in spatial crowdsourcing, where each worker has fine-grained charge for each of their skills, and the goal is to maximize the total utility of the completed tasks based on tasks' budget and requirements on particular skills. The problem is proven to be NP-hard. Thus, we propose two efficient heuristics to solve the problem. Experiments on both synthetic and real datasets demonstrate the effectiveness and efficiency of our solutions.

## 1 Introduction

With the development of mobile Internet and the blossom of sharing economy, all kinds of *spatial crowdsourcing (SC)* platforms become popular, where the *online crowd workers* are employed through their phones to participate in and complete *offline crowdsourcing tasks* in the physical world [9]. Typical SC platforms include Gigwalk[1], TaskRabbit[2] and gMission[3] [2].

One fundamental issue in SC is task assignment, namely assigning crowdsourcing tasks to suitable crowd workers. Generally speaking, there are two kinds of tasks. The first kind is micro tasks which can be completed by any single worker such as taking photos and delivering things. The second kind is specialty-aware tasks such as repairing a house and organizing a party, where crowd workers with different kinds of skills are needed to work collaboratively and finish the task. For micro-task assignment, there are many existing works

---

[1] www.gigwalk.com.
[2] www.taskrabbit.com.
[3] gmission.github.io.

© Springer Nature Switzerland AG 2018
J. Fleuriot et al. (Eds.): AISC 2018, LNAI 11110, pp. 243–254, 2018.
https://doi.org/10.1007/978-3-319-99957-9_19

**Table 1.** Tasks and their lists of skills

| Tasks | Lists of required skills |
|-------|--------------------------|
| $t_1$ | $s_1$(music), $s_2$(drinks) |
| $t_2$ | $s_1$(music), $s_3$(barbecue), $s_4$(lights) |
| $t_3$ | $s_1$(music), $s_2$(drinks), $s_3$(barbecue), $s_4$(lights), $s_5$(stage) |

**Table 2.** Workers' skills and fees

| Workers | Skills and fees |
|---------|-----------------|
| $w_1$ | $(s_1, 3), (s_2, 4), (s_4, 5)$ |
| $w_2$ | $(s_3, 5), (s_5, 3)$ |
| $w_3$ | $(s_4, 2)$ |
| $w_4$ | $(s_1, 5), (s_5, 1)$ |
| $w_5$ | $(s_1, 2), (s_2, 2), (s_3, 3), (s_4, 6)$ |

[6,10,15–18] and we refer the readers to [14] for more details. In this paper we focus on specialty-aware tasks assignment.

Existing works [3,4] on specialty-aware tasks assignment formulate that each crowd worker has multiple skills and will get a unified fee if s/he is employed, which is not very practical as (1) workers often have unbalanced workloads, (2) if more than one workers with the same required skills are employed, they may be confused who should do the job in a task and (3) the payment and the workload do not often match. To solve the above drawbacks, in this paper we propose the Specialty-Aware Task Assignment (SATA) problem where each crowd worker specifies a fee for each of her/his skill to make the payment proportional to the workload.

We then illustrate the SATA problem by a motivation example of organizing a party.

*Example 1.* Suppose we have three tasks of organizing parties, each has different styles and thus different kinds of skills are needed. For example, party 1 is a mini one and only needs music and drinks, while party 3 is ceremonious and requires music, drinks, barbecue, lights and a stage. The skill lists required by the three tasks are shown in Table 1. Besides, we have some workers shown in Table 2, each with different skills and corresponding fees. For example, if $w_1$ is required to be responsible for the music of the party (skill $s_1$), s/he will be paid 3. Besides, each worker will get the transportation fee, which equals the distance from the worker to the assignment task times a global unit price. For example, Fig. 1 shows the locations of tasks and workers, and if the global unit price is 0.5, the transportation fees for assigning $w_1$ to $t_1$ is $\sqrt{5} \times 0.5 \approx 1.12$.

Motivated by the example above, we formalize the SATA problem, which aims to assign crowd workers to specialty-aware tasks, and the goal is to maximize the

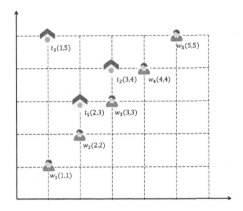

**Fig. 1.** Locations of tasks and workers.

total utility. Note that existing works either focus on assigning workers to micro tasks to optimize different goals, or assume that the workers have a unified fee. Thus, their methods cannot be adopted directly to solve our problem. Briefly, we make the following contributions.

- We formally define a new task assignment problem in spatial crowdsourcing, called the Specialty-Aware Task Assignment (SATA) problem (Sect. 4).
- We prove the SATA problem is NP-hard (Sect. 3), and develop two efficient heuristics to solve it (Sect. 4).
- We verify the effectiveness and efficiency of the proposed methods with extensive experiments on real and synthetic datasets (Sect. 5).

We next review the related works in Sect. 2. The paper is concluded in Sect. 6.

## 2   Related Work

In this section, we review related works from two categories, namely task assignment and team formation problem.

### 2.1   Task Assignment in Spatial Crowdsourcing

The research on task assignment in spatial crowdsourcing mainly includes two parts: micro-task assignment and specialty-aware task assignment.

Micro task refers to the spatial tasks that can be completed by any single worker. [6] is the first work on task assignment in spatial crowdsourcing, whose goal is to maximize the total number of the assignment tasks. [16] is the first work focusing on the online scenario of task assignment, and studies the two-sided online task assignment problem, whose goal is to maximize the total utility score of the assignment. [10] also focuses on the online scenario and considers the influence of work space on task assignment, whose goal is to maximize the total utility score. [15] studies the problem of online minimum weighted bipartite matching, which can be used in online task assignment. [18] considers the problem of flexible online matching where workers can be scheduled if no task is assigned. [11] recommends routes dynamically for workers to deal with online tasks, and the goal is to maximize the total utility. [21] assigns tasks to workers while trading off quality and latency of task completion. [17] proposes a match-based approach to solve the dynamic pricing problem in spatial crowdsourcing. [22] takes the destinations of workers into consideration to perform task assignment. [13] considers performing online task assignment while preserving the privacy of tasks and workers under the circumstance that the server is untrusted. [12,19] propose a real-time framework for task assignment. The difference between our work and the aforementioned works is that they focus on micro tasks which can be completed by a single worker, and we study on the assignment for specialty-aware tasks which have requirements on skills of workers and usually have to be completed by multiple workers collaboratively.

[4,5] recommend top-k teams with the minimum cost to a specialty-aware task. [3] studies assigning workers for specialty-aware tasks to maximize the total utility score. The biggest difference between our work and [3–5] is that in our work workers specify fees for each of their skills, and in [3–5] workers only have a united fee, which is not practical.

## 2.2    Team Formation Problem

A closely related topic is the team formation problem [7], whose goal is to find a team of experts with the minimum cost, according to the skills and social relationships of the users. [1] studies the online version of the team formation problem, where the issue of workload balance is also considered. [8] studies another variant of the team formation problem where the capacity constraint of experts is considered. The difference between our problem and the above works on the team formation problem and its variants is that we do not consider the social relationships between users and focus on task assignment.

## 3    Problem Definition

We first introduce two basic concepts, namely Task and Worker. Then, we introduce how to calculate the reward of worker. Finally, we formally give the definition of the Specialty-Aware Task Assignment (SATA) problem.

**Definition 1 (Worker).** *A worker $w$ is defined as $<L_w, S_w, P_w>$, where $L_w$ is the location of $w$ which can be described by longitude and latitude, $S_w = <s_1^w, s_2^w, \cdots, s_{|S_w|}^w>$ is the list of skills that $w$ masters, and $P_w = <p_1^w, p_2^w, \cdots, p_{|S_w|}^w>$ is the list of fees for each skill in $S_w$.*

**Definition 2 (Task).** *A task $t$ is defined as $<L_t, S_t, B_t>$, where $L_t$ is the location of $t$ which can be described by longitude and latitude, $S_t = <s_1^t, s_2^t, \cdots, s_{|S_t|}^t>$ is the list of skills that are needed to complete $t$ collaboratively, and $B_t$ is the total monetary budget of $t$.*

Briefly, a worker's reward includes two parts: (1) transportation fee, which is directly proportional to the distance between the worker and the task; (2) labor fee, which is the sum of the fees for the skills used to perform a task.

**Definition 3 (Reward of Worker).** *The reward of task $w$ to perform task $t$ equals $r_w = \gamma \cdot dis(L_w, L_t) + \sum_{s \in S'_w} p^w_s$, where $dis(L_w, L_t)$ is the distance between $L_w$ and $L_t$, which can be Euclidean distance or road network distance, $\gamma$ is a global parameter representing the unit transportation fee, and $S'_w$ is the set of skills that $w$ uses to perform the task.*

We define the utility of a task as follows.

**Definition 4 (Utility of Task).** *The utility of task $t$ is defined as $u_t = B_t - \sum_{t \in W_t} r_t$, where $B_t$ is the budget of the task and $\sum_{t \in W_t} r_t$ is the summation of rewards of workers assigned to $t$ if $t$ is completed. If $t$ cannot be finished, the utility is zero.*

We finally define our problem as follows.

**Definition 5 (Specialty-Aware Task Assignment (SATA) Problem).** *Given a set of tasks $T$, a set of workers $W$ and a global unit transportation fee $\gamma$, the problem is to assign workers to tasks to maximize the total utility of the completed tasks and the following constraints should be satisfied:*

- *Specialty Constraint: a task can be completed as long as the workers assigned to it can cover the required skills of the task;*
- *Budget Constraint: the total rewards of workers assigned to a task cannot exceed the task's total budget;*

We then prove the hardness of SATA problem.

**Theorem 1.** *The SATA problem is NP-hard.*

*Proof.* We prove through a reduction from the set cover problem [20]

We first introduce the set cover problem. Given a universe $U = \{s_1, s_2, \cdots, s_n\}$ and its $m$ subsets $S_1, S_2, \cdots, S_m \subseteq U$, $\cup_{i=1}^m S_i = U$. Each $S_i$ is associated with a cost $c_i$. The set cover problem is to find a set $K \subseteq \{1, 2, \cdots, m\}$ to minimize $\sum_{i \in K} c_i$ satisfying $\cup_{i \in K} S_i = U$.

We next show how to transform the set cover problem to an instance of our SATA problem. We only have one task $t$ which requires skills $S_t = U$ and has infinite budget $B_t$. For $m$ workers $\{w_1, w_2, \cdots, w_m\}$, their required fees for skills are all zero, and we adjust their locations and $\gamma$ to make their transportation fee to perform $t$ is $c_i$. For this instance of our SATA problem, we aim to find a set of workers $K$ to maximize the utility of $t$, which equals to minimize $\sum_{i \in K} c_i$. In this way, we reduce set cover problem to our SATA problem. As the set cove problem is known to be NP-hard [20], SATA problem is also NP-hard.

---

**Algorithm 1.** Total Budget Based Algorithm (TBA)

---

    **input**  : set of workers $W$, set of tasks $T$
    **output**: Assignment $M$
**1**  $Q \leftarrow$ sorting tasks in $T$ according to their total budgets in descending order;
**2**  **foreach** $t$ *in* $Q$ **do**
**3**      Assign $w \in W$ to $t$ with minimum $\frac{r_w}{|S'_w \cap S_t|}$;
**4**      Update $M$ and $W$;
**5**      $S_t \leftarrow S_t - S'_w$;
**6**      **if** $S_t$ *is* $\emptyset$ **then**
**7**           Break;

**8**  **return** $M$

---

## 4   Algorithms

In this section, we give two efficient heuristic algorithms to solve the SATA problem.

### 4.1   Total Budget Based Algorithm

Our first algorithm is called the Total Budget Based Algorithm (TBA). The main idea is that we always try to assign workers to the tasks with the largest budget. During the procedure of task assignment, we refer to the greedy algorithm to solve the set cover problem [20].

The procedure of TBA is shown in Algorithm 1. The algorithm takes the set of workers $W$ and set of tasks $T$ as input, and return an assignment $M$ between them. In line 1, the algorithm first sorts the tasks in $T$ in descending order according to their total budgets, and the sorted result is saved in $Q$. In lines 2–7, for each task $t$ in $Q$, we refer to the greedy algorithm to solve the set cover problem [20] to assign workers. Specifically, in lines 3, we find worker with minimum $\frac{r_w}{|S'_w \cap S_t|}$. Notes that here $S'_w$ considers all possible subsets of $S_w$. In lines 4–5, we update $M$, $W$ and $S_t$. In lines 6–7, if $S_t$ is $\emptyset$, which means it can be completed, we break the loop and start to assign workers for the next task.

*Example 2.* Back to our running example in Example 1. TBA first finds the task with the largest total budget, which is $t_3$. The it starts to assign workers for $t_3$. As $w_3$ has the minimal $\frac{r_w}{|S'_w \cap S_t|}$ of 2, we first assign $w_3$ to $t_3$. After assigning $w_3$, $t_3$'s list of skills has not been covered, thus we assign $w_5$ to $t_3$ with $\frac{r_w}{|S'_w \cap S_t|}$ of $\frac{7}{3}$. We finally assign $w_4$ to $t_3$ and the total reward paid to $w_3$, $w_4$ and $w_5$ is $2 + 2 + 2 + 3 + 1 + (\sqrt{2} + 4 + \sqrt{10}) \times 0.5 \approx 15$. Thus, the utility of $t_3$ is $30 - 15 = 15$. Similarly, we assign workers to $t_2$ and $t_3$ successively, and the final utility of TBA is 21.08.

**Complexity.** If taking the maximum number of skills of a worker as a constant, the time complexity of line 1 and lines 2–7 is $O(|T|log|T|)$ and $O(|T||W|)$ respectively. As a result, the time complexity of TBA is $O(max\{|T|log|T|, |T||W|\})$.

---

**Algorithm 2.** $\underline{A}$verage $\underline{B}$udget Based $\underline{A}$lgorithm (ABA)

---

   **input**  : set of workers $W$, set of tasks $T$
   **output**: Assignment $M$
1  $Q \leftarrow$ sorting tasks in $T$ according to their average budgets in descending order;
2  **foreach** $t$ $in$ $Q$ **do**
3      |  Assign $w \in W$ to $t$ with minimum $\frac{r_w}{|S'_w \cap S_t|}$;
4      |  Update $M$ and $W$;
5      |  $S_t \leftarrow S_t - S'_w$;
6      |  **if** $S_t$ $is$ $\emptyset$ **then**
7      |    |  Break;

8  **return** $M$

---

### 4.2 Average Budget Based Algorithm

The TBA algorithm only considers the total budget of tasks. However, a large budget may result from a large number of skills required in the task. Thus, in this subsection, we propose another algorithm, called $\underline{A}$verage $\underline{B}$udget Based Algorithm (ABA). The main idea is that we first measure the average budget (per skill) of all the tasks, and prefer to assign workers to tasks with a larger average budget.

The pseudo codes of ABA is shown in Algorithm 2. The biggest difference between TBA and ABA lies on line 1. In ABA, we first sort tasks in $T$ based on average budget, which is defined as $\frac{B_t}{|S_t|}$. The procedure of how to assign workers to a given task is the same as TBA, which is shown in lines 2–7.

*Example 3.* Back to our running example in Example 1. Different from TBA, ABA first finds the task with the largest average budget, which is $t_1$. Then it assigns workers to $t_1$. As $w_5$ has the minimal $\frac{r_w}{|S'_w \cap S_t|}$ of 2, we first assign $w_5$ to $t_1$. After assigning $w_5$, we find $t_1$'s list of skills has been covered, thus the total utility is $20 - (2 + 2) - (\sqrt{13} \times 0.5 \approx 14.20)$. Similarly, we next assign workers for $t_2$ and $t_3$, and the final utility of TBA is 25.78.

**Complexity.** Similar to the analysis of TBA, if we take the maximum number of skills a worker may have as a constant, the time complexity of ABA is also $O(max\{|T|log|T|, |T||W|\})$.

## 5 Evaluation

### 5.1 Experiment Setup

We use real and synthetic datasets to evaluate our algorithms. Real data comes from CSTO (http://www.csto.com/), which is an outsource task platform. In the

**Table 3.** Synthetic dataset

| Factor | Setting |
|---|---|
| $\|T\|$ | 100 300 **500** 700 900 |
| $\|W\|$ | 1000 3000 **5000** 7000 9000 |
| $\gamma$ | 0.1 0.3 **0.5** 0.7 0.9 |
| $B_t$ | 60 80 **100** 120 140 |
| $P_w$ | 10 15 **20** 25 30 |
| $\|S\|$ | 10 20 **30** 40 50 |

CSTO dataset, each task is associated with a set of skills needed to complete the software development task, and each coder is associated with a set of skills and an average price which can be deduced from the history data. Since the CSTO data is not associated with location information, we generate the distance of each coder from the task following uniform distribution. For synthetic data, based on the observation from real data set, the price of skills owned by a worker and the budget of a task both follow Gaussian distribution. Statistics of the synthetic data are shown in Table 3, where we mark our default settings in bold font.

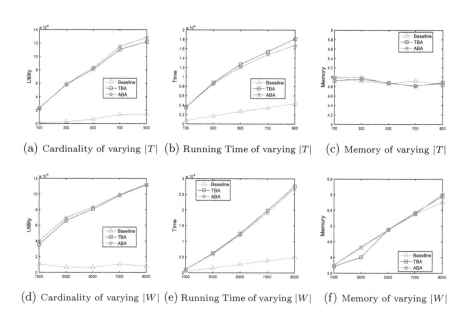

(a) Cardinality of varying $|T|$ (b) Running Time of varying $|T|$ (c) Memory of varying $|T|$

(d) Cardinality of varying $|W|$ (e) Running Time of varying $|W|$ (f) Memory of varying $|W|$

**Fig. 2.** Results on varying $|T|$ and $|W|$.

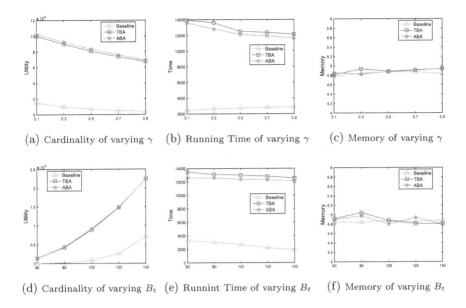

(a) Cardinality of varying $\gamma$   (b) Running Time of varying $\gamma$   (c) Memory of varying $\gamma$

(d) Cardinality of varying $B_t$   (e) Runnint Time of varying $B_t$   (f) Memory of varying $B_t$

**Fig. 3.** Results on varying $\gamma$ and $B_t$.

## 5.2   Experimental Results

In this subsection, we test the performance of our proposed algorithms by setting different parameters. We compare TBA and ABA with a baseline algorithm in terms of total utility score, running time and memory cost, and study the effect of varying parameters on the performance of the algorithms. The baseline algorithm uses a simple random strategy, which assigns workers to tasks randomly. The algorithms are implemented in CodeBlocks 16.1, and the experiments are performed on a machine with Intel(R) Core(TM) i5 2.50 GHZ CPU and 8 GB main memory.

**Effects of the Number of Tasks $|T|$.** The results of varying $|T|$ are presented in Fig. 2a to c. First, we can observe that the utility increases as $|T|$ increases, which is reasonable as more tasks available. Also, we can observe that TBA and ABA are much better than baseline algorithm and TBA has advantages over ABA. The reason may be that the average budget is a better metric for identifying profitable tasks, as tasks with large total budget may result from the requirements on a large number of skills. As for running time, TBA and ABA are slower than the baseline due to sorting tasks and finding more economic schedule, and the running time is acceptable for better performance on utility. Moreover, TBA is faster than ABA for it is easier to find suitable workers for each tasks. The three algorithm do not vary much in memory consumption.

**Effects of the Number of Workers $|W|$.** The results of varying $|W|$ are presented in Fig. 2d to f. We can observe that the utility, running time and memory consumption generally increase as $|W|$ increases, which is reasonable as

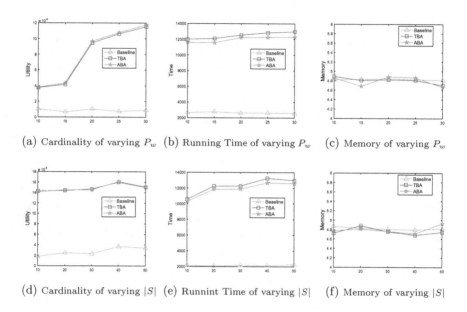

**Fig. 4.** Results on varying $P_w$ and $S$.

more workers need to be assigned. Again, we can see that TBA are better than ABA in terms of Utility and running time.

**Effects of the Global Unit Transportation Fee $\gamma$.** The results of varying $\gamma$ are presented in Fig. 3a to c. We can see that the utility and running time decrease as the $\gamma$ increases for higher transportation fee and less workers that could be assigned to far tasks.

**Effects of the Average Budget of Tasks $B_t$.** The results are presented in Fig. 3d to f. We can first see from the figure that the utility increases as the average budget increases. And there is no large differences of the running time and memory consumption between various $B_t$.

**Effects of the Variance of the Price of Different Skills $P_w$.** The results are presented in Fig. 4a to c. We can see from the figures that TBA and ABA have much better performance than baseline algorithm as the price increases. And the running time and memory consumption do not vary too much in different price.

**Effects of the Total Number of Skills $|S|$.** The results are presented in Fig. 4d to f. First, we can observe that the utility and memory consumption do not change greatly as the number increases. Then, we can see that the running time increases as the number increases, and this is reasonable because it is much harder to find suitable workers to finish the task for more kinds of skills.

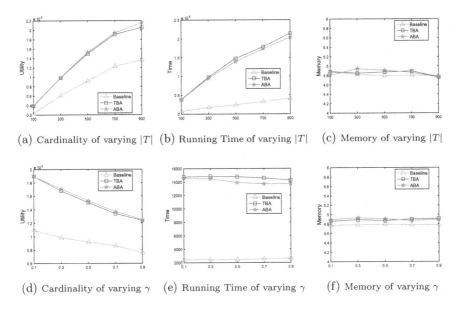

(a) Cardinality of varying $|T|$ (b) Running Time of varying $|T|$ (c) Memory of varying $|T|$

(d) Cardinality of varying $\gamma$ (e) Running Time of varying $\gamma$ (f) Memory of varying $\gamma$

**Fig. 5.** Results of real dataset on varying $|T|$ and $\gamma$.

**Real Dataset.** The results on real dataset are shown in Fig. 5a to f, where we vary $|T|$ and price. We can observe similar patterns as those in Figs. 2a to c and 3a to c.

**Conclusion.** For utility, TBA is better than ABA and baseline algorithm, and both TBA and ABA have a much better performance than baseline algorithm. As for running time, baseline algorithm is fastest, but the speed of TBA and ABA algorithm is acceptable for most circumstances. Moreover, TBA is faster than ABA algorithm.

# 6 Conclusion

In this paper we study the problem of Specialty-Aware Task Assignment (SATA) in spatial crowdsourcing, where the tasks have requirements on skills, and the workers specify fees for each of their skills. The goal is to maximize the total utility of the task assignment between tasks and workers. We prove the SATA problem is NP-hard. Two heuristic algorithms are proposed to solve the problem. We conduct extensive experiments on both synthetic and real-world datasets to evaluate our algorithms. The experimental results show that our solutions are efficient and effective.

**Acknowledgment.** We are grateful to anonymous reviewers for their constructive comments. This work is partially supported by the National Science Foundation of China (NSFC) under Grant No. 61502021 and 61532004.

# References

1. Anagnostopoulos, A., Becchetti, L., Castillo, C., Gionis, A., Leonardi, S.: Online team formation in social networks. In: WWW, pp. 839–848 (2012)
2. Chen, Z.: gMission: a general spatial crowdsourcing platform. PVLDB **7**(14), 1629–1632 (2014)
3. Cheng, P., Lian, X., Chen, L., Han, J., Zhao, J.: Task assignment on multi-skill oriented spatial crowdsourcing. TKDE **28**(8), 2201–2215 (2016)
4. Gao, D., Tong, Y., She, J., Song, T., Chen, L., Xu, K.: Top-$k$ team recommendation in spatial crowdsourcing. In: Cui, B., Zhang, N., Xu, J., Lian, X., Liu, D. (eds.) WAIM 2016. LNCS, vol. 9658, pp. 191–204. Springer, Cham (2016). https://doi.org/10.1007/978-3-319-39937-9_15
5. Gao, D., Tong, Y., She, J., Song, T., Chen, L., Xu, K.: Top-k team recommendation and its variants in spatial crowdsourcing. DSE **2**(2), 136–150 (2017)
6. Kazemi, L., Shahabi, C.: GeoCrowd: enabling query answering with spatial crowdsourcing. In: GIS, pp. 189–198 (2012)
7. Lappas, T., Liu, K., Terzi, E.: Finding a team of experts in social networks. In: SIGKDD, pp. 467–476 (2009)
8. Majumder, A., Datta, S., Naidu, K.: Capacitated team formation problem on social networks. In: SIGKDD, pp. 1005–1013 (2012)
9. Musthag, M., Ganesan, D.: Labor dynamics in a mobile micro-task market. In: CHI (2013)
10. Song, T., et al.: Trichromatic online matching in real-time spatial crowdsourcing. In: ICDE, pp. 1009–1020 (2017)
11. Tao, Q., Zeng, Y., Zhou, Z., Tong, Y., Chen, L., Xu, K.: Multi-worker-aware task planning in real-time spatial crowdsourcing. In: Pei, J., Manolopoulos, Y., Sadiq, S., Li, J. (eds.) DASFAA 2018. LNCS, vol. 10828, pp. 301–317. Springer, Cham (2018). https://doi.org/10.1007/978-3-319-91458-9_18
12. To, H., Fan, L., Tran, L., Shahabi, C.: Real-time task assignment in hyperlocal spatial crowdsourcing under budget constraints. In: PerCom, pp. 1–8 (2016)
13. To, H., Shahabi, C., Xiong, L.: Privacy-preserving online task assignment in spatial crowdsourcing with untrusted server. In: ICDE (2018)
14. Tong, Y., Chen, L., Shahabi, C.: Spatial crowdsourcing: challenges, techniques, and applications. PVLDB **10**(12), 1988–1991 (2017)
15. Tong, Y., She, J., Ding, B., Chen, L., Wo, T., Xu, K.: Online minimum matching in real-time spatial data: experiments and analysis. **9**, 1053–1064 (2016)
16. Tong, Y., She, J., Ding, B., Wang, L., Chen, L.: Online mobile micro-task allocation in spatial crowdsourcing. In: ICDE, pp. 49–60 (2016)
17. Tong, Y., Wang, L., Zhou, Z., Chen, L., Du, B., Ye, J.: Dynamic pricing in spatial crowdsourcing: a matching-based approach. In: SIGMOD (2018)
18. Tong, Y., et al.: Flexible online task assignment in real-time spatial data. PVLDB **10**(11), 1334–1345 (2017)
19. Tran, L., To, H., Fan, L., Shahabi, C.: A real-time framework for task assignment in hyperlocal spatial crowdsourcing. TIST **9**(3), 37 (2018)
20. Vazirani, V.V.: Approximation Algorithms. Springer, Berlin (2013)
21. Zeng, Y., Tong, Y., Chen, L., Zhou, Z.: Latency-oriented task completion via spatial crowdsourcing. In: ICDE (2018)
22. Zhao, Y., Li, Y., Wang, Y., Su, H., Zheng, K.: Destination-aware task assignment in spatial crowdsourcing. In: CIKM, pp. 297–306 (2017)

# Game-Theoretic Analysis on the Number of Participants in the Software Crowdsourcing Contest

Pengcheng Peng[1], Chenqi Mou[1(✉)], and Wei-Tek Tsai[2,3]

[1] LMIB – School of Mathematics and Systems Science /
Beijing Advanced Innovation Center for Big Data and Brain Computing,
Beihang University, Beijing 100191, China
{pengcheng.peng,chenqi.mou}@buaa.edu.cn
[2] School of Computer Science and Engineering, Beihang University,
Beijing 100191, China
[3] School of Computing, Informatics, and Decision Systems Engineering,
Arizona State University, Tempe, USA
wtsai@asu.edu

**Abstract.** In this paper a game theoretic model of multiple players is established to relate the reward from the outsourcer and the number of participants in the software crowdsourcing contest in the winner-take-all mode via Nash equilibria of the game. We show how to construct the payoff function of each participant in this game by computing his expected probability of winning sequential pairwise challenges. Preliminary experimental results with our implementations are provided to illustrate the relationships between the reward and the number of participants for three typical participant compositions.

**Keywords:** Software crowdsourcing · Game theory
Nash equilibrium · Payoff function

## 1 Introduction

Crowdsourcing is a type of activity for accomplishing tasks in an open call to a group of non-specific individuals, usually through the internet [3]. As a successful application, software crowdsourcing has become an influential approach for software development by using collective intelligence of the software developers all over the world [11]. Currently there are many popular software crowdsourcing platforms such as TopCoder[1] [10], Apple's App Store, and Applause[2].

---

This work was partially supported by the National Natural Science Foundation of China (NSFC 61690202, 11401018 and 11771034).

[1] https://www.topcoder.com.
[2] https://www.applause.com.

J. Fleuriot et al. (Eds.): AISC 2018, LNAI 11110, pp. 255–268, 2018.
https://doi.org/10.1007/978-3-319-99957-9_20

The design, analysis, evaluation, and development of software crowdsourcing have been studied from different perspectives [1, 19], among which one kind of popular method is based on the game theory [7, 17]. In particular, a game theoretic model of crowdsourcing contests is proposed in [1] with formal analysis on the asymptotic behaviors of the outcomes of the contests; a framework is proposed in [20] for evaluating the software crowdsourcing projects from many aspects such as quality and cost; the 2-player algorithm challenges on TopCoder are modeled by static games with complete information in [8]; and a crowdsourcing framework is proposed in [16] based on a revised form of $n$-person chicken games.

The ways how the rewards from the oursourcers in software crowdsourcing contests are paid to the participants can be categorized into two modes: the winner-take-all mode and multiple-winners mode [12]. In the former mode all the reward from the outsourcer is taken by the sole winner of the contest, while in the latter multiple or even all the participants get the rewards based on their efforts in the contest. The choice between the two payment modes of the contests has been studied based on the cost functions [15] and the degrees of risk aversion of the participants [1, 9]. It has been shown in [15] that when the cost functions are linear or concave, the choice should be the winner-take-all mode and shown in [1, 9] that the choice should be the winner-take-all mode in the case of risk-neutral participants and the multiple-winners mode in the case of risk-averse participants.

Crowdsourcing contests were first modeled as all-pay auctions in [6], where the behaviors of the participants facing multiple simultaneous winner-take-all contests with different rewards are studied. Then the all-pay auction model has been widely used in the study on crowdsourcing contests [1, 4, 13]. In this paper we focus on the software crowdsourcing contest in the winner-take-all mode using the all-pay auction model and study the influence of the reward on the number of participants in the contest. The study on such influence is helpful for the outsourcers of software crowdsourcing contests to choose an appropriate reward to attract an expected number of participants.

We first model the software crowdsourcing contest by a game of multiple players, each of whom has two pure strategies as joining or quitting the game. Under the assumption that the probability for any player to win a pairwise challenge against any other player is known, we propose a method for constructing the payoff functions of all the players by computing the expected winning probability of each player in this game. With the payoff functions we compute all the Nash equilibria of the game and connect the expected number of participants with the computed Nash equilibria. Experimental results for three typical compositions of candidates are reported based on our implementation for computing payoff functions and the software Gambit [14] for computing all the Nash equilibria.

The paper is organized as follows. After the introduction of all the necessary notions and notations in the game theory in Sect. 2, we present the game theoretic model in which the number of participants is connected to all the Nash equilibria of the game in Sect. 3. Then in Sect. 4 we elaborate how the payoff functions of all the players in the game are constructed. The experimental results are presented in Sect. 5, and we conclude this paper with some remarks in Sect. 6.

## 2   Preliminaries

Let us consider a game of $n$ players $P_1, \ldots, P_n$. Fixed an integer $i$ $(1 \leq i \leq n)$. Let $S_i$ be the set of pure strategies that $P_i$ can choose from and $p_i : \prod_{j=1}^n S_j \to \mathbb{R}$ be the payoff function of $P_i$, where $p_i(s)$ represents $P_i$'s payoff given a profile $s = (s_1, \ldots, s_n)$ of pure strategies $s_j \in S_j$, $j = 1, \ldots, n$. In particular, we denote $s_i = s_i$ for $s = (s_1, \ldots, s_n)$.

Take the game of *matching pennies* between two players $P_1$ and $P_2$ for example. In this game, each player has a penny and they show their choices of turning the penny to head or tail simultaneously. Here $S_1 = S_2 = \{H, T\}$, where $H$ and $T$ denote the pure strategies of turning the penny to head and tail respectively, and therefore $S_1 \times S_2 = \{(H, H), (H, T), (T, H), (T, T)\}$. Two possible payoff functions $p_1$ and $p_2$ are shown below.

$$
\begin{array}{ll}
\quad\quad p_1 & \quad\quad p_2 \\
(H, H) \mapsto 1 & (H, H) \mapsto -1 \\
(H, T) \mapsto -1 & (H, T) \mapsto 1 \\
(T, H) \mapsto -1 & (T, H) \mapsto 1 \\
(T, T) \mapsto 1 & (T, T) \mapsto -1
\end{array}
$$

A *mixed strategy* $\sigma_i$ of $P_i$ is a probability distribution over the pure strategies in $S_i$, where $\sigma_i(s)$ is the probability for $P_i$ to choose a pure strategy $s \in S_i$. Denote the space of mixed strategies of $P_i$ by $\Sigma_i$ and $\Sigma_{-i} := \prod_{j \in \{1, \ldots, n\} \setminus \{i\}} \Sigma_j$. For a profile $\sigma = (\sigma_1, \ldots, \sigma_n)$ of mixed strategies $\sigma_j \in \Sigma_j$, $j = 1, \ldots, n$, denote $\sigma_i = \sigma_i$ and $\sigma_{-i} = (\sigma_1, \ldots, \sigma_{i-1}, \sigma_{i+1}, \ldots, \sigma_n) \in \Sigma_{-i}$. The payoff of $P_i$ for a mixed strategy profile $\sigma$, denoted by $\tilde{p}_i(\sigma)$, is defined as

$$
\tilde{p}_i(\sigma) = \sum_{s \in \prod_{j=1}^n S_j} \left( \prod_{k=1}^n \sigma_k(s_k) \right) p_i(s), \tag{1}
$$

which is essentially the expected payoff of $P_i$ for the probability distribution $\sigma$. For the sake of simplicity, we write $\tilde{p}_i(\sigma)$ as $\tilde{p}_i(\sigma_i, \sigma_{-i})$ and as $\tilde{p}_i(s, \sigma_{-i})$ when $\sigma_i(s) = 1$ for some $s \in S_i$.

For the game of matching pennies, consider the mixed strategies $\sigma_1$ of $P_1$ with $\sigma_1(H) = 0.3$ and $\sigma_1(T) = 0.7$ and $\sigma_2$ of $P_2$ with $\sigma_2(H) = 0.6$ and $\sigma_2(T) = 0.4$. Then for the mixed strategy profile $\sigma = (\sigma_1, \sigma_2)$, the payoffs $\tilde{p}_1(\sigma)$ of $P_1$ and $\tilde{p}_2(\sigma)$ of $P_2$, computed by (1), are

$$
\tilde{p}_1(\sigma) = 0.3 \times 0.6 \times 1 + 0.3 \times 0.4 \times (-1) + 0.7 \times 0.6 \times (-1) + 0.7 \times 0.4 \times 1 = -0.08,
$$
$$
\tilde{p}_2(\sigma) = 0.3 \times 0.6 \times (-1) + 0.3 \times 0.4 \times 1 + 0.7 \times 0.6 \times 1 + 0.7 \times 0.4 \times (-1) = 0.08
$$

respectively.

A mixed strategy profile $\sigma^* \in \prod_{j=1}^n \Sigma_j$ is called a *Nash equilibrium* of a game if for each $i = 1, \ldots, n$, $\tilde{p}_i(\sigma_i^*, \sigma_{-i}^*) \geq \tilde{p}_i(s_i, \sigma_{-i}^*)$ for any pure strategy $s_i \in S_i$. A Nash equilibrium is a state such that each player in the game maximizes his expected payoff under the condition that the mixed strategies of the other players are fixed, and thus anyone attempting to change his mixed strategy from the Nash equilibrium will face a reduced payoff. For a finite non-cooperative game, there always exists at least one Nash equilibrium [17].

It is easy to verify that $\sigma^* = ((1/2, 1/2), (1/2, 1/2))$ is a Nash equilibrium of the game of matching pennies:

$$\tilde{p}_1(\sigma^*) = 0 \geq \tilde{p}_1(H, \sigma^*_{-1}) = \tilde{p}_1(T, \sigma^*_{-1}) = 0,$$
$$\tilde{p}_2(\sigma^*) = 0 \geq \tilde{p}_2(H, \sigma^*_{-2}) = \tilde{p}_2(T, \sigma^*_{-2}) = 0,$$

and that $\tilde{\sigma} = ((2/3, 1/3), (1/2, 1/2))$ is not a Nash equilibrium:

$$\tilde{p}_2(\tilde{\sigma}) = 0 < \tilde{p}_2(T, \tilde{\sigma}_{-2}) = 1/3.$$

In fact, for any $\sigma \neq \sigma^*$, there exist some $i = 1$ or $2$ and $s \in S_i$ such that $\tilde{p}_i(\sigma_i, \sigma_{-i}) < \tilde{p}_i(s, \sigma_{-i})$, and thus $\sigma^*$ is the only Nash equilibrium of the game of matching pennies.

## 3   A Multi-Player Game-Theoretic Model

This paper studies the relationship between the reward and the expected number of participants in the software crowdsourcing contest by using Nash equilibria of a corresponding game of multiple players with two pure strategies. In this context, the players in this game are called *candidates*. Each candidate has two optional pure strategies as joining or quitting the contest, denoted by $J$ and $Q$ respectively. A candidate who chooses to join the contest is further called a *participant*.

Based on the discussions above, we model the software crowdsourcing contest as a game of $n$ candidates $C_1, \ldots, C_n$. For each $i = 1, \ldots, n$, let $S_i = \{J, Q\}$ be the set of pure strategies for $C_i$ and let $p_i : \prod_{j=1}^n S_j \to \mathbb{R}$ be the payoff function of $C_i$. Let $R$ be the reward paid to the sole winner of the software crowdsourcing contest and $c_i$ be the cost of $C_i$ for finishing the contest. The payoff function $p_i$ of $C_i$ is defined as

$$p_i(s) = \begin{cases} R \cdot W_i(s) - c_i, & \text{if } s_i = J \\ 0, & \text{if } s_i = Q \end{cases} \tag{2}$$

for any $s \in \prod_{j=1}^n S_j$, where $W_i(s)$ is the probability for $C_i$ to win the contest with participants specified by $s$ (note that $s_j \in S_j = \{J, Q\}$ for $j = 1, \ldots, n$ and thus the participants of the contest are known once $s$ is given). As one may find clearly from (2), the reward $R$ of the contest appears in the payoff function of each candidate. The explicit computation of $W_i(s)$ (and thus the construction of all the payoff functions) is elaborated in Sect. 4.

Assume that the payoff functions $p_1, \ldots, p_n$ are known. Then all the Nash equilibria of the game described above can be computed, say by the methods proposed in [2,5]. Each Nash equilibrium is a mixed strategy profile $\sigma = (\sigma_1, \ldots, \sigma_n) \in \prod_{i=1}^n \Sigma_i$, where $\sigma_i(J)$ describes the probability for $C_i$ to choose to join the game. Then the expected number of participants $N(\boldsymbol{\sigma})$ for this Nash equilibrium is equal to $\sum_{i=1}^{n} \sigma_i(J)$. For example, suppose that $\boldsymbol{\sigma} = (\sigma_1, \ldots, \sigma_4)$ is a Nash equilibrium of a 4-player game described above such that $\sigma_1(J) = 1$, $\sigma_2(J) = 0.7$, $\sigma_3(J) = 0$, and $\sigma_4(J) = 0.2$. Then the expected number of participants in the game for this Nash equilibrium is $1 + 0.7 + 0 + 0.2 = 1.9$. In the case of multiple Nash equilibria $\boldsymbol{\sigma}^{(1)}, \ldots, \boldsymbol{\sigma}^{(m)}$ of the game, the expected number of participants of the contest is defined to be $\sum_{i=1}^{m} N(\boldsymbol{\sigma}^{(i)})/m$.

The game above of multiple candidates is the underlying theory for the study in this paper on the influence of the reward of the software crowdsourcing contest, which appears in the payoff functions as shown in (2), on the number of participants in this contest, which can be computed via the Nash equilibria as discussed above. In particular, for any given value $R = R_0$, by computation of all the Nash equilibria we are able to obtain the expected number of participants corresponding to $R_0$. Furthermore, the changes in the value of $R$ lead to changes in the expected number of participants, and this correspondence is what we are interested in for the influence of the reward on the number of the participants.

## 4    Construction of Payoff Functions

For the game described above of $n$ candidates $C_1, \ldots, C_n$, assume that the probability for any candidate $C_i$ to win a pairwise challenge against another candidate $C_j$ ($j \neq i$) is known and denoted by $P_{ij}$. Next we show how to compute the probability $W_i(\boldsymbol{s})$ in (2) for a given pure strategy profile $\boldsymbol{s}$ in a model of sequential pairwise challenges of all participants specified by $\boldsymbol{s}$. Note that in software crowdsourcing contests like those held in TopCoder, the information regarding the participants in the contests is usually available in the host platform, and it can be further analyzed and processed to furnish such a winning probability in a pairwise challenge.

Assume that for a pure strategy profile $\boldsymbol{s}$, there are $m$ out of $n$ candidates who choose to join the contest. Without loss of generality, we further assume that the $m$ participants are $C_1, \ldots, C_m$.

Let $\boldsymbol{i} = (i_1, i_2, \ldots, i_m)$ be a permutation of $\{1, 2, \ldots, m\}$. Then $\boldsymbol{i}$ defines an order for sequential pairwise challenges in the following way: the winner of the pairwise challenge $C_{i_1}$ and $C_{i_2}$ goes to the next round of pairwise challenge with $C_{i_3}$, whose winner goes further to the next round. This process is repeated and ended with one final winner. Obviously two distinct permutations $\boldsymbol{i}^{(1)} = (i_1^{(1)}, i_2^{(1)}, \ldots, i_m^{(1)})$ and $\boldsymbol{i}^{(2)} = (i_1^{(2)}, i_2^{(2)}, \ldots, i_m^{(2)})$ define the same order for sequential pairwise challenges if and only if

$$i_1^{(1)} = i_2^{(2)}, i_2^{(1)} = i_1^{(2)}, i_3^{(1)} = i_3^{(2)}, \ldots, i_m^{(1)} = i_m^{(2)}.$$

Denote

$\mathbb{P}(m) := \{\text{permutations of } \{1, \ldots, m\} \text{ such that no pair defines the same order}\}.$

The permutations in $\mathbb{P}(m)$ define all the possible orders of $C_1, \ldots, C_m$ for sequential pairwise challenges, and it is straightforward that the cardinality of $\mathbb{P}(m)$ is

$\#\mathbb{P}(m) = \binom{m}{2} \cdot (m-2)!$, where $\binom{m}{2}$ denotes the number of 2-combinations in a set of $m$ elements and $(m-2)!$ is the factorial of $m-2$.

For an arbitrary $\boldsymbol{i} = (i_1, \ldots, i_m) \in \mathbb{P}(m)$, next we study the probability for $C_i$ $(1 \le i \le m)$ to be the final winner in the sequential pairwise challenges in the order defined by $\boldsymbol{i}$, denoted by $W_i(\boldsymbol{i})$. Let $i = i_k$ for some $k$ $(1 \le k \le m)$. Then $C_i$ is the final winner if and only if $C_i$ wins the pairwise challenge against any possible winner of the sequential pairwise challenges in the order $(i_1, \ldots, i_{k-1})$ and $C_i$ wins successive pairwise challenges against $C_{i_{k+1}}, \ldots, C_{i_m}$. This observation leads to

$$W_i(\boldsymbol{i}) = \sum_{r=1}^{k-1} [P_{i_k,i_r} \cdot W_{i_r}((i_1, \ldots, i_{k-1}))] \cdot \prod_{s=k+1}^{m} P_{i_k,i_s}, \tag{3}$$

where $P_{ij}$, as defined at the beginning of this section, is the probability for $C_i$ to win a pairwise challenge against $C_j$ $(j \ne i)$ and $W_{i_r}((i_1, \ldots, i_{k-1}))$ is the probability for $C_{i_r}$ to be the final winner of the sequential pairwise challenges in the order $(i_1, \ldots, i_{k-1})$. Since the length of the permutation $(i_1, \ldots, i_{k-1})$ is strictly smaller than that of the permutation $(i_1, \ldots, i_m)$, the probability $W_i(\boldsymbol{i})$ can be effectively computed in a recursive way.

Obviously, for a given permutation $\boldsymbol{i} = (i_1, \ldots, i_m)$, the probability $W_i(\boldsymbol{i})$ is heavily dependent on the position of $i$ in $\boldsymbol{i}$. Assume that all the permutations in $\mathbb{P}(m)$ follow an equiprobable distribution. Then the expected probability for $C_i$ $(1 \le i \le m)$ to be the final winner of a sequential pairwise challenge, which is $W_i(\boldsymbol{s})$ in (2) by our notation, is

$$W_i(\boldsymbol{s}) = \frac{1}{\binom{m}{2} \cdot (m-2)!} \cdot \sum_{\boldsymbol{i} \in \mathbb{P}(m)} W_i(\boldsymbol{i}). \tag{4}$$

At this point, for each $i = 1, \ldots, n$, we are able to compute $W_i(\boldsymbol{s})$ in (2) for any pure strategy profile $\boldsymbol{s}$, and thus the payoff function $p_i$ can be constructed immediately with (2).

### 4.1   Algorithm Description

The procedure to construct all the payoff functions $p_1, \ldots, p_n$ described above is formulated as Algorithm 1 below. In this algorithm, $A$ cat $B$ for $A = [A_1, \ldots, A_r]$ and $B = [B_1, \ldots, B_s]$ returns $[A_1, \ldots, A_r, B_1, \ldots, B_s]$ and a payoff function $p_i$ is stored as $p_i = \{[\boldsymbol{s}, p] : \boldsymbol{s} \in \prod_{j=1}^{n} \{J, Q\}, p \in \mathbb{R}\}$, where in each pair $[\boldsymbol{s}, p]$, $p$ is the payoff of $C_i$ for the pure strategy profile $\boldsymbol{s}$. The subroutine in Algorithm 1 for computing the probability for some participant to be the final winner in a sequential pairwise challenge is formulated as Algorithm 2 and is essentially a recursive function.

### 4.2   An Illustrative Example

Let us consider a game of seven candidates $C_1, \ldots, C_7$ with the following matrix $P$ of pairwise winning probability such that $P_{ij}$ is the probability for $C_i$ to win a pairwise challenge against $C_j$ for $1 \le i \ne j \le 7$:

---

**Algorithm 1.** Computation of payoff functions $(p_1, \ldots, p_n) :=$ Payoff$(n, \mathsf{P}, R, (c_1, \ldots, c_n))$

---

**Input:** $n$: number of candidates; $\mathsf{P}$: matrix of pairwise winning probability such that $\mathsf{P}_{ij}$ $(i \neq j)$ is the probability for $C_i$ to win a pairwise challenge against $C_j$; $R$: reward; $(c_1, \ldots, c_n)$: costs of $n$ candidates $C_1, \ldots, C_n$

**Output:** $(p_1, \ldots, p_n)$, payoff functions of $C_1, \ldots, C_n$

1   $p_i := \{\ \}, i = 1, \ldots, n;$
2   **for** $s \in \prod_{i=1}^{n}\{J, Q\}$ **do**
3     **for** $i = 1, \ldots, n$ **do**
4       **if** $s_i = Q$ **then**
5         $p_i := p_i \cup \{[s, 0]\};$
6       **else**
7         $\tilde{s} = [\ ];$             [$\tilde{s}$ records the participants]
8         **for** $j = 1, \ldots, n$ **do**
9           **if** $s_j = J$ **then**
10            $\tilde{s} := \tilde{s}$ cat $[j]$ ;
11         $\mathbb{P}(\tilde{s}) := \{$permutations of $\tilde{s}$ with no pair defining the same order$\};$
12         $W_i := 0;$
13         **for** $i \in \mathbb{P}(\tilde{s})$ **do**
14           Suppose that $i_k = i;$
15           $W_i^{(2)} := \prod_{j=k+1}^{\#\tilde{s}} \mathsf{P}_{i, i_j};$     [$W_i^{(2)}$: second part of probability in (3)]
16           $W_i^{(1)} := 0;$                    [$W_i^{(1)}$: first part of probability in (3)]
17           $i(k-1) := (i_1, \ldots, i_{k-1});$
18           **for** $j = 1, \ldots, k-1$ **do**
19             $W_i^{(1)} := W_i^{(1)} + \mathsf{P}_{i, i_j} \cdot \text{WinPairwise}(i_j, i(k-1), \mathsf{P});$
20           $W_i := W_i + W_i^{(1)} \cdot W_i^{(2)};$
21         $p_i := p_i \cup \{[s, R \cdot \frac{W_i}{\binom{\#\tilde{s}}{2} \cdot (\#\tilde{s}-2)!} - c_i]\};$        [as in (4)]

22 **return** $(p_1, \ldots, p_n);$

---

$$\mathsf{P} = \begin{pmatrix} * & 0.90 & 0.92 & 0.91 & 0.93 & 0.94 & 0.96 \\ 0.10 & * & 0.52 & 0.50 & 0.53 & 0.56 & 0.60 \\ 0.08 & 0.48 & * & 0.49 & 0.50 & 0.52 & 0.57 \\ 0.09 & 0.50 & 0.51 & * & 0.52 & 0.53 & 0.55 \\ 0.07 & 0.47 & 0.50 & 0.48 & * & 0.50 & 0.54 \\ 0.06 & 0.44 & 0.48 & 0.47 & 0.50 & * & 0.52 \\ 0.04 & 0.40 & 0.43 & 0.45 & 0.46 & 0.48 & * \end{pmatrix}.$$

Let $s = (J, Q, J, J, Q, Q, J)$ be a pure strategy profile. Next we show how the payoff $p_1(s)$ of $C_1$ is computed with Algorithm 1. First with $s_2 = s_5 = s_6 = Q$ and $s_1 = s_3 = s_4 = s_7 = J$, we construct the list $\tilde{s} = [1, 3, 4, 7]$ of indexes for the participants as in Line 10 for the profile $s$.

---

**Algorithm 2.** Probability to win sequential pairwise challenges $W :=$ WinPairwise$(i, \boldsymbol{s}, \mathsf{P})$

---

**Input:** $i$: an integer; $\boldsymbol{s}$: a sequence; $\mathsf{P}$: matrix of pairwise winning probability such that $\mathsf{P}_{ij}$ $(i \neq j)$ is the probability for $C_i$ to win a pairwise challenge against $C_j$

**Output:** $W$, the probability for $C_i$ to be the final winner in the sequential pairwise challenges in the order $\boldsymbol{s}$

1  **if** $s_1 = i$ **then**
2  $\quad$ **return** $\prod_{j=2}^{\#s} \mathsf{P}_{i,s_j}$;
3  **else if** $\#s = 2$ **then**
4  $\quad$ **return** $\mathsf{P}_{i,s_1}$;
5  **else**
6  $\quad$ Suppose that $s_k = i$;
7  $\quad$ $W_i^{(2)} := \prod_{j=k+1}^{\#s} \mathsf{P}_{i,s_j}$; $\qquad\qquad$ [$W_i^{(2)}$: second part of probability in (3)]
8  $\quad$ $W_i^{(1)} := 0$ $\qquad\qquad\qquad\qquad$ [$W_i^{(1)}$: first part of probability in (3)]
9  $\quad$ $s(k-1) := (s_1, \ldots, s_{k-1})$;
10 $\quad$ **for** $j = 1, \ldots, k-1$ **do**
11 $\quad\quad$ $W_i^{(1)} := W_i^{(1)} + \mathsf{P}_{i,s_j} \cdot$ WinPairwise$(s_j, s(k-1), \mathsf{P})$; $\quad$ [a recursive call]
12 $\quad$ **return** $W_i^{(1)} \cdot W_i^{(2)}$;

---

One example of all the permutations of $\tilde{\boldsymbol{s}} = [1, 3, 4, 7]$ such that no pair defines the same order is shown as follows.

$$(4,7,3,1), \quad (3,7,4,1), \quad (3,4,7,1), \quad (4,7,1,3), \quad (1,7,4,3), \quad (1,4,7,3),$$
$$(3,7,1,4), \quad (1,7,3,4), \quad (1,3,7,4), \quad (3,4,1,7), \quad (1,4,3,7), \quad (1,3,4,7).$$

Note that there are $\binom{4}{2} \cdot (4-2)! = 12$ possible permutations in total.

Take the permutation $\boldsymbol{i} = (4, 7, 1, 3)$ for example. In $\boldsymbol{i}$ we find that $i_3 = 1$. Then $W_1^{(2)}$, the second part of winning probability as in (3), is equal to $\mathsf{P}_{13} = 0.92$, and $W_1^{(1)}$, the first part as in (3) is

$$W_1^{(1)} = \mathsf{P}_{14}\mathsf{P}_{47} + \mathsf{P}_{17}\mathsf{P}_{74} = 0.91 \times 0.55 + 0.96 \times 0.45 = 0.9325.$$

Therefore the probability for $C_1$ to be the final winner in the sequential pairwise challenges in the order $(4, 7, 1, 3)$ is $0.9325 \times 0.92 = 0.8579$.

The probabilities for $C_1$ to be the final winners for all the 12 possible permutations are listed in Table 1 below.

The expected probability for $C_1$ to be the final winner in the contest of the four participants $C_1$, $C_3$, $C_4$, and $C_7$ is therefore

$$\frac{0.924935 + 0.922468 + 0.934803 + 0.8579 + 0.852852 + 0.878304 + 0.803712 \times 6}{12} \approx 0.8495.$$

Then with any given $R$ and $c_1$, the payoff $p_1(\boldsymbol{s})$ can be computed immediately with (2).

**Table 1.** Probability for $C_1$ to be the final winner for different permutations

| Permutation | Probability | Permutation | Probability | Permutation | Probability |
|---|---|---|---|---|---|
| (4, 7, 3, 1) | 0.924935 | (3, 7, 4, 1) | 0.922468 | (3, 4, 7, 1) | 0.934803 |
| (4, 7, 1, 3) | 0.8579 | (1, 7, 4, 3) | 0.803712 | (1, 4, 7, 3) | 0.803712 |
| (3, 7, 1, 4) | 0.852852 | (1, 7, 3, 4) | 0.803712 | (1, 3, 7, 4) | 0.803712 |
| (3, 4, 1, 7) | 0.878304 | (1, 4, 3, 7) | 0.803712 | (1, 3, 4, 7) | 0.803712 |

## 5    Experimental Results

In this section, the experimental results on the influence of the reward of the software crowdsourcing contest on the expected number of participants are reported for three types of candidates based on our implementations of Algorithms 1 and 2 and the software Gambit for computing all Nash equilibria.

We consider a software crowdsourcing contest of 7 candidates $C_1, \ldots, C_7$, each of whom has a fixed cost $c_i = 20$ for $i = 1, \ldots, 7$. The three types of candidates we consider are the following: the first is such that all the candidates have average capabilities, the second is such that one candidate is super strong while the others have average capabilities, and the third is such that the capabilities of the candidates follow a normal distribution.

### 5.1    Candidates of Average Capabilities

In this case, the matrix of pairwise winning probability is set as

$$
P_{average} = \begin{pmatrix}
* & 0.49 & 0.50 & 0.53 & 0.54 & 0.58 & 0.60 \\
0.51 & * & 0.51 & 0.52 & 0.52 & 0.55 & 0.56 \\
0.50 & 0.49 & * & 0.52 & 0.53 & 0.56 & 0.57 \\
0.47 & 0.48 & 0.48 & * & 0.50 & 0.53 & 0.54 \\
0.46 & 0.48 & 0.47 & 0.50 & * & 0.52 & 0.54 \\
0.42 & 0.45 & 0.44 & 0.47 & 0.48 & * & 0.51 \\
0.40 & 0.44 & 0.43 & 0.46 & 0.46 & 0.49 & *
\end{pmatrix}.
$$

For each $R = 10, 20, \ldots, 700$, we construct the payoff functions $p_1, \ldots, p_7$ with Algorithm 1 and then compute the expected number $N(R)$ of participants by using the Nash equilibria of the corresponding game. The relationship between $R$ and $N(R)$ is shown in Fig. 1.

For example, at $R = 70$, in total there are 25 Nash equilibria returned by the Gambit software. For example, one Nash equilibrium $\sigma = (\sigma_1, \ldots, \sigma_7)$ is shown as follows.

$$
\sigma_1(J) = 1, \quad \sigma_1(Q) = 0, \quad \sigma_2(J) = 0.0655276, \quad \sigma_2(Q) = 0.934472,
$$
$$
\sigma_3(J) = 1, \quad \sigma_3(Q) = 0, \quad \sigma_4(J) = 0.568249, \quad \sigma_4(Q) = 0.431751,
$$
$$
\sigma_5(J) = 0, \quad \sigma_5(Q) = 1, \quad \sigma_6(J) = 0, \quad \sigma_6(Q) = 1, \quad \sigma_7(J) = 0.993354,
$$
$$
\sigma_7(Q) = 0.00664647.
$$

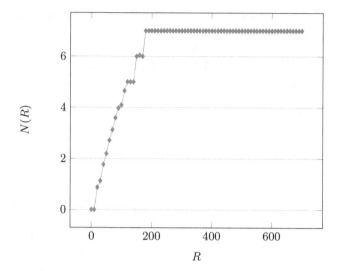

**Fig. 1.** Relationship between the reward $R$ and the expected number of participants $N(R)$: the case of candidates of average capabilities

Therefore the expected number of participants is $\sum_{i=1}^{7} \sigma_i(J) = 3.6271306$ for this Nash equilibrium $\sigma$.

As can be seen from Fig. 1, the expected number of participants $N(R)$ increases almost at a constant speed with respect to the reward $R$ and reaches its maximum value $N(R) = 7$ when $R = 180$, which means that all the candidates choose to join the contest at this point.

## 5.2   One Super Strong Candidate

In this case, the matrix of pairwise winning probability is set as

$$
P_{strong} = \begin{pmatrix}
* & 0.90 & 0.92 & 0.91 & 0.93 & 0.94 & 0.96 \\
0.10 & * & 0.52 & 0.50 & 0.53 & 0.56 & 0.60 \\
0.08 & 0.48 & * & 0.49 & 0.50 & 0.52 & 0.57 \\
0.09 & 0.50 & 0.51 & * & 0.52 & 0.53 & 0.55 \\
0.07 & 0.47 & 0.50 & 0.48 & * & 0.50 & 0.54 \\
0.06 & 0.44 & 0.48 & 0.47 & 0.50 & * & 0.52 \\
0.04 & 0.40 & 0.43 & 0.45 & 0.46 & 0.48 & *
\end{pmatrix}.
$$

The relationship between $R$ and $N(R)$ is shown in Fig. 2.

As can be found in Fig. 2, the function curve of $N(R)$ is in the shape of a ladder with respect to $R$. Compared to the slope of the curve in Fig. 1, the expected number of participants in this case increases at a slower speed than that in the case of candidates of average capabilities. In particular, the expected number of participants $N(R)$ reaches 6 at $R = 510$ and 7 at $R = 670$ respectively. This relationship between $R$ and $N(R)$ reflects the fact that in a software crowdsourcing

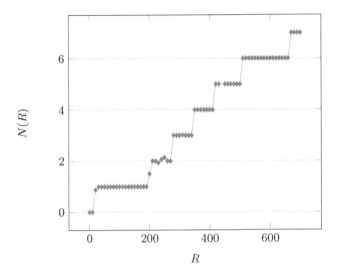

**Fig. 2.** Relationship between the reward $R$ and the expected number of participants $N(R)$: the case of one super strong candidate

contest, there are fewer participants when there is one super strong candidate, for the other candidates are reluctant to join a contest in which he tends to lose unless the reward from the contest is great enough. In some sense, the width of each step in the ladder in the curve of $N(R)$ reflects the difference in the reward to draw an additional participant.

## 5.3   Normal Distribution

In this case, we choose the capabilities $r_1, \ldots, r_7$ of the 7 candidates randomly from a normal distribution $N(\mu, \sigma^2)$ with the mean value $\mu = 5$ and the standard deviation $\sigma = 5/3$ so that the probability for the capability to fall in the interval $[\mu - 3\sigma, \mu + 3\sigma] = [0, 10]$ is about 99.74%. Then, based on their capabilities, the probability $P_{ij}$ for $C_i$ to win a pairwise challenge against $C_j$ for $1 \leq i \neq j \leq 7$ is set to $P_{ij} = r_i/(r_i + r_j)$.

For example, one instance of randomly chosen capabilities of 7 candidates is $[4.13, 6.03, 5.54, 4.11, 8.35, 2.81, 2.18]$, and the corresponding matrix of pairwise winning probability is constructed as follows. The relationship between $R$ and $N(R)$ is shown in Fig. 3.

$$
P_{normal} = \begin{pmatrix}
* & 0.41 & 0.43 & 0.50 & 0.33 & 0.60 & 0.65 \\
0.59 & * & 0.52 & 0.59 & 0.42 & 0.68 & 0.73 \\
0.57 & 0.48 & * & 0.57 & 0.40 & 0.66 & 0.72 \\
0.50 & 0.41 & 0.43 & * & 0.33 & 0.59 & 0.65 \\
0.67 & 0.58 & 0.60 & 0.67 & * & 0.75 & 0.79 \\
0.40 & 0.32 & 0.34 & 0.41 & 0.25 & * & 0.56 \\
0.35 & 0.27 & 0.28 & 0.35 & 0.21 & 0.44 & *
\end{pmatrix}.
$$

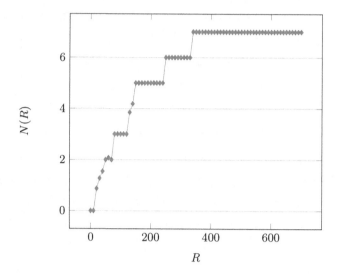

**Fig. 3.** Relationship between the reward $R$ and the expected number of participants $N(R)$: the case of normal distribution

As can be found in Fig. 3, the slope of the curve is between those of the curves in Fig. 1 and 2, and the expected number of participants $N(R)$ reaches 7 at $R = 340$. We have tested 5 instances of randomly generated capabilities from the normal distribution mentioned above, and the least rewards for the numbers of participants to reach 7 are presented in the column labeled $R_7$ in Table 2 below.

**Table 2.** Least rewards for 7 participants in the case of normal distribution

| No | Capabilities | $R_7$ |
|----|--------------|-------|
| 1 | [4.13, 6.03, 5.54, 4.11, 8.35, 2.81, 2.18] | 340 |
| 2 | [6.64, 3.96, 3.68, 5.22, 6.57, 3.23, 3.87] | 220 |
| 3 | [5.83, 2.05, 4.95, 3.69, 5.65, 4.76, 4.66] | 340 |
| 4 | [7.39, 7.14, 7.86, 4.69, 6.48, 5.59, 5.35] | 200 |
| 5 | [8.91, 4.91, 6.86, 4.38, 6.40, 5.51, 5.23] | 210 |

## 6   Concluding Remarks

A game theoretic model of the software crowdsourcing contest is proposed in this paper, and in this model the number of participants in the contest is connected to the Nash equilibria of the game. Based on the winning probabilities of pairwise challenges, we show how to construct the payoff functions of all the players in the game and thus the Nash equilibria of the game can be computed with these payoff functions. Experimental results illustrate different behaviors of the influence of the reward on the expected number of participants for three types of candidates.

As regards the efficiency of our algorithms, computation of all the Nash equilibria of the game is the current bottleneck. In fact, the study on and implementation of efficient algorithms for computing Nash equilibria of different kinds of games remain a highly non-trivial problem of common interest in game theory and computer science [18].

The proposed method in this paper for computing the payoff functions are based on the winning probabilities of pairwise challenges, but these probabilities may not be directly available in actual software crowdsourcing contests. In this case, one can try to estimate these winning probabilities of pairwise challenges by analyzing the information disclosed by the contest platform or establish a new model for computing the payoff function of each player in the game.

In this paper we only study software crowdsourcing contests in the winner-take-all mode. The study on the influence of allocation of rewards on the number of participants in contests in the multiple-winners mode by extending our algorithms is our future work.

**Acknowledgements.** The first author wishes to thank his supervisor, Professor Dongming Wang, for his support and encouragement.

# References

1. Archak, N., Sundararajan, A.: Optimal design of crowdsourcing contests. In: Proceedings of International Conference on Information Systems 2009, p. 200 (2009)
2. Avis, D., Rosenberg, G.D., Savani, R., Von Stengel, B.: Enumeration of Nash equilibria for two-player games. Econ. Theor. **42**(1), 9–37 (2010)
3. Brabham, D.C.: Crowdsourcing as a model for problem solving: an introduction and cases. Convergence **14**(1), 75–90 (2008)
4. Chawla, S., Hartline, J.D., Sivan, B.: Optimal crowdsourcing contests. Games Econ. Behav. (2015, in press)
5. Datta, R.S.: Finding all Nash equilibria of a finite game using polynomial algebra. Econ. Theor. **42**(1), 55–96 (2010)
6. DiPalantino, D., Vojnovic, M.: Crowdsourcing and all-pay auctions. In: Proceedings of the 10th ACM Conference on Electronic Commerce, pp. 119–128. ACM (2009)
7. Fudenberg, D., Tirole, J.: Game Theory. MIT Press, Cambridge, Massachusetts (1991)
8. Hu, Z., Wu, W.: A game theoretic model of software crowdsourcing. In: Proceedings of IEEE 8th International Symposium on Service Oriented System Engineering, pp. 446–453. IEEE (2014)
9. Kalra, A., Shi, M.: Designing optimal sales contests: a theoretical perspective. Mark. Sci. **2**(20), 170–193 (2001)

10. Lakhanih, K., Garvin, D.A., Lonstein, E.: Topcoder (A): developing software through crowdsourcing. Harvard Business School General Management Unit Case No. 610–032 (2010)
11. Li, W., Huhns, M.N., Tsai, W.-T., Wu, W. (eds.): Crowdsourcing: Cloud-Based Software Development. Springer, Heidelberg (2015). https://doi.org/10.1007/978-3-662-47011-4
12. Liang, X., Yan, Z.: A survey on game theoretical methods in human-machine networks. Future Gener. Comput. Syst. (2017, in press)
13. Liu, T.X., Yang, J., Adamic, L.A., Chen, Y.: Crowdsourcing with all-pay auctions: a field experiment on Taskcn. Manage. Sci. **60**(8), 2020–2037 (2014)
14. McKelvey, R.D., McLennan, A.M., Turocy, T.L.: Gambit: Software tools for game theory (2006)
15. Moldovanu, B., Sela, A.: The optimal allocation of prizes in contests. Am. Econ. Rev. **3**(91), 542–558 (2001)
16. Moshfeghi, Y., Rosero, A.F.H., Jose, J.M.: A game-theory approach for effective crowdsource-based relevance assessment. ACM Trans. Intell. Syst. Technol. **7**(4), 55 (2016)
17. Nash, J.: Non-cooperative games. Ann. Math. **54**(2), 286–295 (1951)
18. Nisan, N., Roughgarden, T., Tardos, E., Vazirani, V.V.: Algorithmic Game Theory. Cambridge University Press, Cambridge (2007)
19. Wu, W., Tsai, W.-T., Li, W.: Creative software crowdsourcing: from components and algorithm development to project concept formations. Int. J. Creative Comput. **1**(1), 57–91 (2013)
20. Wu, W., Tsai, W.-T., Li, W.: An evaluation framework for software crowdsourcing. Front. Comput. Sci. **7**(5), 694–709 (2013)

# Author Index

Printed in the United States
By Bookmasters